✦

Solving

Problems with

NMR Spectroscopy

Solving Problems with NMR Spectroscopy

Atta-ur-Rahman

and

Muhammad Iqbal Choudhary

H.E.J. Research Institute of Chemistry
University of Karachi, Karachi-75270, Pakistan

ACADEMIC PRESS

San Diego ✦ New York ✦ Boston ✦ London ✦ Sydney ✦ Tokyo ✦ Toronto

This book is printed on acid-free paper. ∞

Copyright © 1996 by ACADEMIC PRESS, INC. PHYSICS

All Rights Reserved.
No part of this publication may be reproduced or transmitted in any form or by any means, electronic or mechanical, including photocopy, recording, or any information storage and retrieval system, without permission in writing from the publisher.

Academic Press, Inc.
A Division of Harcourt Brace & Company
525 B Street, Suite 1900, San Diego, California 92101-4495

United Kingdom Edition published by
Academic Press Limited
24-28 Oval Road, London NW1 7DX

Library of Congress Cataloging-in-Publication Data

Rahman, Atta-ur-, date.
   Solving problems with NMR spectroscopy / by Atta-ur-Rahman,
Muhammad I. Choudhary.
       p.   cm.
   Includes index.
   ISBN 0-12-066320-1 (paper : alk. paper)
   1. Nuclear magnetic resonance spectroscopy.   I. Choudhary,
Muhammad I. (Muhammad Iqbal)   II. Title.
QC762.R34   1995
543'.0877--dc20   95-15392
                                                                   CIP

PRINTED IN THE UNITED STATES OF AMERICA
95   96   97   98   99   00   BB   9   8   7   6   5   4   3   2   1

Contents

CHAPTER 2

Spin-Echo and Polarization Transfer

CHAPTER 3

The Second Dimension

CHAPTER 4

Nuclear Overhauser Effect

CHAPTER 5

Important 2D NMR Experiments

Foreword

The past twenty years have witnessed the development of nuclear magnetic resonance (NMR) into one of the most powerful analytical tools of natural science. Applications range from solid state physics to chemistry, biology, and medicine. NMR takes advantage of the presence of nuclear spins inserted by nature at strategic points in virtually all materials and biological organisms. NMR has become not only one of the most useful but also one of the most sophisticated techniques for exploring nature. An enormous number of ingenious pulse techniques have been developed by the experts to enhance its applicability. An in-depth understanding of this advanced technology is necessary for a fruitful application of NMR to a still-growing palette of important practical issues in industry, science, and health care.

Solving Problems with NMR Spectroscopy is a very welcome addition to the existing literature. It fulfills a real need for an up-to-date and authoritatively written introduction for students and practitioners of NMR. The vast experience of Professor Atta-ur-Rahman and Dr. Muhammad Iqbal Choudhary in the field of structural natural-product chemistry combined with a profound understanding of the concepts and techniques of NMR has led to a very useful and reliable treatise of practical NMR that is useful both for graduate students and for research workers. Professor Atta-ur-Rahman and Dr. Muhammad Iqbal Choudhary are congratulated for their admirable achievement.

PROF. DR. RICHARD R. ERNST
Nobel Prize in Chemistry, 1991

Foreword

Since its inception as a method for investigating structures of organic molecules, NMR spectroscopy has been characterized by its ever-increasing complexity and diversity. This has presented organic chemists of all levels of experience with some special difficulties. Despite the considerable successes of manufacturers in designing instruments and software that are user friendly, the modern spectrometer is far from being a "black box." The effectiveness of NMR as a tool for solving structures is still greatly enhanced by the practitioner's understanding of the fundamental concepts of the experiments. Furthermore, unlike with other physical methods for structure determination, the user is confronted with a veritable galaxy of NMR experiments that may be applied to particular problems. How then is the subject best presented to organic chemists, many of whom may lack the basic physics and mathematics necessary for adopting a truly rigorous treatment? Professor Atta-ur-Rahman and Dr. Choudhary have approached this problem by liberally interspersing the text with formal problems plus their solutions, thereby reinforcing an understanding of the basic principles.

It is my opinion that this approach has considerable merit, provided that the questions posed in the problems are wisely selected, as indeed they are in this text. The authors themselves are well versed in natural-product chemistry, an area that presents a wide array of "small molecule" structural problems. They are therefore concerned that the reader reach the practical goal of applying the full power of NMR spectroscopy to problems of this type. To this end they have selected problems that address methods for solving structures as well as those that pertain to basic theory. The authors have wisely made a point of treating the more widely used 1D and 2D experiments in considerable detail. Nevertheless, they also introduce the reader to many of the less common techniques.

Organic chemists who read this book and *do the problems* as they occur in the text will be rewarded with a functional understanding of NMR spectroscopy at a level that will allow them to make full use of this most versatile spectroscopic method for investigating the structures, stereochemistries, and conformations of organic molecules.

LLOYD M. JACKMAN
The Pennsylvania State University
University Park

Preface

The explosive growth in the area of structural chemistry is based mainly on new and sensitive spectroscopic techniques developed in the past few decades, particularly in the field of NMR spectroscopy. The advent of faster and affordable computer systems with sophisticated microprocessors and large and powerful magnets has led to the development of many extremely sensitive and precise NMR spectrometers. The advent of microcomputers that enable one to control the durations of the pulse as well as the time interval between pulses (or sets of pulses) has made it possible to manipulate nuclear spins in prescribed fashions. The superconducting magnets now available have provided increased sensitivity and field dispersions and, more importantly, very high field stabilities so that scans can be accumulated over long time periods with little or no change in the recording conditions. The use of two-dimensional NMR spectroscopy, triggered by Jeener's original experiment, has transformed the field to such an extent that 2D experiments such as COSY, NOESY, and hetero COSY can now be routinely performed on modern NMR spectrometers.

Chemists, biochemists, biotechnologists, and physicists now routinely use NMR spectroscopy as a powerful research tool. The effective application of 1D and 2D NMR experiments depends largely on the skill and innovation of the user. This book is intended to provide practical knowledge to research workers in the use of NMR spectroscopic techniques to elucidate the structure of organic molecules. Every attempt has been made to prevent the book from becoming too technical, and the underlying principles behind many of the experiments have been described nonmathematically.

Special emphasis has been given to the more important techniques for solving practical problems related to the interpretation of the spectral data obtained from one- and two-dimensional NMR techniques. An introductory treatment of the underlying principles is also included. We hope that the book will be useful to a large community of students of NMR spectroscopy who wish to acquire practical knowledge on how to use this important technique to solve structural problems.

We are most grateful to Mr. Mohammad Ashraf for his constant and invaluable help in the preparation of the manuscript. We are also grateful to Mr. Ahmedullah for typing the manuscript and Mr. Mahmood Alam for secretarial work.

ATTA-UR-RAHMAN
M. IQBAL CHOUDHARY
Karachi, Pakistan

CHAPTER 1

The

Basics of

Modern NMR Spectroscopy

1.1 WHAT IS NMR?

Nuclear magnetic resonance (NMR) spectroscopy is the study of molecules by recording the interaction of radiofrequency (Rf) electromagnetic radiation with the nuclei of molecules placed in a strong magnetic field. Zeeman first observed the strange behavior of certain nuclei subjected to a strong magnetic field at the end of the last century, but practical use of the so-called "Zeeman effect" was made only in the 1950s when NMR spectrometers became commercially available.

Like all other spectroscopic techniques, NMR spectroscopy involves the interaction of the material being examined with electromagnetic radiation. The simplest example of electromagnetic radiation is a ray of light, which occurs in the visible region of the electromagnetic spectrum and has a wavelength of 380 nm to 780 nm. Each ray of light can be thought of as a sine wave. However, this wave can actually be considered to be made up of two mutually perpendicular waves that are exactly in phase with each other; i.e., they both pass through their maxima and minima at exactly the same point of time. One of these two perpendicular waves represents an oscillatory electric field (E) in one plane, while the second wave, oscillating in a plane perpendicular to the first wave, represents an oscillating magnetic field, B.

Cosmic rays, which have a very high frequency (and a short wavelength), fall at the highest-energy end of the known electromagnetic spectrum and involve frequencies greater than 3×10^{20} Hz. Radiofrequency (Rf) radiation, which is the type of radiation that concerns us in NMR spectroscopy, occurs at the other (lowest-energy) end of the electromagnetic spectrum and involves energies of the order of 100 MHz (1 MHz = 10^6 Hz). Gamma rays, X rays, ultraviolet rays, visible light, infrared rays, and microwaves fall between these two extremes. The various types of radiation and the corresponding ranges of wavelength, frequency, and energy are presented in Table 1.1.

Electromagnetic radiation also exhibits behavior characteristic of particles, in addition to its wavelike character. Each quantum of radiation is called a *photon,* and each photon exhibits a discrete amount of energy, which is directly proportional to the frequency of the electromagnetic radiation. The strength of a chemical bond is typically around 400 kJ mol^{-1}, so that only radiation above the visible region will be capable of breaking bonds. But infrared rays, microwaves, and radio-frequency radiation will not be able to do so.

Let us now consider how electromagnetic radiation can interact with a particle of matter. Quantum mechanics (the field of physics dealing with

Table 1.1
The Electromagnetic Spectrum

Radiation	Wavelength (nm) λ	Frequency (Hz) ν	Energy (kJ mol^{-1})
Cosmic rays	$<10^{-3}$	$>3 \times 10^{20}$	$>1.2 \times 10^8$
Gamma rays	10^{-1} to 10^{-3}	3×10^{18} to 3×10^{20}	1.2×10^6 to 1.2×10^8
X rays	10 to 10^{-1}	3×10^{16} to 3×10^{18}	1.2×10^4 to 1.2×10^6
Far ultraviolet rays	200 to 10	1.5×10^{15} to 3×10^{16}	6×10^2 to 1.2×10^4
Ultraviolet rays	380 to 200	8×10^{14} to 1.5×10^{15}	3.2×10^2 to 6×10^2
Visible light	780 to 380	4×10^{14} to 8×10^{14}	1.6×10^2 to 3.2×10^2
Infrared rays	3×10^4 to 780	10^{13} to 4×10^{14}	4 to 1.6×10^2
Far infrared rays	3×10^5 to 3×10^4	10^{12} to 10^{13}	0.4 to 4
Microwaves	3×10^7 to 3×10^5	10^{10} to 10^{12}	4×10^{-3} to 0.4
Radiofrequency (Rf) waves	10^{11} to 3×10^7	10^6 to 10^{10}	4×10^{-7} to 4×10^{-3}

energy at the atomic level) stipulates that in order for a particle to absorb a photon of electromagnetic radiation, the particle must first exhibit a uniform periodic motion with a frequency that exactly matches the frequency of the absorbed radiation. When these two frequencies exactly match, the electromagnetic fields can "constructively" interfere with the oscillations of the particle, and the system is then said to be "in resonance" and absorption of Rf energy can take place. Nuclear magnetic resonance involves the immersion of nuclei in a magnetic field, and then matching the frequency at which they are precessing with electromagnetic radiation of exactly the same frequency so that energy absorption can occur.

1.1.1 The Birth of a Signal

Certain nuclei, such as ^{1}H, ^{13}C, ^{19}F, ^{2}H, and ^{15}N, possess a spin angular momentum and hence a corresponding magnetic moment μ, given by:

$$\mu = \frac{\gamma h[I(I+1)]^{1/2}}{2\pi},\tag{1}$$

where h is Planck's constant and γ is the magnetogyric ratio. When such nuclei are placed in a magnetic field B_o applied along the z-axis, they can adopt one of $(2I+1)$ quantized orientations, where I is the spin quantum number of the nucleus (Fig. 1.1). Each of these orientations corresponds to a certain energy level:

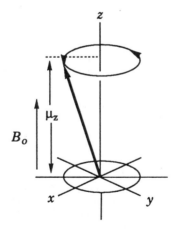

Figure 1.1 Representation of the precession of the magnetic moment about the axis of the applied magnetic field, B_0. The magnitude μ_z of the vector corresponds to the Boltzmann excess in the lower energy (α) state.

$$E = -\mu_z B_0 = -\frac{m_I \gamma h B_0}{2\pi}, \qquad (2)$$

where m_I is the magnetic quantum number of the nucleus. In the lowest energy orientation, the magnetic moment of the nucleus is most closely aligned with the external magnetic field (B_0), while in the highest energy orientation it is least closely aligned with the external field. Organic chemists are most frequently concerned with ^{1}H and ^{13}C nuclei, both of which have a spin quantum number (I) of ½, and only two quantized orientations are therefore allowed, in which the nuclei are either aligned parallel to the applied field (lower energy orientation) or antiparallel to it (higher energy orientation). Transitions from the lower energy level to the higher energy level can occur by absorption of radiofrequency radiation of the correct frequency. The energy difference ΔE between these energy levels is proportional to the external magnetic field (Fig. 1.2) as defined by the equation $\Delta E = \gamma h B_0 / 2\pi$. In frequency terms, this energy difference corresponds to:

$$\upsilon_0 = \frac{\gamma B_0}{2\pi}. \qquad (3)$$

Before being placed in a magnetic field, the nucleus is spinning on its axis, which is stationary. The external magnetic field (like that generated

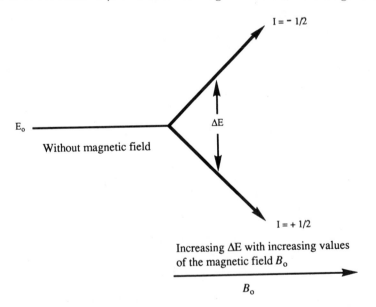

Figure 1.2 The energy difference between two energy states ΔE increases with increasing value of the applied magnetic field B_0, with a corresponding increase in sensitivity.

by the NMR magnet) causes the spinning nucleus to exhibit a characteristic wobbling motion (precession) often compared to the movement of a gyroscopic top before it topples, when the two ends of its axis no longer remain stationary but trace circular paths in opposite directions (Fig. 1.3). If a radiofrequency field is now applied in a direction perpendicular to the external magnetic field and at a frequency that exactly matches the precessional frequency ("Larmor" frequency) of the nucleus, absorption of energy will occur and the nucleus will suddenly "flip" from its lower energy orientation (in which its magnetic moment was precessing in a direction aligned with the external magnetic field) to the higher energy orientation, in which it is aligned in the opposite direction. It can then relax back to the lower energy state through *spin–lattice relaxation* T_1 by transfer of energy to the assembly of surrounding molecules ("lattice"), or by *spin–spin relaxation* (T_2), involving transfer of energy to a neighboring nucleus. The change in the impedance of the oscillator coils caused by the relaxation is measured by the detector as a signal in the form of a decaying beat pattern, known as a *free induction decay* (FID) (Fig. 1.4), which is stored in the computer memory and converted by a mathematical operation known as Fourier transformation to the conventional NMR spectrum.

Thus, excitations caused by absorption of radiofrequency energy cause nuclei to migrate to a higher energy level, while relaxations cause them to flip back to the lower energy level, and an equilibrium state is soon established. In nuclei with positive magnetogyric ratios, such as ^{1}H or ^{13}C, the lower energy state will correspond to the $+\frac{1}{2}$ state, and the higher energy

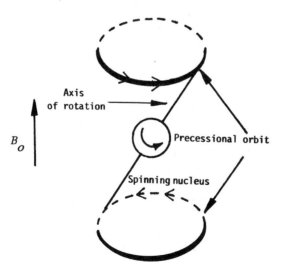

Figure 1.3 Precessional motion of an NMR active nucleus in magnetic field B_0.

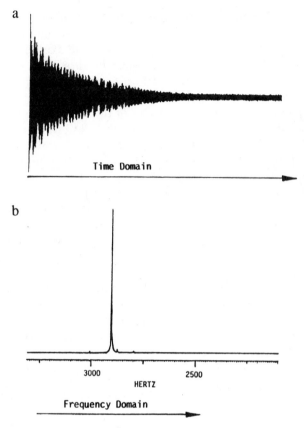

Figure 1.4 (a) Free induction decay (FID) in the time domain. (b) Fourier transformation of the time domain signal yields the conventional frequency domain spectrum.

state to the $-\frac{1}{2}$ state; but in nuclei with negative magnetogyric states, e.g., ^{29}Si or ^{15}N, the opposite will be true. The magnetogyric ratios of some important nuclei are given in Table 1.2 (Harris, 1989).

  If the populations of the upper and lower energy states were equal, then no energy difference between the two states of the nucleus in its parallel and antiparallel orientations would exist and no NMR signal would be observed. However, at equilibrium there is a slight excess ("Boltzmann excess") of nuclei in the lower energy (α) state as compared to the upper energy (β) state, and it is this difference in the populations of the two levels that is responsible for the NMR signal (Fig. 1.5). The ratio of the populations between the two states is given by the Boltzmann equation:

Table 1.2

Magnetogyric Ratios of Some Important NMR-Active Nuclei

Nucleus	Magnetogyric ratio, γ $(10^7 \text{ rad } T^{-1}s^{-1})$	Larmor frequency, v_o (MHz) for $B_o = 7.0461\,T$
^{1}H	26.7520	300.000
^{13}C	6.7283	75.435
^{15}N	−2.712	30.410
^{19}F	25.181	282.282
^{29}Si	−5.3188	59.601
^{31}P	10.841	121.442

$$\frac{N_\beta}{N_\alpha} = \exp\left(\frac{-\Delta E}{kt}\right), \tag{4}$$

where N_α is the population of the lower energy state and N_β is the population of the upper energy state. On a 100-MHz instrument, if there are a million nuclei in the lower energy level, there will be 999,987 in the upper energy level, yielding only a tiny excess of 13 nuclei in the lower energy state. It is this tiny excess that is detected by NMR spectrometers as a signal. Since the signal intensity is dependent on the population difference between

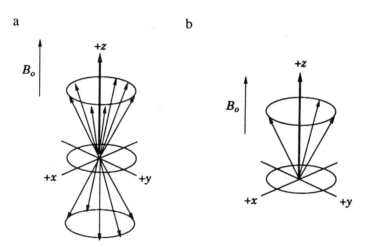

Figure 1.5 (a) Vector representation displaying a greater number of spins aligned with the magnetic field B_0. (b) Excess spin population (Boltzmann distribution excess) aligned with B_0 results in a bulk magnetization vector in the +z direction.

nuclei in the upper and lower energy states, and since the population difference depends on the strength of the applied magnetic field, the signal intensities will be significantly higher on instruments with more powerful magnets. Nuclear Overhauser enhancement (Section 4.1) or polarization transfer techniques (Section 2.2) can also be employed to enhance the population of the ground state over that of the higher energy state to obtain a more intense signal.

✦ *PROBLEM 1.1*

What will happen if the radiofrequency pulse is applied for an unusually long time?

── ✦

✦ *PROBLEM 1.2*

From the discussion in Section 1.1, can you summarize the factors affecting the population difference between the lower energy state (N_α) and the upper energy state (N_β) and how is the population difference related to the NMR signal strength?

── ✦

✦ *PROBLEM 1.3*

As mentioned in the text, there is only a slight excess of nuclei in the ground state (about 13 in a million protons at 100 MHz). Would you expect in the case of a ^{13}C-NMR experiment for the same population difference to prevail?

── ✦

✦ *PROBLEM 1.4*

Explain what is meant by the *Larmor frequency*. What is its importance in an NMR experiment?

── ✦

✦ *PROBLEM 1.5*

What is the magnetogyric ratio, and how does it affect the energy difference between two states and the nuclear species sensitivity to the NMR experiment?

── ✦

1.2 INSTRUMENTATION

NMR spectrometers have improved significantly, particularly in the present decade, with the development of very stable superconducting magnets and of minicomputers that allow measurements over long time periods under homogeneous field conditions. Repetitive scanning and signal accumulation allow ^{1}H-NMR spectra to be obtained with very small sample quantities.

There are two types of NMR spectrometers—continuous wave (CW) and pulsed Fourier transform (FT). In the CW instruments, the oscillator frequency is kept constant while the magnetic field is changed gradually. The value of the magnetic field at which a match is reached between the oscillator frequency and the frequency of nuclear precession depends on the shielding effects that the protons experience. Different protons will therefore sequentially undergo transitions between their respective lower and upper energy levels at different values of the changing applied magnetic field as and when the oscillator frequency matches exactly their respective Larmor frequencies during the scan, and corresponding absorption signals will be observed.

One limitation of this procedure is that at any given moment, only protons resonating at a particular chemical shift can be subjected to excitation at the appropriate value of the magnetic field, and it is therefore necessary to *sequentially* excite the protons that have differing precessional frequencies in a given molecule. A given set of protons will therefore be scanned for only a small fraction of the total scan time, with other protons or base line noise being scanned for the rest of the time. Moreover, since it is often necessary to distinguish splitting that is only a fraction of a hertz in width, a serious time constraint is introduced. This is because, in order to distinguish spectral features separated by Δv Hz, we need to spend $1/\Delta v$ s measuring each element of the spectrum of width Δv. If the spectral range is W, then the minimum time required to record this spectrum would be $W/\Delta v$ s. Thus, on a 500-MHz instrument with a spectral width of 12 ppm (6000 Hz), to achieve a resolution of 0.1 Hz would require $6000/0.1 = 60,000$ s (almost 17 h) if the spectrum were recorded in the continuous-wave mode! Clearly this imposes a severe time limitation in CW NMR spectroscopy, making it impractical at higher field strengths.

Fortunately, an alternative method of excitation exists, instead of the sequential excitation of nuclei by the slow variation of the magnetic field. This involves the application of a short but intense radiofrequency pulse extending over the entire bandwidth of frequencies in which the nuclei to be observed resonate, so that all the nuclei falling within the region are excited *simultaneously*. As a result the total scan time is made independent of the sweep width W and becomes reduced to $1/\Delta\mu$. The relaxations that occur immediately after this excitation process are measured as exponentially decaying waves (FID) and converted to NMR spectra by Fourier transformation.

Such instruments, often called *pulse Fourier transform* (PFT) NMR spectrometers, have largely replaced the CW instruments. The NMR measurements on the earlier CW instruments were in the *frequency domain,* involving the measurement of the signal amplitude as a function of frequency. The sample in such experiments was subjected to a *weak* field, and the energy *absorbed* was measured. In pulse NMR, the sample is subjected to a *strong* burst of radiofrequency energy; when the pulse is switched off, the energy *emitted* by the relaxing nuclei is measured. Thus the CW NMR experiment may be considered as providing an absorption spectrum, while the pulse NMR experiment affords an emission spectrum. A 500-MHz PFT NMR spectrometer is shown in Fig. 1.6.

NMR experimenters need to be aware of some basic features of NMR spectrometers, to be presented briefly next.

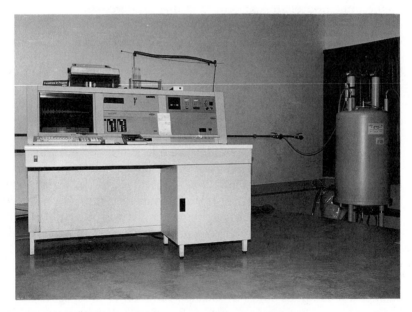

Figure 1.6 A 500-MHz NMR spectrometer (Bruker AMX 500). The console is the computer-controlled recording and measuring system; the superconducting magnet and NMR probe assemblies are on the right.

1.2.1 The Magnet

The heart of the NMR spectrometer is the magnet. Modern high-field NMR spectrometers have oscillators with frequencies of up to 750 MHz, and intensive efforts are under way to develop magnets with higher oscillator frequencies. The solenoid in these magnets is made of a niobium alloy wire dipped in liquid helium ($-269°C$). At this temperature the resistance to electron flow is zero, so, once charged, the "superconducting" magnets become permanently magnetized without consuming any electricity. The liquid helium is housed in an inner container, with liquid nitrogen in an outer container to minimize the loss of helium by evaporation. A large balloon can be connected to the magnet to collect the evaporated helium gas, for subsequent liquefication and recycling. In places where liquid helium is not readily available, it is advisable to order special larger magnet Dewars with the instrument, with longer helium hold times. Fitted with such special Dewars, 300-MHz instruments need to be refilled only about once a year; on 500-MHz instruments the refill period is about once in 4–5 months. Superconducting magnets are very stable, allowing measurements to be made over long periods with little or no variation of the magnetic field.

✦ *PROBLEM 1.6*

Which of the following conditions would be desirable?

 (a) More sample—measurement on a lower-MHz NMR spectrometer.
 (b) Less sample—use of a higher-MHz NMR spectrometer.

── ✦

✦ *PROBLEM 1.7*

Describe the effect of the magnet's power B_0 on the separation of the nuclei in the frequency spectrum. Does it also affect the coupling constant?

── ✦

1.2.2 The Probe

The probe, situated between the field gradient coils in the bore of the magnet, consists of a cylindrical metal tube that transmits the pulses to the sample and receives the resulting NMR signals. The glass tube containing the sample solution is lowered gently onto a cushion of air from the top of the magnet into the upper regions of the probe. The probe, which is inserted into the magnet from the bottom of the cryostat, is normally kept at room temperature, as is the sample tube. The sample is spun on its axis in a stream of air to minimize the effects of any magnetic field inhomogeneities. The gradient coils are also kept at room temperature. If only ^{1}H-NMR spectra are to be recorded, then it is advisable to use a dedicated ^{1}H probe that affords maximum sensitivity. If, however, ^{13}C-NMR spectra are also to be recorded frequently, as is usually the case, then a dual ^{1}H/^{13}C probe is recommended, which, although having a somewhat (10–20%) lower sensitivity than the dedicated probe, has the advantage of avoiding frequent changing of the probe, retuning, and reshimming. If other nuclei (e.g., ^{19}F, ^{31}P) are to be studied, then broadband multinuclear probes can be used, although the sensitivity of such probes is lower by a factor of about 2 than that of dedicated probes.

We also need to choose the probe diameter to accommodate 5-, 10-, or 15-mm sample tubes. In wide-bore magnets, the probes can be several centimeters in diameter, allowing insertion of larger sample tubes and even small animals, such as cockroaches and mice. Normally, the 5-mm probe is used, unless sample solubility is a critical limitation, when it may become necessary to use a larger quantity of sample solution to obtain a sufficiently strong signal. The usual limitation is, however, that of sample quantity rather than sample solubility, and it is often desirable to be able to record good spectra with very small sample quantities. In such situations we should use the smallest-diameter probe possible that affords stronger signals than larger-diameter probes with the same amount of sample. A microprobe of diameter 2.5 mm with special sample tubes is particularly useful in such cases, and special NMR tubes are used with it. If, however, the amount of sample available is not a limiting factor, then it may be preferable to use a larger-diameter probe to subject as much sample as possible to the NMR experiment so as to obtain a good spectrum in the shortest possible measuring time. Such a situation may arise, for instance, in INADEQUATE spectra (Section 5.2.1.7) in which ^{13}C–^{13}C couplings are being observed, and it may be necessary to scan for several days to obtain an acceptable spectrum.

The significant improvements in sensitivity achieved during the last 5 years have been because of improved probe design and radiofrequency circuits. Since the probe needs to be located very close to the sample, it must be made of a material with a low magnetic susceptibility, for otherwise it would cause distortions of the static magnetic field B_0, thereby adversely affecting line shape and resolution. Much research has therefore been undertaken by NMR spectrometer manufacturers to develop materials that have low magnetic susceptibilities suitable for use in probes. The probe must also have a high field (B_1) homogeneity; i.e., it must be able to receive and transmit radiofrequency signals from different regions of the sample solution in a uniform manner. A typical probe assembly is shown in Fig. 1.7.

✦ *PROBLEM 1.8*

Recommend the most suitable probe for each of the following laboratories:

(i) A laboratory involved in biochemical work or in analytical studies on natural products.

(ii) A laboratory involved in the synthesis of phosphorus compounds and organometallic complexes.

(iii) A laboratory where large-scale synthesis of organic compounds is carried out.

(iv) A laboratory where various nitrogenous bases are prepared and studied.

✦

✦ *PROBLEM 1.9*

What properties should an "ideal" NMR probe have?

✦

1.2.3 Probe Tuning

Inside the probe is a wire coil that surrounds the sample tube. This wire transmits the radiofrequency pulses to the sample and then receives

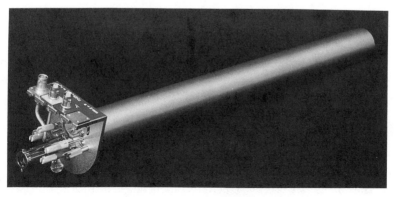

Figure 1.7 A typical probe assembly.

the NMR signals back from the sample. It is vital that the impedance of the wire be identical to those of the transmitter and receiver to properly perform the dual function of pulse transmitter and signal receiver. To match the impedance correctly, two capacitors deep inside the probe resonant circuit can be adjusted by a long screw driver (Fig. 1.8). Adjusting one of the capacitors changes the resonant frequency of the circuit, and this adjustment is carried out so the circuit resonant frequency matches precisely the precessional frequency of the observed nucleus. The other capacitor controls the impedance of the circuit, and it is adjusted to match the probe impedance.

Normally it is not necessary to adjust these capacitors. But if a very high-quality NMR spectrum is wanted, then it may be necessary to tune the probe by adjusting these two capacitors, since the inductance of the coil will vary from sample to sample. The two capacitors are adjusted in

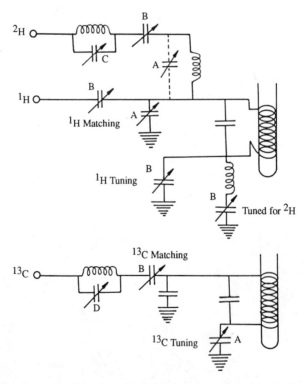

Figure 1.8 A schematic representation of a typical resonant circuit for a dual ^{1}H/^{13}C probe. The capacitors A, B, C, and D perform various functions, such as symmetry and matching resonance.

conjunction with one another, since adjustment of one tends to affect the other and an optimum combination of settings is required. This process is facilitated by employing a *directional coupler* that is inserted between the probe and the transmitter output (Fig. 1.9). The power of the pulse transmitter reflected from the probe is measured by the directional coupler, and the probe is tuned so the reflected power is kept to a minimum, to obtain the best performance. The power level is displayed on a separate directional power meter or Voltage Standing Wave Ratio (VSWR). This meter may be built-in in some spectrometers.

✦ *PROBLEM 1.10*

How does probe tuning affect the quality of the NMR spectrum?

─────────────────────────────────────── ✦

1.2.4 Shimming

Modern superconducting magnets have a set of superconducting gradient coils that are never adjusted by the user. There is, however, another set of printed coils at room temperature that are wrapped around the magnet cylinder. The weak magnetic fields produced by these coils can be adjusted to simplify any errors in the static field, a process known as *shimming*. The shim assembly contains many different coils, which have their respective fields aligned with the *x*-, *y*-, and *z*-axes. The NMR probe lies in between the shim assembly, with the sample tube being located in the

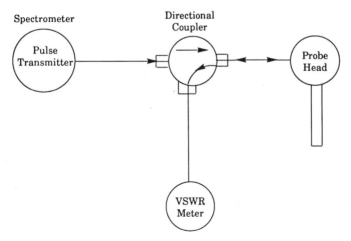

Figure 1.9 Use of a directional coupler for probe tuning.

center of the z-gradient coil. The static field in superconducting magnets lies along the z-axis (in an iron magnet it is aligned horizontally). The proper adjustment of the vertical z- and z^2-gradients is most important, particularly since most of the field inhomogeneities along the x- and y-axes are eliminated by the rapid spinning of the sample tube along the z-axis. It is therefore necessary to correct the x- and y-gradients only to the third order (x, x^2, x^3, y, y^2, y^3), while the z-gradients need to be corrected to the fourth or fifth order, particularly on high-field instruments.

Since it is the z- and z^2-gradients that have to be adjusted most frequently, the operator has to become proficient in the rapid and optimum adjustment of these gradients each time the sample is changed. The adjustments afford maximum lock levels, which in turn lead to higher resolution and improved line shape. The intensity of the lock signal (Section 1.2.5) displayed on the lock-level meter or on some other gradient device indicates the field homogeneity, and it is therefore used to monitor the shimming process.

One feature of the shimming process is the interdependability of the gradients; i.e., changing one set of gradients alters others, so that an already optimized gradient will need to be readjusted if other gradients have been subsequently altered. Good shimming therefore requires patience and perseverance, since there are several gradients to be adjusted and they affect each other.

However, shimming of the various gradients is not done randomly, since certain gradients affect others more or others less. The NMR operator soon recognizes these pairs or small groups of interdependent gradients that need to be adjusted together. The adjustment of x- and y-gradients corresponds to first-order shimming, changes in xy-, xz-, yz-, and x^2-y^2-gradients represents second-order shimming, while optimization of xz^2- and yz^2-gradients is called third-order shimming. It is normally not necessary to alter the xy-, xz-, yz-, xy-, or x^2-y^2-gradients.

Adjustment of the z-gradients affects the line widths, with changes in z-, z^3-, and z^5-gradients altering the symmetrical line broadening and adjustments of z^2- and z^4-gradients causing unsymmetrical line broadening. Changes in the lower-order gradients, e.g., z or z^2, cause more significant effects than changes in the higher-order gradients (z^3, z^4, and z^5). The height and shape of the spinning side bands is affected by changing the horizontal x- and y-gradients, adjustments to these gradients normally being carried out without spinning the sample tube, since field inhomogeneity effects in the horizontal (xy) plane are suppressed by spinning the sample tube. A recommended stepwise procedure for shimming is as follows:

1. First optimize the z-gradient to maximum lock level. Note the maximum value obtained.

2. Then adjust the z^2-gradient, and note carefully the direction in which the z^2-gradient is changed.
3. Again adjust the z-gradient for maximum lock level.
4. Check if the strength of the lock level obtained is greater than that obtained in step 1. If not, then readjust z^2, changing the setting in a direction opposite to that in step 2.
5. Readjust the z-gradient for maximum lock level, and check if the lock level obtained is greater then that in steps 1 and 3.
6. Repeat the preceding adjustments till an optimum setting of z/z^2-gradients is achieved, adjusting the z^2-gradient in small steps in the direction so that maximum lock level is obtained after subsequent adjustment of the z-gradient.
7. If x^2-, y^2-, or z^2-gradients require adjustment, then follow this by readjustment of the x- and y-gradients, making groups of three (x^2, x, y; y^2, x, y; z^2, x, y).
8. If a shim containing a z-gradient is adjusted (e.g., xz, yz, xz^2), then it should be followed by readjustment of the z-gradient.

The main shim interactions are presented in Table 1.3. Note that since adjustments are made for maximum lock signal corresponding to the *area* of the single solvent line in the deuterium spectrum, a high lock signal will correspond to a high intensity of the NMR lines but will not represent improvement in the *line shape*. The duration and shape of the FID is a better indication of the line shape. Shimming should therefore create an exponential decay of the FID over a long time to produce correct line shapes.

The duration for which an FID is acquired also controls the resolution obtainable in the spectrum. Suppose we have two signals, 500.0 Hz and 500.2 Hz away from the tetramethylsilane (TMS) signal. To observe these two signals separately, we must be able to see the 0.2-Hz difference between them. This would be possible only if these FID oscillations were collected long enough that this difference became apparent. If the FID was collected for only a second, then 500 oscillations would be observed in this time, which would not allow a 0.2-Hz difference to be seen. To obtain a resolution of signals separated by n Hz, we therefore need to collect data for $0.6/n$ s. Bear in mind, however, that if the intrinsic nature of the nuclei is such that the signal decays rapidly, i.e. if a particular nucleus has a short T_2^*, then the signals will be broad irrespective of the duration for which the data is collected. As already stated, FIDs that decay over a long time produce sharp lines, whereas fast-decaying FIDs yield broad lines. Thus, to obtain sharp lines, we should optimize the shimming process so the signal decays slowly.

Table 1.3

Main Shimming Interactions

Gradient adjusted*	Main interactions	Subsidiary interactions
z	—	—
z^2	z	—
x	y	z
y	x	z
xz	x	z
yz	y	z
xy	x,y	—
z^3	z	z^2
z^4	z^2	z,z^3
z^5	z,z^3	z^2z^4
x^2-y^2	xy	x,y
xz^2	xz	x,z
yz^2	yz	y,z
zxy	xy	x,y,z
$z(x^2-y^2)$	x^2-y^2	x,y,z
x^3	x	—
y^3	y	—

*Alteration in any gradient in the first column will affect the gradients in the second column markedly, while those in the third column will be affected less.

FIDs have to be accumulated and stored in the computer memory, often over long periods, to obtain an acceptable signal-to-noise ratio. During this time there may be small drifts in the magnetic field due to a slight electrical resistance in the magnet solenoid, variations in room temperature, and other outside influences, such as the presence of nearby metal objects. It is therefore desirable to lock the signal on to a standard reference to compensate for these small changes.

The deuterium line of the deuterated solvent is used for this purpose, and, as stated earlier, the intensity of this lock signal is also employed to monitor the shimming process. The deuterium lock prevents any change in the static field or radiofrequency by maintaining a constant ratio between the two. This is achieved via a lock feedback loop (Fig. 1.10), which keeps a constant frequency of the deuterium signal. The deuterium line has a dispersion-mode shape; *i.e.,* its amplitude is zero at resonance (at its center), but it is positive and negative on either side (Fig. 1.11). If the receiver reference phase is adjusted correctly, then the signal will be exactly on resonance. If, however, the field drifts in either direction, the detector will

a   b

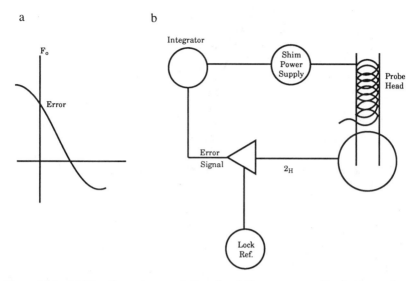

Figure 1.10   (a) The dispersion mode line should have zero amplitude at resonance. (b) The deuterium lock keeps a constant ratio between the static magnetic field and the radiofrequency. This is achieved by a lock feedback loop, which keeps the frequency of the deuterium signal of the solvent unchanged throughout the experiment.

experience a positive or negative signal (Fig. 1.12), which will be fed to a coil lying coaxially with the main magnet solenoid. The coil will generate a field that will be added or subtracted from the main field to compensate for the effect of the field drift. The deuterium lock therefore comprises a simple deuterium spectrometer operating in parallel to the nucleus being observed (Fig. 1.10b).

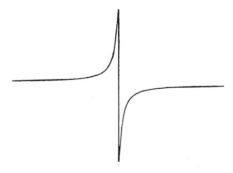

Figure 1.11   The dispersion-mode line shape showing the zero amplitude at the center of the peak but nonzero amplitude on each side.

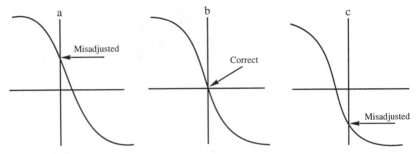

Figure 1.12 (a and c). The reference phase of the receiver is not correctly adjusted. (b) Zero amplitude is achieved by accurate receiver reference phase setting.

✦ *PROBLEM 1.11*

How would you expect the NMR spectrum to be affected if the instrument is poorly shimmed?

_____ ✦

1.2.5 Deuterium Lock

Two other parameters that need to be considered in the operation of the lock channel, besides adjusting the receiver reference phase as just described, are (1) Rf power and (2) the gain of the lock signal. If too much Rf power is applied to the deuterium nuclei, a state of saturation will result, since they will not be able to dissipate the energy as quickly via relaxation processes, producing line broadening and variation of the signal amplitude. It is desirable to achieve the highest transmitter power level that is just below the saturation limit to obtain a good lock signal amplitude. The gain of the lock signal should also be optimized, since too high a lock gain will result in overamplification of the lock signal, thereby causing excessive noise.

✦ *PROBLEM 1.12*

What physical changes would you expect in the shape of the NMR signal if the deuterium lock is not applied during data acquisition?

_____ ✦

1.3 CREATING NMR SIGNALS

As stated earlier, when placed in a magnetic field, the hydrogen nuclei adopt one of two different orientations, aligned either with or against the

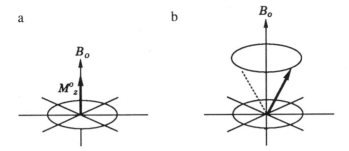

Figure 1.13 (a) Bulk magnetization vector, M_z^o, at thermal equilibrium. (b) Magnetization vector after the application of a radiofrequency pulse.

applied magnetic field B_0. There is slight excess of nuclei in the lower energy (α) orientation, in which the nuclei are aligned with the external field, and there is a bulk magnetization M_z^o corresponding to this difference that points in the direction of the applied field (z-axis, by convention) and that represents the sum of the magnetizations of the individual spins. As long as the bulk magnetization remains aligned along the z-axis, the precessional motion of the nuclei, though present, will not become apparent. This is the situation prevailing at the thermal equilibrium.

The NMR experiment is aimed at unveiling and measuring this precessional motion of the nuclei. One way to do this is to displace the magnetization from its position along the z-axis by application of a radiofrequency pulse. Once the magnetization M is displaced from its position along the z-axis, it experiences a force due to the applied magnetic field B_0, causing it to "precess" about the z-axis at a frequency of γB_0 radians per second (or $\gamma B_0/2\pi$ Hz) (Fig. 1.13), a motion called *Larmor precession*. The object of the NMR experiment is to measure this motion. When the magnetization M is bent away from the z-axis, it will have a certain component in the xy-plane; it is this component that is measured as the NMR signal.

✦ PROBLEM 1.13

Deuterium (^{2}H) is the most common lock signal. ^{1}H, ^{7}Li, and ^{19}F have also been employed for this purpose. Can we use a nucleus as lock signal while recording a spectrum of the same nuclei?

——✦

For a transition to occur from the lower energy state to the upper energy state, the precessing nuclei must absorb radiofrequency energy. This can happen only if the radiofrequency of the energy being emitted from the oscillator matches exactly the precessional frequency of the nuclei. Here's a simple analogy: Imagine that you are holding a sandwich (an "energy packet," if you like) in your right hand while you are constantly rotating your left hand at a particular angular velocity. To pass this sandwich from your right hand to your left without stopping the rotation of your left hand, your right hand must also rotate with exactly the same angular velocity. The transfer of the sandwich (energy) between the two hands can take place only if they rotate *synchronously*. Similarly, only when the oscillator frequency *matches exactly* the nuclear precessional frequency can the nuclei absorb energy and flip to the higher energy state. Via spin–spin or spin–lattice relaxations, they can then relax back to their lower energy state. The resulting change in the impedance of the magnet coils during the relaxation process is recorded in the form of an NMR signal.

The z-component of the magnetization (M_z) corresponds to the population difference between the lower and upper energy states. Any change in this population difference—for instance, by nuclear Overhauser enhancement—will result in a corresponding change in the magnitude of M_z, the NMR signal being strong when the magnitude of M_z is large.

1.3.1 Effects of Pulses

A pulse is a burst of radiofrequency energy that may be applied by switching on the Rf transmitter. As long as the pulse is on, a constant force is exerted on the sample magnetization, causing it to precess about the Rf vector.

The slight Boltzmann excess of nuclei aligned with the external magnetic field corresponds to a net magnetization pointing towards the $+z$-axis. We can bend this magnetization in various directions by applying a pulse along one of the axes (e.g., along the $+x$-, $-x$-, $+y$-, or $-y$-axis). If a pulse is applied along the x-axis, a linear field is generated along the y-axis that is equivalent to two vectors rotating in opposite directions in the xy-plane. However, interaction with the precessing nuclear magnetization occurs only with the vector that rotates in the same direction and with exactly the same frequency.

We can control the extent by which the $+z$-magnetization is bent by choosing the duration for which the pulse is applied. Thus the term "90° pulse" actually refers to the *time period* for which the pulse has to be applied to bend the magnetization by 90°. If it takes, say, t μs to bend a pulse by 90°, it would require half that time to bend the magnetization by 45°, i.e., $t/2$ μs. A 180° pulse, on the other hand, will require double that time, i.e., $2t$ μs and cause the z-magnetization to become inverted so that it comes to lie along the $-z$-axis (Fig. 1.14). There will then be a Boltzmann excess of nuclei in the upper energy state, and the spin system is said to possess a "negative spin temperature." A subscript such as x or y is usually placed after the pulse angle to designate the direction in which the pulse is applied. A "90°_x" pulse therefore refers to a pulse applied along the $+x$-axis that bends the nuclear magnetization by 90°, while a 40°_{-y} pulse is a pulse applied along the $-y$-axis (i.e., in the direction $-y$ to $+y$) for a duration just enough to bend the nuclear magnetization 40° from its previous position.

The duration for which the pulse is applied is inversely proportional to the bandwidth; i.e., if we wish to stimulate nuclei in a large frequency range, then we must apply a pulse of a short duration. Nuclear excitation will of course only occur if the magnitude of the B_1 field is large enough to produce the required tip angle. Typically, if the transmitter power is adjusted to 100 W on a high field instrument, then a 90° pulse width would have a duration of a few microseconds, and it would have a bandwidth of tens of kilohertz over which nuclei could be uniformly excited. A "*soft*" *pulse* is one that has low power or a long duration (milliseconds rather than microseconds), and such pulses can be used to excite nuclei selectively in specific regions of the spectrum.

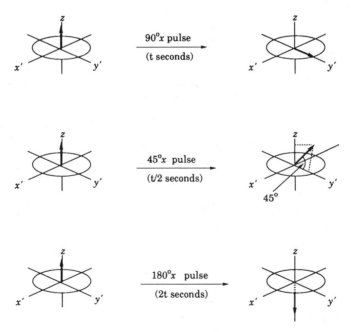

Figure 1.14 Effect of radiofrequency pulses of different durations on the position of the magnetization vector.

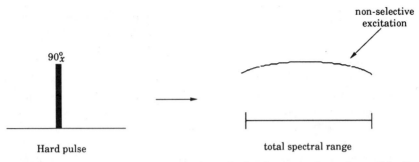

Hard pulse   total spectral range

Figure 1.15 Time domain representation of a hard rectangular pulse and its frequency domain excitation function. The excitation profile of a hard pulse displays almost the same amplitude over the entire spectral range.

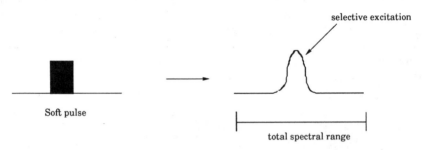

Soft pulse

total spectral range

Figure 1.16 Time domain representation and frequency excitation function of a soft pulse. The soft pulse selectively excites a narrow region of a spectral range and leads to a strong offset-dependent amplitude of the excitation function.

Gaussian pulses are frequently applied as soft pulses in modern 1D, 2D, and 3D NMR experiments. The power in such pulses is adjusted in milliwatts. *"Hard" pulses,* on the other hand, are short-duration pulses (duration in microseconds), with their power adjusted in the 1–100 W range. Figures 1.15 and 1.16 illustrate schematically the excitation profiles of hard and soft pulses, respectively. Readers wishing to know more about the use of shaped pulses for frequency-selective excitation in modern NMR experiments are referred to an excellent review on the subject (Kessler *et al.*, 1991).

✦ PROBLEM 1.14

The bandwidth is inversely proportional to the pulse duration. Following is a computer simulation of a short pulse (*left*) and its calculated spectrum (*right*). Can you predict the shape of the pulse and its spectrum if a long pulse were employed?

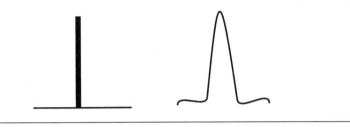

The direction in which the nuclear magnetization is bent by a particular pulse is controlled by the direction in which the pulse is applied. *A pulse applied along a certain axis causes the magnetization to rotate about that axis in a plane defined by the other two axes.* For instance, a pulse along the x-axis will rotate the magnetization *around the x-axis in the yz-plane.* Similarly a pulse along the y-axis will rotate the magnetization in the xz-plane. The final position of the magnetization will depend on the time t_1 for which the Rf pulse is applied (usually a few microseconds). The angle (flip angle) by which the magnetization vector is bent may be calculated as $B = \gamma B_1 t_p$, where B_1 is the applied field and t_p is the duration of the pulse. Since the spectrometers are normally set up to detect the component lying along the y-axis, only this component of the total magnetization will be recorded as signals. The pulse, therefore, serves to convert longitudinal magnetization, or z-magnetization, to transverse, or detectable, magnetization along the y-axis. The magnitude of the magnetization component along the y-axis is given by $B_0 \sin \theta$, where θ is the angle by which the magnetization is bent away from the z-axis (Fig. 1.17).

The direction in which the magnetization is bent by a particular pulse may be predicted by a simple "right-hand thumb" rule: If the thumb of your right hand represents the direction along which a pulse is applied, then the partly bent fingers of that hand will show you the direction in which the magnetization will be bent. Let us consider three different situations and see if we can predict the direction of bending of the magnetization vector in each case.

First let us imagine that the net magnetization lies along the +z-axis and that we apply a 90°_x pulse. If you point your right-hand thumb along the direction $+x$ to $-x$, then the bent fingers of that hand indicate that the magnetization will be bent away from the z-axis and toward the +y-axis, adopting the position shown in Fig. 1.18. If we continue to apply this pulse for a second, identical time duration—i.e., if we apply a second 90°_x

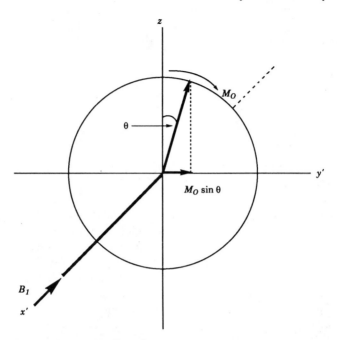

Figure 1.17 A cross-sectional view of the $y'z$-plane (longitudinal plane) after application of pulse B_1 along the x'-axis (perpendicular to this plane). The pulse B_1 applied along the x'-axis causes the equilibrium magnetization M_0 to bend by an angle θ. The magnitude of the component M_y along the y'-axis is $M_0 \sin \theta$.

pulse immediately after the first 90°_x pulse (the two pulses actually constituting a 180°_x pulse)—then the magnetization vector that has come to lie along the $+y$-axis will be bent further by another 90° so it comes to lie

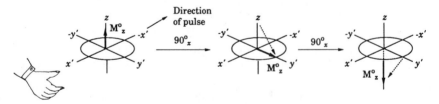

Figure 1.18 Effect of applying a 90°_x pulse on the equilibrium magnetization M°_z. Continuous application of a pulse along the x'-axis will cause the magnetization vector (M°_z) to rotate in the $y'z$-plane. If the thumb of the right hand points in the direction of the applied pulse, then the partly bent fingers of the right hand point in the direction in which the magnetization vector will be bent.

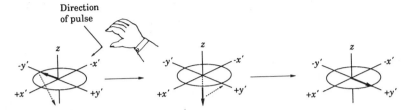

Figure 1.19 Applying a $90°_x$ pulse will bend the magnetization vector lying along the $-y'$-axis to the $-z$-axis, while continued application of the pulse along the x'-axis will cause the magnetization to rotate in $y'z$-plane.

along the $-z$-axis. Continuous application of this pulse effectively rotates this vector continuously around the x-axis in the $y'z$-plane.

Now assume the vector lies along the $-y$-axis and a $90°_x$ pulse is applied (i.e., a pulse in the $-x$ to $+x$ direction along the x-axis). This pulse will now cause the magnetization vector to bend from the $-y$-axis to the z-axis (Fig. 1.19). Applying another $90°_x$ pulse without any intervening delay will rotate the magnetization vector by a further 90° so it comes to lie along the $+y$-axis. Continuous application of this pulse will cause the magnetization vector to rotate around the x-axis in the $y'z$-plane.

Finally, let us assume that the magnetization vector lies along the x-axis and that the magnetization pulse is also applied along the x-axis (i.e., in the $+x$ to $-x$ direction, Fig. 1.20). There is now no magnetization component to be bent in the yz-plane, since the entire magnetization component lies along the x-axis. In other words, a pulse applied along a particular axis will have no effect on any magnetization component vector *lying along that axis.*

1.3.2 Rotating Frame of Reference

As stated earlier, when placed in a magnetic field, the bulk magnetization of the nuclei will precess about the applied magnetic field with a

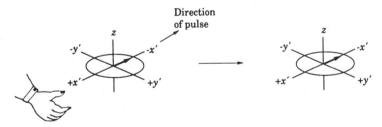

Figure 1.20 Applying a pulse along the x'-axis will not change the position of the magnetization vector lying along that axis.

frequency γB_0. If this precessing magnetization is exposed to a second magnetic field B_1 produced by an oscillator perpendicular to B_0, then this causes it to tilt away from the z-axis and to exhibit an additional precessional motion *around* the applied Rf field at an angular frequency γB_1. This simultaneous precession about two axes (B_0 and B_1) makes it quite difficult to visualize the movement of the magnetization. To simplify the picture, we can eliminate the effects of the precession about B_0 by assuming that the plane defined by the x- and y-axes is itself rotating about the z-axis at a frequency γB_0.

✦ *PROBLEM 1.15*

Based on the right-hand thumb rule described in the text, and assuming that only an equilibrium magnetization directed along the z-axis exists, draw the positions of the magnetization vectors after the application of:

(a) a 90°_x pulse
(b) a 180°_x pulse
(c) a 90°_{-y} pulse
(d) a 90°_y pulse and then a 90°_x pulse.

―― ✦

✦ *PROBLEM 1.16*

Describe what is meant by "hard" and "soft" pulses.

―― ✦

To understand this simplification, imagine a man sitting on a merry-go-round with a large placard on which several sentences are written. As long as the merry-go-round is rotating rapidly, a stationary observer may not be able to read the words on the placard as it whirls round. However, if the observer were to jump onto the merry-go-round, then the apparent circular motion of the man with the placard would disappear to the observer since the two would now have the same angular velocity, and the words on the placard would be readable.

Another analogy is that of communication satellites in geostationary orbits. Imagine a satellite in the sky directly above us that is moving with an angular velocity that matches exactly the rotation of the Earth. From our rotating frame of reference, the satellite will appear stationary to us, although it will be rotating with a much higher angular velocity than our own rotation on the Earth, since it will be in a larger orbit. If we could somehow change the angular velocity of the satellite, say, by firing its rockets so it starts to travel faster, then it would appear to move slowly away from us; if it were made to travel more slowly, then it would appear to move away from us, in the opposite direction. Only the *relative* motion between us and the satellite would be visible in such a rotating frame of reference.

Considering the motions of nuclear magnetization in a "rotating frame" greatly simplifies their description. By rotating the plane defined by the xy-axes at the frequency B_0, we have effectively switched off the precession due to B_0, so that in the rotating frame only the precession due to the field B_1 of the oscillator will be evident. If the field B_1 is applied along the x-axis, it causes the bulk magnetization that was originally lying along the z-axis to precess about the x-axis (i.e., about B_1) in the yz-plane. A $\pi/2$ (90°) pulse along the x-axis would bend the magnetization from the z-axis to the y-axis. Let us assume that the plane defined by the x- and y-axes is rotating at a frequency exactly equal to the Larmor frequency of the methyl protons of TMS. When a 90° pulse is applied along the x-axis, the magnetization vector (M) of the TMS protons will flip away from the z-axis by 90° and come to rest along the y-axis. Since the y-axis and the magnetization vector M are now rotating with identical frequencies, the vector M will appear static along the y-axis in this "rotating frame."

Now let us consider a second case, in which we record the NMR spectrum of TMS in $CHCl_3$ and the xy-plane is again assumed to be rotating at the precession frequency of the TMS methyl protons. Application of a 90° pulse would superimpose both vectors (the one for the TMS methyl protons as well as the one for the $CHCl_3$ methine proton) on the y-axis. Because the TMS vector is rotating with the same frequency as the y-axis, the movement of the TMS vector becomes concealed in this rotating frame of reference, and it appears static along the y-axis. The frequency of precession of the $CHCl_3$ methine proton is, however, somewhat greater than that of the TMS vector, so in the time interval immediately after the pulse, it appears to move away clockwise with a *differential* angular velocity (Fig. 1.21). Similarly, if there are protons in a compound that resonate upfield from TMS (such as those in an organometallic substance), then after the 90° pulse their respective magnetization vectors would move away from TMS in the opposite (counter-clockwise) direction in the delay immediately after the 90° pulse. By convention, the x- and y-axes in the rotating frame of reference are differentiated by adding primes: x' and y'.

After the $90°_x$ pulse is applied, all the magnetization vectors for the different types of protons in a molecule will initially come to lie together along the y'-axis. But during the subsequent time interval, the vectors will separate and move away from the y'-axis according to their respective precessional frequencies. This movement now appears much slower than that apparent in the laboratory frame since only the *difference* between the a   b   c

Figure 1.21   (a) Position of the magnetization vector at thermal equilibrium. (b) After the $90°_x$ pulse, the magnetization vector $M°_z$ will come to lie along the y'-axis on the $x'y'$-plane. (c) During the subsequent delay, the sample magnetization vector will move away from the TMS vector due to its (generally) higher precessional frequency on account of its chemical shift. Since the $x'y'$-plane is itself precessing at the precessional frequency of the TMS vector, the TMS vector appears to be stationary along the y'-axis in this rotating frame.

reference frequency of the rotating frame and the precessional frequencies of the nuclei is observable in the rotating frame. If we consider the relative movement of two different vectors in the $x'y'$-plane, the angle between the faster vector and the slower vector will grow with time till it reaches 180°, then start decreasing, the two vectors thus repetitively coming in phase and then out of phase. The faster-moving vector corresponds to a downfield proton, while the slower-moving vector corresponds to an upfield proton. *Chemical shifts can therefore be considered in terms of differences between the angular velocities of the magnetization vectors of the nuclei as compared to the angular velocity of the TMS vector.*

1.3.3 Free Induction Decay

As long as the bulk magnetization lies along the z-axis, there is no signal. Once a pulse is applied along the x'-axis, the magnetization M is bent away from the z-axis and a component is generated along the y'-axis. Since a pulse (a "hard" one) produces a wide band of frequencies with about the same amplitude, all the nuclei lying within the entire chemical shift range are *simultaneously* tipped by the same angle. If a $90°_x$ pulse is applied, then the total magnetization M_0 will "flip" to the y'-axis; but if a pulse of a smaller pulse angle θ is applied, then the component aligned along the y'-axis will be correspondingly smaller ($M_0 \sin \theta$).

✦ PROBLEM 1.17

Following is a pictorial vector representation of a doublet. It shows the evolution of the components of a signal. Draw vector positions in the fourth and fifth frames, along with their directions of rotation.

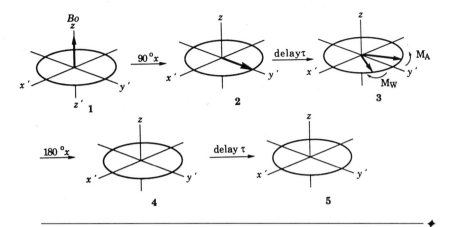

✦

✦ *PROBLEM 1.18*

The effect of pulses on magnetization vectors is much easier to understand in the rotating frame than in the fixed frame. How do we arrive
at the chemical-shift frequencies in the rotating frame?

✦

Since the detector is conventionally regarded as being set up to detect
signals along the y'-axis, the signal intensity will be at a maximum immediately after the 90_x° pulse. During the subsequent delay, the magnetization
vector will precess away from the y'-axis and toward the x'-axis, and the
signal will decay to a zero value when it reaches the x'-axis. As it moves
further toward the $-y'$-axis, a negative signal appears that will reach its
maximum negative amplitude when the vector points directly toward the
$-y'$-axis. As the vector then moves toward the $-x'$ axis, the negative signal
amplitude will again proceed to become zero and then grow to a maximum
positive amplitude as it completes one full circle and reaches the y'-axis.
A sinusoidal variation of signal amplitude will thus occur. The relationship
of the magnetization vector's position with the signal amplitude and signal
phase is shown in Fig. 1.22.

The transverse magnetization and the applied radiofrequency field will
therefore periodically come in phase with one another, and then go out
of phase. This causes a continuous variation of the magnetic field, which
induces an alternating current in the receiver. Furthermore, the intensity
of the signals does not remain constant but diminishes due to T_1 and T_2
relaxation effects. The detector therefore records both the exponential
decay of the signal with time and the interference effects as the magnetization vectors and the applied radiofrequency alternately dephase and re-

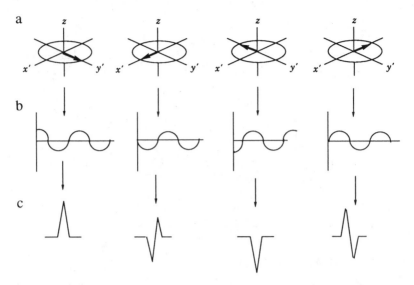

Figure 1.22 Relationship of (a) the magnetization vector's position with (b) the signal time-domain NMR signal and (c) the frequency-domain NMR signal.

phase in the form of a decaying beat pattern, the *pulse interferogram* or the *free induction decay* (indicating that it is *free* of the influence of the Rf field, that it is *induced* in the coils, and that it *decays* to a zero value).

✦ *PROBLEM 1.19*

Why in a decaying signal (FID) does the amplitude decay asymptotically toward zero while the precessional frequency remains unchanged?

─── ✦

✦ *PROBLEM 1.20*

In practice we sample the FID for a duration of about 2–3 seconds till most of it has been recorded. This means that a small portion of the "*tail*" of the FID will not be recorded. What effect would you expect this to have on the quality of the spectrum?

─── ✦

1.3.4 Fourier Transformation

The data on modern NMR spectrometers are obtained in the *time* domain (FIDs); i.e., they are collected and stored in the computer memory as a function of time. However, in order for such data to be used, they must be converted into the *frequency* domain, since the NMR spectrum required for interpretation must show characteristic line resonances at specific frequencies. The conversion of the data from the time domain to the frequency domain is achieved by *Fourier transformation,* a mathematical process that relates the time-domain data $f(t)$ to the frequency-domain data $f(w)$ by the Cooley-Tukey algorithm:

$$f(w) = \sum_{-t}^{+t} f(t) \exp\{iwt\}\, dt. \qquad (5)$$

Apparently, the time-domain and frequency-domain signals are interlinked with one another, and the shape of the time-domain decaying exponential will determine the shape of the peaks obtained in the frequency domain after Fourier transformation. A decaying exponential will produce a Lorentzian line at zero frequency after Fourier transformation, while an exponentially decaying cosinusoid will yield a Lorentzian line that is offset from zero by an amount equal to the frequency of oscillation of the cosinusoid (Fig. 1.23).

Fourier transformation of Rf pulses (which are in the time domain) produces frequency-domain components. If the pulse is long, then the Fourier components will appear over a narrow frequency range (Fig. 1.24); but if the pulse is narrow, the Fourier components will be spread over a wide range (Fig. 1.25). The time-domain signals and the corresponding frequency-domain partners constitute *Fourier pairs.*

Protons in different environments in a molecule will normally exhibit different chemical shifts (i.e., they will resonate at different frequencies) and different multiplicities (i.e., the individual components of each multiplet will resonate at their characteristic frequencies). Each of the individual components will contribute its decaying beat pattern to the FID, which will therefore represent a *summation* of the various decaying beat patterns. Fourier transformation produces the NMR spectrum, in which the *positions* at which the individual signals appear will depend on the precessional frequencies of the respective nuclei, while the *signal widths* will depend on the life span of the decaying transverse magnetization. If the transverse magnetization decays slowly (i.e., if there is a long FID due to a long effective transverse relaxation time, T_2,) then a sharp signal will be observed; but if there is a short FID, then a broad signal results (Fig. 1.26). If a molecule contains only one type of proton, then only a single peak will be observed after Fourier transformation. Such a simple FID will have only one decaying sinusoidal pattern. However, in a molecule containing several different types of nuclei, the FID obtained will be more complex, since

a

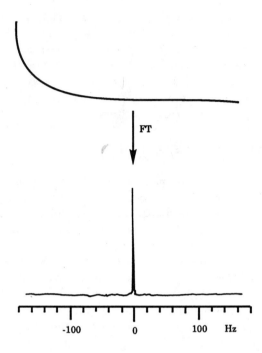

b

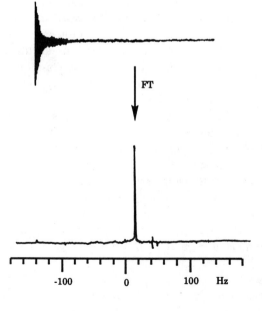

it will be a summation of the various individual decaying beat patterns (Fig. 1.27).

✦ *PROBLEM 1.21*

Why is Fourier transformation essential in pulse NMR spectroscopy?

─── ✦

1.3.5 Data Acquisition and Storage

Although the idea of two-dimensional NMR was introduced by Jeener as far back as 1971 (Jeener, 1971), it was largely the limitations in the data systems available at the time that prevented the experiments from becoming popular in the early 1970s (the data systems typically needed 8K words of core memory). Only in the early 1980s did sufficiently large data systems become available that were capable of handling the large amounts of data required in 2D NMR experiments, and the NMR spectrometers began to be fitted with sophisticated pulse programmers that made it possible to operate multipulse sequences in a highly flexible manner. This was accompanied by the development of both reliable 160-MB or larger sealed Winchester disk drives for storing all the data generated in 2D NMR experiments and array processors for fast Fourier transformations. Thus, a 512×512-word data matrix that required several hours for transformation on the instruments available in 1970s can now be processed in seconds!

The free induction decay obtained in the NMR experiment takes the form of high-frequency electrical oscillations. These must be converted into numerical form before being stored in computer memory. The electrical input is converted into binary output, representing the magnitude of the electrical voltage, by an analog-to-digital converter (ADC). To simplify calculations, the reference frequency is subtracted from the observed frequency before the data are processed mathematically. For instance, on a 300-MHz (i.e., 300,000,000-Hz) instrument, the protons would normally resonate in the region between 0 and 12 ppm (0–3600 Hz), so the frequencies to be processed would be between 300,000,000 Hz and

───

Figure 1.23 (a) The Fourier transform of an exponentially decaying FID yields a Lorentzian line at zero. (b) The FT of an exponentially decaying consinusoid FID gives a Lorentzian line offset from zero frequency. The offset from zero is equal to the frequency of oscillation of the consinusoid. (Reprinted from S. W. Homans, *A dictionary of concepts in NMR,* copyright © 1990, p. 127–129, by permission of Oxford University Press, Walton Street, Oxford OX2 6DP, U.K.)

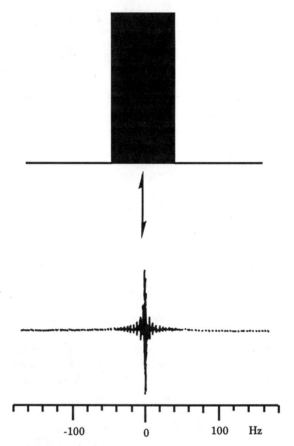

Figure 1.24 Fourier components of a long Rf pulse ("soft" pulse) are spread over a relatively narrow frequency range. (Reprinted from S. W. Homans, *A dictionary of concepts in NMR*, copyright © 1990, pp. 127–129, by permission of Oxford University Press, Walton Street, Oxford OX2 6DP, U.K.)

300,003,600 Hz. Since it is only the *difference* between these two frequencies that concerns us, and since the use of such large frequencies would unnecessarily occupy computer memory, it is convenient first to subtract the reference frequency (300,000,000 Hz in this case) from the observed frequency, and to store the remainder (0–3600 Hz) in computer memory.

After each pulse, the digitizer (ADC) converts the FID into digital form and stores it in the computer memory. Ideally, we should keep *sampling* i.e., acquiring data till each FID has decayed to zero. This would require about $5T_1$ seconds, where T_1 is the spin-lattice relaxation time of the slowest-relaxing protons. Since this may often take minutes, it is more convenient

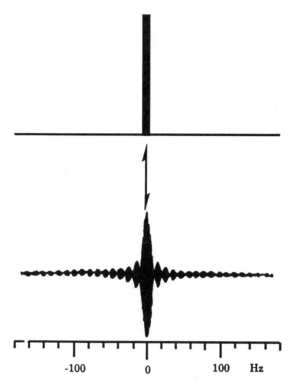

Figure 1.25 A short ("hard") Rf pulse has Fourier components spread over a relatively wide frequency range. (Reprinted from S. W. Homans, *A dictionary of concepts in NMR*, copyright © 1990, pp. 127–129, by permission of Oxford University Press, Walton Street, Oxford OX2 6DP, U.K.)

to sample each FID for only a few seconds. This means that the data obtained are from the larger-volume "head" of the FID, while a small portion of its "tail" is cut off. Since most of the required information is present in the "head" of the FID, we can thus avoid the time-wasting process of waiting for the "tail" to disappear before recollecting data from a new FID.

The next issue to consider is the *rate* at which the FIDs must be sampled. To draw a curve unambiguously, we need to have many data points per cycle. This becomes clear if we consider two different situations: In the first case, the signal is oscillating much faster than the rate at which the data are being collected; in the second case, the data are being collected faster than the rate of signal oscillation. Clearly, it is in the second case that we will obtain several data points per cycle, allowing a proper sinusoidal wave shape to be drawn. In the first case, the data collected will not correspond to any particular wave shape, and many different curves may be drawn through the data points.

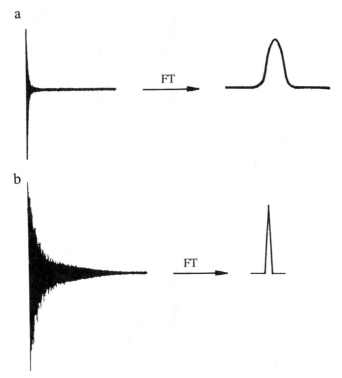

Figure 1.26 Free induction decay and corresponding frequency-domain signals after Fourier transformations. (a) Short-duration FIDs result in broader peaks in the frequency domain. (b) Long-duration FIDs yield sharp signals in the frequency domain.

To represent a frequency with accuracy, we must have *at least* two data points per cycle, so that a unique wave can be drawn that would pass through *both* data points. Thus, in Fig. 1.28a, only one data point has been registered per cycle, and the resulting ambiguity is apparent, since many curves can be drawn through these points. In Fig. 1.28b, however, two data points have been registered per cycle, so only one curve can be drawn through them. This causes a problem: On a 500-MHz instrument, the signal will have to be sampled at twice its oscillator frequency i.e., at 1000 MHz. In other words, we will need to collect one data point every $1/(10^6 \times 10^3)$ s, i.e., every 1 ns, for a second or longer.

Collecting such a large amount of data would require highly sophisticated and expensive instrumentation and is unnecessary, since we are really interested in the signals appearing in a rather narrow frequency range.

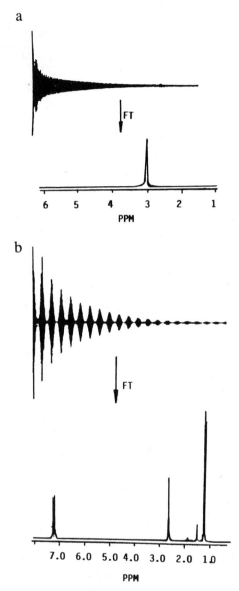

Figure 1.27 FIDs and Fourier-transformed frequency spectra. (a) Single-proton system displaying only one decaying sinusoidal pattern. (b) Several different types of nuclei in a complex molecule yield a more complex FID pattern.

a

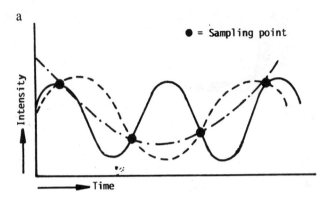

b

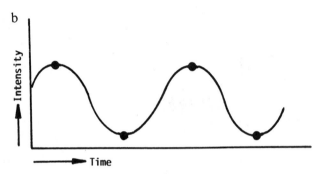

Figure 1.28 (a) If fewer than two data points are sampled per cycle, then more than one curve can be drawn through the points. This will result in an ambiguous wave shape. (b) Sampling of at least two data points per cycle removes the uncertainty and yields a particular wave shape.

For instance, on a 500-MHz instrument, a 15-ppm range would represent 7500 Hz, and because we are interested only in the *difference* between the frequencies at which the signals of one sample appear and the frequencies of a reference standard, such as TMS, the sampling can be carried out at low frequencies (which can be digitized easily) even though detection is being done at much higher frequencies (radiofrequency range). In the earlier example, the minimum number of words of data storage per second required to cover the entire spectral width of 7500 Hz (0–15 ppm) would be $2 \times 7500 = 15,000$, and the *sampling rate* would need to be $1/15,000$ s (0.066 ms).

The frequency at which the signals must be sampled is determined by the maximum separation between the signals in a spectrum, i.e., by the highest frequency to be characterized. If this is S Hz, then according to the Nyquist theorem, the signals will need to be sampled every $\frac{1}{2}S$ s. Moreover, to distinguish spectral features separated by $\Delta\delta$ Hz, the sampling must be continued for $1/\Delta\delta$ s. The total number of points that are sampled, stored in computer memory, and subjected to Fourier transformation is given by $2S/\Delta\delta$. In practice we actually set the *dwell time* between the data points on the computer (which is equal to $\frac{1}{2}SW$, where SW is the sweep width) and hence indirectly set the required sweep width. The design of the timers installed in NMR spectrometers does not allow them to be set at all possible time durations, so the computer program will select the dwell time that is slightly shorter than the required dwell time, and the spectral window will therefore be slightly larger than the required window.

✦ PROBLEM 1.22

Why it is necessary to subtract the reference frequency from the observed frequency before data storage and processing?

———————————————————————————————— ✦

1.3.6 Digital Resolution

Even when the NMR spectrometer has been adjusted for optimum resolution, the spectrum may still have a poor resolution if enough computer memory was not allocated for the purpose, i.e., it will have poor *digital resolution*. A large allocation of computer memory (i.e., a greater number of words of data storage) requires a corresponding increase in acquisition time. The maximum acquisition time, in seconds, required after each pulse is given by:

$$\frac{\text{No. of words of data storage}}{2 \times \text{spectral width, in Hz}} . \tag{6}$$

In the example given in the preceding section, the number of words of data storage was 15,000 and the spectral width was 7500 Hz, so an acquisition time of $15,000/(2 \times 7500) = 1.0$ s was required after each pulse.

The resolution is controlled not only by such factors as field homogeneity and intrinsic nature of the compound (such as the presence of exchange or restricted rotation), but also by the *sampling rate*. With a spectral width of 7500 Hz on a 500-MHz instrument, the sampling rate was $1/(2(7500))$ s, i.e., 0.066 ms. An acquisition time of 1.0 s requires 1.0 s$/0.066$ ms $\cong 15,000$ data points of computer memory.

Suppose this gave a digital resolution of 1.0 Hz when the desired resolution was 0.2 Hz. A fivefold increase in digital resolution would now necessitate a corresponding fivefold increase in the number of data points (i.e., 75,000 instead of 15,000) to improve the line resolution from 1.0 Hz to 0.2 Hz. Alternatively, we could maintain the same number of data points (15,000) but use a five-times smaller spectral width, i.e., scan in a 3-ppm region rather than the original spectral width of 15-ppm, to improve digital resolution from 1.0 Hz to 0.2 Hz. As already indicated, there are other limiting factors that control the maximum achievable resolution, such as field inhomogeneity, and increase in the sampling rate would only improve the resolution to a certain point.

Apparently, the digital resolution (DR) depends on the amount of computer memory used in recording the spectrum. If the memory size is M, then there will be $M/2$ real and $M/2$ imaginary data points in the frequency spectrum, and the separation (in hertz) between these data points (DR) will be given by:

$$DR = \frac{2SW}{M},\qquad(7)$$

where SW represents the sweep width. In a spectrum recorded with 16K ($2^{14} = 16,384$) data points, there will be 8K (8,192) real and 8K imaginary data points. For a spectral width of 6000 Hz (12 ppm on a 500-MHz instrument), the digital resolution per point will be 6000/8192 = 0.73 Hz (i.e., the accuracy of measurement will be ±0.73 Hz), so it will be impossible to differentiate between lines separated by less than 1.46 Hz.

If we wish to measure at a higher digital resolution, then the same spectrum could be recorded at, say, 32K (16,384 real and 16,384 imaginary data points), giving a digital resolution of 6000/16,384 = 0.365 Hz. The same improvement could be achieved by halving the spectral width to 3000 Hz and measuring the spectrum at the same 16K digital resolution (3000/8192 = 0.365 Hz). Working at high digital resolutions increases the computing time, so spectra are normally not recorded beyond 32K or 64K. If increase in digital resolution does not improve resolution, then field homogeneity, instrumental factors, or the intrinsic nature of the compound may be the limiting considerations.

We should decide in advance the digital resolution at which we wish to acquire a spectrum and then set the acquisition time accordingly. The acquisition time AT (that is, the product of the number of data points to be collected and the dwell time between the data points) is calculated as simply the reciprocal of the digital resolution:

$$AT = \frac{1}{DR}.\qquad(8)$$

✦ PROBLEM 1.23

A poor digital resolution will result in loss of some of the fine structure of an NMR signal. To increase the digital resolution, we need either to maintain the same number of data points but reduce the spectral width or, alternatively, to maintain the spectral width but increase the number of data points. Which method would you prefer for achieving a better signal-to-noise ratio?

─── ✦

Figure 1.29 shows the effect of digital resolution on the appearance of signals. It is important to remember that *each* data point in the FID contains information about *every* peak in the spectrum, so as more data points are accumulated, and the FIDs acquired for a longer period, a finer digital resolution is achieved.

1.3.7 Peak Folding

If too small a spectral width is chosen, the signal lying outside the selected region will still appear in the spectrum, but at the wrong frequency. This is because the peaks lying outside the spectral width *fold over* and become superimposed on the spectrum (Fig. 1.30). The folded (or *aliased*) peaks can usually be recognized easily, since they show different phasing than the other peaks. To check if a peak is indeed a folded one, the spectral window may be shifted (say, by 500 Hz) to one side. All the "normal" peaks will then shift in the same direction by exactly this value, but the folded peaks will either shift in the wrong direction or will shift by the wrong value.

If too large a spectral width is chosen to avoid the folding of speaks, much of the computer memory will be wasted in storing noise data lying beyond the frequency range of the spectrum. The spectral width should therefore be chosen to be just wide enough to cover the region in which the signals are likely to appear. Even if the correct spectral width is chosen, the noise lying within the spectral region would still be digitized and processed. The magnitude of this problem can be reduced by filtering off the noise before digitization. Modern NMR spectrometers are fitted with filters that remove most of the noise, but they also somewhat reduce signal intensities within the spectral range while cutting them off outside the spectral range.

1.3.8 The Dynamic Range Problem

The analog-to-digital converter receives the FID signals in the form of electrical voltages and converts them into binary numbers proportional to

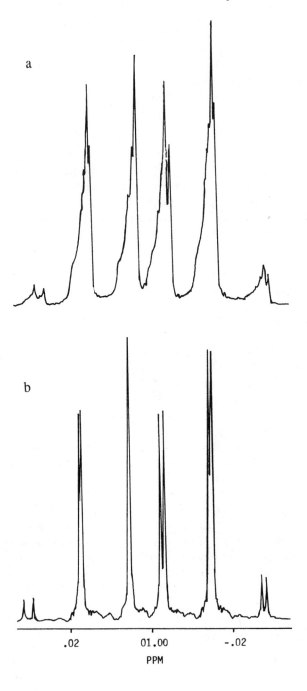

the voltages stored in computer memory. To a great extent, the *digitizer resolution* (which is different from the *digital resolution* discussed earlier) determines the efficiency of this process. The digitizer resolution is dependent on the *word length* used in the ADC, i.e., the number of bits (binary 1 or 0) used to sample the electrical voltages. An 8-bit digitizer can process numbers from 00000000 binary (which is 0 in decimal) to 11111111 (which is 255 in decimal, or $2^8 - 1$). Thus, an 8-bit digitizer would accept the largest signal with an output of 255 and the smallest signal with an output of 1. The *dynamic range* of the 8-bit digitizer would then be described as 255 : 1.

The NMR signal should fit into the dynamic range properly; i.e., the gain of the instrument must be adjusted correctly. If the gain is too high, then the largest signal will not fall within the dynamic range of the digitizer and an ADC overflow will result, causing baseline distortions. If, however, some of the signals are too small, then they will be below the detection limits of the digitizer and will not be observed. To cope with this problem, modern NMR spectrometers are fitted with 12-bit digitizers with a dynamic range of 4095 : 1 or 16-bit digitizers with a dynamic range of 65,535 : 1.

✦ PROBLEM 1.24

If the spectral width is inadequate to cover every peak in the spectrum, then some peaks in the downfield or upfield region may fold over and appear superimposed on the spectrum. How can you identify these folded signals?

─── ✦

The dynamic range problems become acute if we are trying to record a spectrum in which there is a large size difference between the largest and the smallest signals, such as in the spectra of proteins in the presence of a signal for the hydroxylic protons in monodeuterated water (HOD), present as an impurity. Attempts to fit the large hydroxylic signal within the dynamic range may make the signals for the protein molecule fall below the digitizer detection limit. Careful adjustment of the gain and special water-suppression pulse sequences may be used to solve this problem.

Figure 1.29 The effect of increased digital resolution (DR) on the appearance of the NMR spectrum. (a) The spectrum of *o*-dichlorobenzene recorded at a digital resolution of 0.1 Hz per point, allowing the spectral lines to be seen at their natural line width. (b) The spectrum of the same molecule recorded at a digital resolution of 0.4 Hz per point.

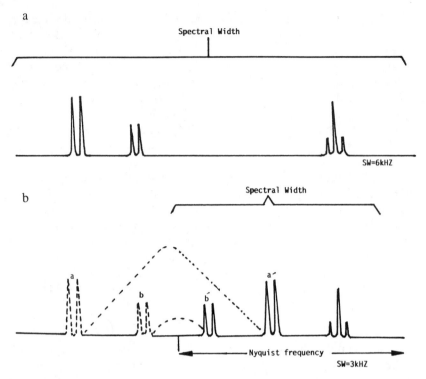

Figure 1.30 (a) Normal NMR spectrum resulting from the correct selection of spectral width. (b) When the spectral width is too small, the peaks lying outside the spectral width can fold over. Thus a' and b' represent artifact peaks caused by the fold-over of the a and b signals.

One way to solve the dynamic range problem might be simply to adjust the gain in order to bring the smaller signals within the dynamic range and allow the intense FID signals to overflow, i.e., cut off a part of the "head" of the FID that contains the most intense signals. However, as stated earlier, *each* data point in the FID contains information about all parts of the spectrum. Cutting off the top and bottom sections of the "head" of the FID will therefore reduce the signal-to-noise (S/N) ratio, distort the peaks (with negative peaks appearing on the sides of the signals), and give the appearance of new artifact peaks (Fig. 1.31).

With larger-capacity digitizers, the digitization rate is slower and the spectral width is also reduced. Thus, a 12-bit digitizer operates at a maximum speed of 300 kHz, requiring 3 ms for each sampling to characterize a spectral range of 150 kHz, while a 16-bit digitizer would reduce the spectral

a

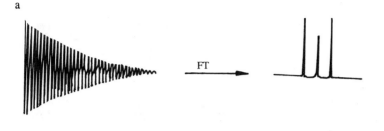

b

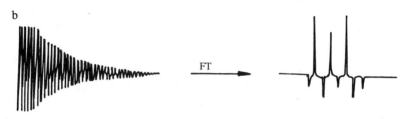

Figure 1.31 (a) When the entire FID falls within the dynamic range of the digitizer, a regular-shaped signal results. (b) Dynamic range overflow (nonrecording of the top and bottom of the head of the FID) reduces S/N ratio and distorts signal shapes.

range to 50 kHz. In spite of the slower digitization rate, it is still preferable to use larger-capacity digitizers to detect the smaller signals, which would otherwise be lost in the quantization noise caused by digitization at smaller word lengths (Fig. 1.32).

After digitization, the data have to be readded in the computer memory. It is convenient to have a larger-capacity computer (24- or 32-bit) to avoid a memory overflow that would cause loss of information and spectral distortions. The software checks for such memory overflow and automatically compensates for such problems by reducing the digitizer precision. Alternatively, we can collect the data at full digitizer resolution in blocks, and store them on disks. The blocks of data are then subjected to Fourier transformation and the frequency spectra coadded. Such a "block averaging" procedure allows the larger signals to overflow the memory limit without causing spectral distortion.

✦ *PROBLEM 1.25*

The dynamic range of the digitizer is very important. The NMR signal should fit into the digitizer appropriately; i.e., the gain must be adjusted

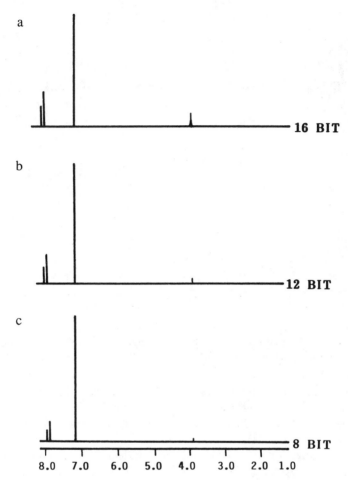

Figure 1.32 Increase in sensitivity via increase in ADC resolution. (a) A tiny signal of impurity at δ 3.90 appears clearly with a 16-bit digitizer. (b) The signal at δ 3.90 is smaller when a 12-bit digitizer is used. (c) The signal at δ 3.90 is hard to see when the spectrum is recorded with an 8-bit digitizer.

properly. What will happen if there is a large solvent signal that makes the gain too high?

── ✦

1.3.9 Quadrature Detection

So far we have been concerned with a single detector measuring only the *y*-component of the magnetization. In such a single-detection system we can measure only the *frequency difference* between the reference frequency and the frequency of the signal, but not its *sign;* i.e., it is impossible to tell whether the signal frequency is greater or less than the reference frequency. If the Rf pulse is placed at one end of the spectrum, then all the signals will appear on one side of the pulse frequency while only noise will lie on the other side. Because the positive and negative frequencies cannot be distinguished, this noise will fold over and come to lie on top of the spectrum, thereby decreasing the signal-to-noise ratio by a factor of $\sqrt{2}$, or 1.4.

Clearly, this is not acceptable. In modern NMR spectrometers, the irradiating frequency is therefore placed in the middle of the spectrum rather than at one end. To distinguish between the signals lying above and below the reference frequency, the so-called *quadrature detection* method is employed. This involves detecting not only the magnetization component along the *y*-axis, but also those along the *x*-axis by employing *two detectors* to detect the signals from the same coil, with their reference phases differing by 90°. This allows us to discriminate the sign of the detected frequencies. The *x*-magnetization M_x results in a signal of magnitude cos Ωt in the *x*-channel, while the *y*-magnetization M_y creates a signal of magnitude sin Ωt in the *y*-channel.

Two different methods have been employed for quadrature detection. In the first method, the two signals along the *x*- and *y*-axes are collected *simultaneously;* in the other method, they are collected *sequentially.* This has implications in two-dimensional spectra in which the spectral widths may be defined differently in the two dimensions. The sequence of 90° phase shifts used in the quadrature phase-cycling routine (also known as *cyclops*) results in the magnetization being successively generated along the *y*-, *x*-, $-y$-, and $-x$-axes, corresponding to the 0°, 90°, 180°, and 270° phase shifts, respectively. The data from the four separate phase shifts are coadded in the computer.

The underlying principle of this Redfield technique is illustrated in Fig. 1.33. Since the spectral width is now reduced, the noise lying outside the spectral region cannot fold back onto the spectrum, resulting in an improved signal-to-noise ratio (Fig. 1.34). For simplicity, we can assume that one of the two phase-sensitive detectors is set up correctly to detect the cosine (odd) component of the magnetization while the other detects the sine (even) component (actually each detector detects both components). The two signals, corresponding to the real and imaginary parts of a complex spectrum, are digitized separately and subjected to Fourier transformation so that one line in the frequency domain is reinforced while the other is cancelled (Fig. 1.35). This allows us to distinguish the signs of

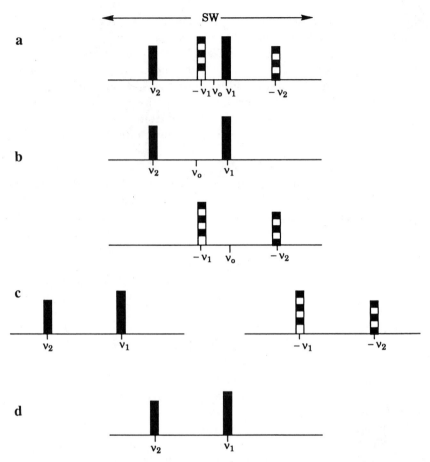

Figure 1.33 The underlying principle of the Redfield technique. Complex Fourier transformation and single-channel detection gives spectrum (a), which contains both positive and negative frequencies. These are shown separately in (b), corresponding to the positive and negative single-quantum coherences. The overlap disappears when the receiver rotates at a frequency that corresponds to half the sweep width (SW) in the rotating frame, as shown in (c). After a real Fourier transformation (involving folding about $n_1 = 0$), the spectrum (d) obtained contains only the positive frequencies.

the signals, i.e., whether they are lying at higher or lower frequencies than the reference frequency.

Since quadrature detection involves the cancellation of an unwanted component by adding two signals that have been processed through different parts of the detection system, they will cancel out completely only if their phases differ by precisely 90° and if their amplitudes are exactly equal.

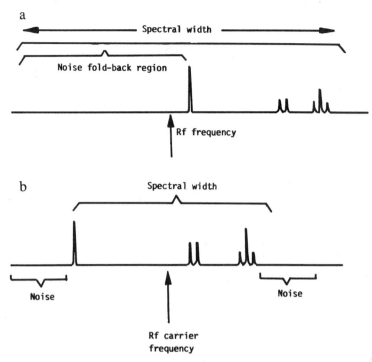

Figure 1.34 (a) Reduced S/N ratio resulting from noise folding. If the Rf carrier frequency is placed outside the spectral width, then the noise lying beyond the carrier frequency can fold over. (b) Better S/N ratio is achieved by quadrature detection. The Rf carrier frequency in quadrature detection is placed in the center of the spectrum. Due to the reduced spectral width, noise cannot fold back on to the spectrum.

In practice this may not be achieved perfectly, so weak *quad images* may be produced. They can be readily recognized as they show different phases than the rest of the spectrum.

Suppose we wish to collect signals in the range of 4000 Hz involving 2000 time domain points, with 1000 data points from each of the quadrature signals. In the simultaneous data-collection mode, the spectral window will be defined as 2000 Hz (i.e., ±2000 Hz) and the dwell time will be ½(2000) = 0.25 ms (DW = ½SW) and the acquisition time will be 1000 data points × 0.25 ms/point = 0.25 s. The digital resolution will be 4000/1000 = 4 Hz per point. If, however, we acquire the quadrature signals alternately, the spectral window will be 4000 Hz and the dwell time will be 1/(2 × 4000) = 0.125 ms, the acquisition time will remain at 0.25 s (2000 points × 0.125 ms), and the digital resolution will remain at 4 Hz per point.

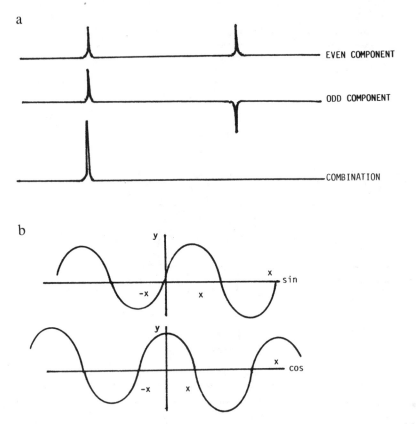

Figure 1.35 (a) Combination of odd and even components yields a spectrum with cancellation of signals through the addition of opposite-phased signals. (b) Sine waves are odd, and cosine waves are even. (Reprinted from E. Derome, *Modern NMR techniques for chemistry research,* copyright © 1987, pp. 78–79, with permission from Pergamon Press Ltd., Headington Hill Hall, Oxford OX3 OBW, U.K.)

1.3.10 Signal-to-Noise Ratio

As stated earlier, the spectral width of the frequencies is determined by the rate at which the data are collected; so in order to have a spectral width of W Hz, the data must be sampled at $2W$ Hz. In other words, to collect data over wider spectral widths, we must collect data more *slowly.* However, because the *accuracy* with which the frequencies are measured depends on the length of time spent in collecting data points *in each FID,* to obtain a high resolution spectrum we need to collect data points for a greater *length of time.* Fourier transformation converts the FID, which is in the time domain, to the spectrum, which is in the frequency domain, $2N$ data points in the time domain giving 2^{N-1} points in the frequency domain.

✦ *PROBLEM 1.26*

A serious problem associated with quadrature detection is that we rely on the cancellation of unwanted components from two signals that have been detected through different parts of the hardware. This cancellation works properly only if the signals from the two channels are exactly equal and their phases differ from each other by exactly 90°. Since this is practically impossible with absolute efficiency, some so-called "image peaks" occasionally appear in the center of the spectrum. How can you differentiate between genuine signals and image peaks that arise as artifacts of quadrature detection?

─── ✦

However, many FIDs have to be recorded sequentially and the FIDs coadded in the memory before Fourier transformation is carried out to afford a spectrum with an acceptable signal-to-noise ratio. On the continuous-wave instruments, improvement in the signal-to-noise ratio could be achieved by incorporating a computer of averaged transients (CAT) and coadding successive scans. As we scan slowly from one end of the spectrum to the other, different nuclei come to resonance *sequentially.* However, since at any one time only nuclei in a specific region of the spectrum are being subjected to excitation while for the remainder of the scan time only the baseline noise or other spectral regions are being scanned, acquisition of spectra on continuous-wave instruments, even if fitted with a CAT, involves the time-wasting process of scanning one region at a time.

Fourier transform NMR spectroscopy overcame this problem because all the nuclei in the spectral range of interest are excited simultaneously by a short, sharp burst of radiofrequency energy (pulse). On the modern PFT instruments, the FIDs can therefore be obtained much more rapidly, with only a certain relaxation delay being inserted between successive scans. Since the sample signals appear at exactly the same frequency while the noise signals vary in their positions, the noise tends to grow at a slower rate. If n scans are accumulated, the signal will grow n times while the noise will grow by a factor of $\sqrt{n}$, resulting in an improvement of the signal-to-noise ratio by $\sqrt{n}$. Thus, if a certain signal-to-noise ratio is achieved with 64 scans ($\sqrt{64} = 8$), then to double this ratio we would need to accumulate 256 scans ($\sqrt{256} = 16$), which would take four times longer than did accumulating the original 64 scans. Simply doubling the scan time, therefore, does not double the signal-to-noise ratio, a common misconception. With small sample quantities, a point is soon reached beyond which it is no longer feasible to devote more instrument time to accumulate additional scans for improving the signal-to-noise ratio. It is also possible to obtain an improved signal-to-noise ratio by manipulating the spectrum through digital filtering and other apodization techniques that will be discussed later.

The maximum signal intensity is obtained if the magnetization vectors point directly to the y'-axis, i.e., if a 90° pulse is employed. However, the nuclei would require a certain time, depending on the relaxation rate R, to relax back to their equilibrium state before a new pulse is applied. If the pulses are applied too rapidly so the nuclei cannot relax sufficiently, then a state of saturation would soon be reached and the signal will disappear. Moreover, if all the nuclei do not have the same relaxation rates (as is usually the case), and if the spectrum is recorded before they have relaxed to their respective equilibrium states, the signal intensities will not correspond with their respective integrations. If, however, we wait long enough between the pulses so the z-magnetization of all the nuclei is fully restored (normally, $5 T_1$ of the slowest-relaxing nuclei in the molecule), then so much time would be spent between successive scans that an insufficient number of scans may be accumulated in a given time. A compromise between these two extremes is therefore desirable. It is more time-efficient to tip the nuclei by a smaller angle so they take less time to relax back to their respective equilibrium states, allowing the spectra to be recorded with smaller time delay intervals between successive scans. In general, the most efficient combination of pulse angle θ, delay interval t between successive pulses, and relaxation rate R is given by the equation

$$\cos \theta = \exp(-tR). \tag{8}$$

The relaxation rates of the individual nuclei can be either measured or estimated by comparison with other related molecules. If a molecule has a very slow-relaxing proton, then it may be convenient *not* to adjust the delay time with reference to that proton and to tolerate the resulting inaccuracy in its intensity but adjust it according to the *average* relaxation rates of the other protons. In 2D spectra, where 90° pulses are often used, the delay between pulses is typically adjusted to $3 T_1$ or $4 T_1$ (where T_1 is the spin–lattice relaxation time) to ensure no residual transverse magnetization from the previous pulse that could yield artifact signals. In 1D proton NMR spectra, on the other hand, the tip angle θ is usually kept at 30°–40°.

✦ *PROBLEM 1.27*

What is signal-to-noise (S/N) ratio, and how can it be improved by acquiring a large number of FIDs?

─── ✦

1.3.11 Apodization

Having recorded the FID, it is possible to treat it mathematically in many ways to make the information more useful by a process known as *apodization* (Ernst, 1966; Lindon and Ferrige, 1980). By choosing the right "window function" and multiplying the digitized FID by it, we can improve either the signal-to-noise ratio or the resolution. Some commonly used apodization functions are presented in Fig. 1.36.

✦ *PROBLEM 1.28*

The signal-to-noise ratio can be increased by treating the data so as to bias the spectrum in favor of the signals and against the noise. This can be done by multiplying the FIDs by the proper apodization functions. What would happen if the spectrum is recorded without apodization?

─── ✦

✦ *PROBLEM 1.29*

Apodization is likely to change the relative intensities of signals with different line widths. Can it also affect the chemical shifts of the signals?

─── ✦

1.3.11.1 SENSITIVITY ENHANCEMENT

To increase the signal-to-noise ratio, we need to multiply the FIDs by a window function that will reduce the noise and lead to a relative increase in signal strength. Since most of the signals lie in the "head" of the FID while its "tail" contains relatively more noise, we multiply the FID by a mathematical function that will emphasize the "head" of the FID and suppress its "tail."

Most simply this can be done by multiplying each data point in the FID by an exponential decay term that starts at unity but decays to a negligible value at its end. Such an exponential multiplication (EM) is a simple and effective way to increase the signal-to-noise ratio at the expense of added line broadening (LB). The LB term in this function can be altered

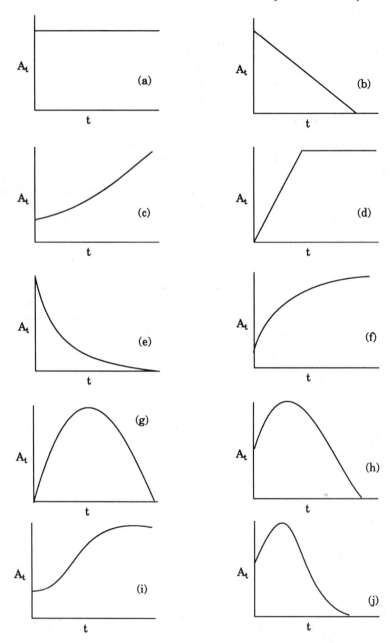

Figure 1.36 Various selected apodization window functions: (a) an unweighted FID;
(b) linear apodization; (c) increasing exponential multiplication; (d) trapezoidal
multiplication; (e) decreasing exponential multiplication; (f) convolution differ-

by the operator, the larger the value of LB, the more rapid will be the apodized FID decay, leading after Fourier transformation to some broadening of lines (i.e., reduction in resolution) but often to significant improvement in the signal-to-noise ratio. If the lines are already rather broad, then LB should be chosen to correspond to the existing line widths; if the lines are narrow, then LB may be set to correspond to the digital resolution. If a negative LB value is chosen, then an opposite effect will be observed after Fourier transformation, and resolution will be improved; i.e., the lines will be sharper but there will also be more noise. Such a "resolution-enhanced" spectrum will have a lower signal-to-noise ratio. Figure 1.37 shows the effects of multiplying an FID when various LB values are included in the exponential multiplication.

1.3.11.2 RESOLUTION ENHANCEMENT

The process of exponential multiplication just described produces a rapid decay of the FID and the production of broad lines; suppressing the decay of the FID gives narrow lines and better resolution, with increased noise level. An alternative approach to resolution enhancement is to reduce the intensity of the earlier part of the FID. Ideally, we should use a function that reduces the early part of the FID, to give sharper lines, as well as reduces the tail of the FID, to give a better signal-to-noise ratio.

A simple way to do this is to multiply by a symmetrical shaping function, such as the *sine-bell* function (Marco and Wuethrich, 1976), which is zero in the beginning, rises to a maximum, and then falls to zero again, resembling a broad inverted cone (Fig. 1.36g). One problem with this function is that we cannot control the point at which it is centered, and its use can lead to severe distortions in line shape. A modification of the function, the *phase-shifted sine bell* (Wagner *et al.*, 1978) (Fig. 1.36h), allows us to adjust the position of the maximum. This leads to a lower reduction in the signal-to-noise ratio and improved line shapes in comparison to the sine-bell function. The *sine-bell squared* and the corresponding *phase-shifted sine-bell squared* functions have also been employed (see Section 3.2.2. also).

Gaussian multiplication (Ernst, 1966; Marco and Wuethrich, 1976) has been used widely for resolution enhancement without significant loss of sensitivity in 1D NMR spectra. There are two parameters altered by the

---

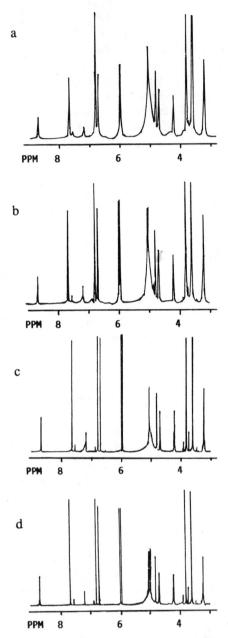

Figure 1.37 The effect of line broadening (LB) multiplication on the appearance of ¹H-NMR spectra. (a:LB = 10) and (b:LB = 5) ¹H-NMR spectra recorded after multiplying the FIDs by positive LB values. (c:LB = 0). The same ¹H-NMR spectrum recorded without line broadening. (d:LB = −2) Sharper signals are obtained when the FID is multiplied by negative LB values.

operator: Lorentzian Broadening (LB) controls the line widths (negative values lead to sharper lines), and Gaussian Broadening (GB) determines the position of the maximum of the function. LB initially may be set at -2 or at 3DR (where DR is digital resolution), and GB should be located at the point where the FID begins to merge into the noise. Increasingly negative values of LB lead, up to a point, to sharper lines, but with corresponding decreases in the signal-to-noise ratio; smaller values of GB tend to prevent the lowering of the signal-to-noise ratio.

The existence of a high signal-to-noise ratio in the collected data therefore allows the use of resolution enhancement functions with greater freedom, since it offers the operator greater flexibility in sacrificing some of the signal-to-noise ratio to obtain the desired resolution. If the signal-to-noise ratio in the original data is low to start with, then multiplication by a resolution enhancement function may lead to weakening of the signals to the extent that they become indistinguishable from the noise.

✦ *PROBLEM 1.30*

Define *sensitivity* and *resolution* in NMR spectroscopy.

─── ✦

✦ *PROBLEM 1.31*

What changes in the line shape of an NMR spectrum occur after resolution enhancement?

─── ✦

✦ *PROBLEM 1.32*

Why are pulse Fourier transform (PFT) NMR experiments preferred over continuous wave (CW) NMR techniques?

─── ✦

✦ *PROBLEM 1.33*

How does saturation affect the sensitivity of an NMR experiment?

─── ✦

✦ *PROBLEM 1.34*

Summarize the common methods for enhancing sensitivity in NMR spectroscopy.

─── ✦

1.3.12 Pulse Width Calibration

It is important that the pulse widths be calibrated accurately, since the NMR experiments cannot be performed properly if the wrong flip angles are chosen. The flip angle may not be critical in simple one-dimensional NMR experiments, but it is still important to know the duration of the 180° pulse. In two-dimensional NMR experiments, however, accurate knowledge of pulse widths is essential. In some heteronuclear 2D NMR experiments in which a proton or X-band decoupler is used as a pulse transmitter, it may be necessary to determine the pulse widths of the decoupler amplifiers. In 2D multiple quantum coherence NMR experiments, even a slight deviation from the required pulse angle can adversely affect the outcome of the experiment, making the accurate determination of the relationship between spin flip angle and pulse duration even more critical. Also, the pulse calibration must be repeated regularly. Since pulse widths depend on the power output of the transmitter and decoupler, and since power levels can vary with time, pulse widths determined on one day may not be valid on the following day. In certain 2D NMR experiments it is also advisable to calibrate the pulse width immediately before the start of the experiment.

It is important to avoid saturation of the signal during pulse width calibration. The Bloch equations predict that a delay of $5*T_1$ will be required for complete restoration to the equilibrium state. It is therefore advisable to determine the T_1 values; an approximate determination may be made quickly by using the inversion-recovery sequence (see next paragraph). The protons of the sample on which the pulse widths are being determined should have relaxation times of less than a second, to avoid unnecessary delays in pulse width calibration. If the sample has protons with longer relaxation times, then it may be advisable to add a small quantity of a relaxation reagent, such as $Cr(acac)^3$ or $Gd(FOD)_3$, to induce the nuclei to relax more quickly.

The inversion-recovery pulse sequence used to determine T_1 values is shown in Fig. 1.38. The first 180° pulse causes the magnetization of the protons to be inverted so that it comes to rest along the $-z$-axis. During the subsequent evolution period τ, the magnetization would relax from the $-z$-axis back toward its original equilibrium position along the $+z$-axis. The return to equilibrium is not instantaneous, but usually takes place with a first-order rate constant R_1. In the study of nuclear relaxations involving time-dependent measurements, however, it is more convenient to consider the relaxation-time constant $T_1 (= 1/R_1)$ rather than the rate constant R_1. To convert the longitudinal magnetization into a signal, it must first be converted into transverse magnetization (i.e., brought in the $x'y'$-plane)

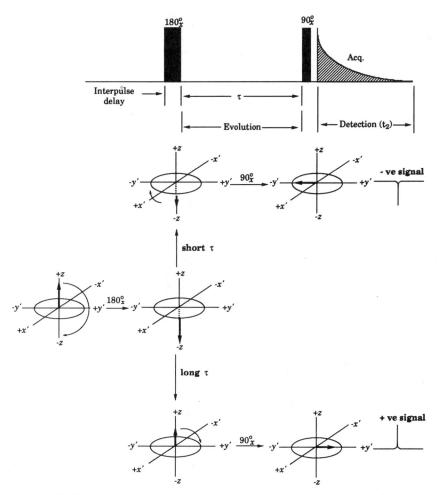

Figure 1.38   The effect of the duration of the evolution delay τ on the spin-lattice relaxation in an inversion-recovery experiment is shown. A short duration produces only minimal relaxation along the z-axis. Thus, when a 90° pulse is applied to the evolved magnetization, the z-magnetization vector (which points to the $-z$-axis) will be rotated to the $-y'$-axis, ultimately giving rise to a *negative* signal in the spectrum (assuming the detector lies along y'). On the other hand, a longer duration results in a restoration of the $+z$-magnetization, so the 90° pulse will cause the magnetization to rotate to the $+y'$-axis, producing a *positive* signal.

by the application of a subsequent 90°_x pulse. If this 90°_x pulse is applied very soon after the original 180°_x pulse (i.e., if the evolution period or delay τ is very short), then the 90°_x pulse will "catch" the magnetization while it is still on the $-z$-axis and cause it to rotate to the $-y'$-axis, thereby giving a negative signal. As the evolution period τ is increased progressively, the 90°_x pulses will "encounter" the z-magnetization as it recedes along the $-z$-axis to a zero value; at still longer τ intervals, it would grow along the $+z$-axis toward its original equilibrium value, the one that existed before application of the 180°_x pulse. This would produce a series of corresponding spectra in which the signals initially have decreasing negative amplitudes and then, after going through a null point, increasingly positive amplitudes. The null points may sometimes appear as tiny out-of-phase signals and are readily recognized (Fig. 1.39).

The whole sequence of successive pulses is repeated n times, with the computer executing the pulses and adjusting automatically the values of the variable delays between the 180°_x and 90°_x pulses as well as the fixed relaxation delays between successive pulses. The intensities of the resulting signals are then plotted as a function of the pulse width. A series of "stacked plots" are obtained (Fig. 1.40), and the point at which the signals of any particular proton pass from negative amplitude to positive is determined. This zero transition time τ_0 will vary for different protons in a molecule,

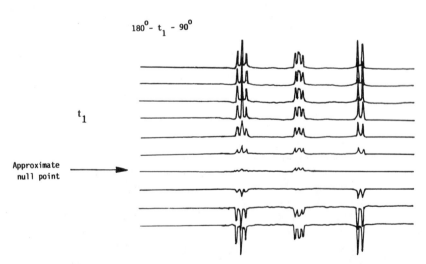

Figure 1.39 Representation of an inversion-recovery T_1 experiment. (Reprinted from W. R. Croasmun and R. M. K. Carlson, *Two-dimensional NMR spectroscopy applications for chemists and biochemists,* copyright © 1987, p. 13, with permission, from VCH Publishers Inc., 220 East, 23rd Street, New York 10010-4606.)

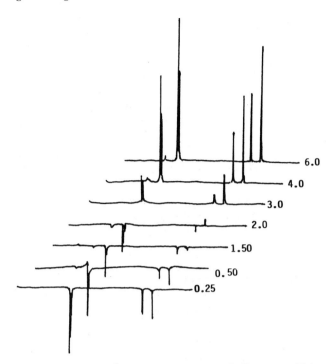

Figure 1.40 Stacked plots of ¹H-NMR spectra for ethylbenzene. This experiment can be used to measure the spin-lattice relaxation time, T_1.

depending on their respective spin-lattice relaxation times, and it is related to T_1 by the equation $T_1 = \tau_0/\ln 2 = \tau_0/0.693$. Determining τ, in seconds, at which the amplitude of a particular nucleus is zero and dividing it by 0.693 thus gives its spin-lattice relaxation time T_1. There are other ways to determine the spin-lattice relaxation times, but this is the most convenient.

As already indicated, determination of the T_1 times helps in setting the interpulse delays to be used during pulse calibrations. Having derived the T_1 values, we next acquire an FID with a very short pulse (say, 250 µs). Fourier transformation and phasing produces a spectrum. The process is repeated to give a series of FIDs with the appropriate intervening relaxation delays, and the pulse widths are incremented successively in small steps (say, 5 µs). The spectra thus collected are arranged as a stacked plot, and each signal is seen to undergo an oscillation that reaches its maximum positive amplitude at 90° and its maximum negative amplitude at 270°. The 180° pulse width (zero amplitude or very weak signal) can thus be determined.

It is convenient to use a sample of *p*-dioxane in $CDCl_3$ having enough concentration to give an acceptable signal-to-noise ratio in one acquisition. The 90° pulse width is not read directly, since it is often difficult to see when the peaks reach their maximum amplitude but easier to see the null point. The 90° pulse is therefore calculated by halving the duration determined for the 180° pulse. We should check that inadequate intervening delays between successive pulses are not producing an erroneous value due to saturation. We can do this by repeating the experiment, with increased intervals between successive FIDs and checking that the signal intensity does not increase. If it does, then the plot should be repeated, with longer delays between successive FIDs. Once the approximate null point is known, to get a more accurate reading the experiment is repeated, with smaller incrementation of pulse durations, say, 1 or 2 μs.

A closely related procedure involves setting the pulse width to the approximate value of the 180° pulse and then recording many spectra. Before doing this, the approximate phase constants must first be determined. The transmitter frequency is placed close to the signals being used for calibration. We should obtain a reasonable signal-to-noise ratio with a single transient. But if the signal obtained with a single transient is too weak, then we can accumulate several transients, with a relaxation delay of at least $5T_1$ s between successive transients. The transmitter offset is placed close to the signal in the sample of interest, and the FID is acquired with a small flip angle (10°–20°). The sensitivity is enhanced by exponential line broadening, the FID subjected to Fourier transformation in the absolute intensity mode, and the phase corrected for the pure absorption mode across the spectrum. The phase correction constants thus defined are carefully noted.

The pulse width is next adjusted to the expected value of the 180° pulse, and a new FID recorded. This is again subjected to exponential multiplication, Fourier transformation, and phase correction using the phase constants defined in the earlier experiment. The phasing of the signal (i.e., positive or negative) depends on whether the pulse is longer or shorter than 180°. Many spectra are thus recorded, each with a slightly different value of the pulse width, till the 180° condition is reached, when the residual signal will have a symmetric dispersive (positive/negative) phasing with minimum amplitude. A broadened positive hump often also remains, due to poor shimming, but only the positions of the central narrow dispersive region should be used for judging the 180° pulse width.

There are many other methods known for accurate calibration of pulse widths, but such discussion is beyond the scope of this text (see: Thomas *et al.*, 1981; Lawn and Jones, 1982; Bax, 1983; Wesener and Gunther, 1985; Nielsen *et al.*, 1986).

In practice it is usually unnecessary to determine exact pulse widths for each sample; we can use approximate values determined for each probehead, except in certain 2D experiments in which the accuracy of pulse widths employed is critical for a successful outcome. Proper tuning of the probehead is advisable, since pulse widths will normally not vary beyond $\pm 10\%$ with well-tuned probeheads.

1.3.13 Composite Pulses

Since there is a slight delay between when a pulse is switched on and when it reaches full power, an error may be introduced when measuring $90°$ or smaller pulses directly. If the $90°$ pulse width is required with an accuracy of better than ± 0.5 μs, then it may be determined more accurately by using self-compensating *pulse clusters* that produce accurate flip angles even when there are small ($<10\%$) errors in the setting of pulse widths.

For example, to bend the equilibrium magnetization from the z-axis to the $-z$-axis, we need to apply a $180°_x$ pulse. Suppose the "$180°_x$" pulse is wrongly adjusted so that in practice it rotates the z-magnetizations by $170°$ instead of the desired $180°$. One simple way to compensate for this error is to apply a cluster of pulses comprising a $90°_x$–$180°_y$–$90°_x$ *composite pulse*. The first "$90°_x$" pulse in this cluster actually produces a similar error factor, so it will in reality be an $85°_x$ pulse, causing the magnetization to come to lie $5°$ above the y'-axis in the $y'z$-plane. The $180°y$-pulse causes the magnetization vector to jump across the y'-axis and adopt a mirror image position* on the other side of the y'-axis, i.e., $5°$ below the y'-axis in the $y'z$-plane. Noted that a $180°$ pulse about any axis would cause the magnetization vector lying on one side of that axis to adopt a position that is the exact mirror image of its earlier position, across the axis about which the pulse was applied. The angle traversed may *appear* to be smaller than $180°$ if viewed in a particular plane. For instance, in the example given, the vector has traveled only about $10°$ in the $y'z$-plane but has actually moved by a *full semicircle* (Fig. 1.41). The magnetization now lies as far below the y'-axis as it was above the y'-axis before the application of the "$180°_y$" pulse; i.e., an equal and opposite error has been produced by the "$180°_y$" pulse. Applying the final "$90°_x$" (actually, $85°_x$) pulse now moves the vector an additional $85°$, so it comes to lie fairly accurately on the $-z$-axis.

---

*Actually, it will not be quite the mirror image position: The $180°_y$ pulse now applied will really be only a $170°_y$ pulse. But since the vector is now only moving a very short distance on the circular surface in the $y'z$-plane (about $10°$), the error thus produced will be significantly less.

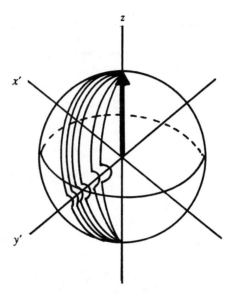

Figure 1.41   Applying the first incorrectly adjusted "90_x°" pulse (actually, 85_x° pulse) bends the z-magentization vector 5° above the y'-axis. The 180_y° pulse at this stage will bring the magnetization vector 5° below the y'-axis (to the mirror image position). Applying another similarly maladjusted "90_x°" pulse causes a further bending of the magnetization vector precisely to the $-z$-axis. The composite pulse sequence (i.e., $90_x^\circ–180_y^\circ–90_x^\circ$) is thus employed to remove imperfections in the 90_x° pulse.

✦ *PROBLEM 1.35*

What problems would you expect to encounter in the case of incorrect alignment of the pulse width during an NMR experiment?

─── ✦

Composite pulses are usually described in abbreviated notations. For instance, a $\pi_x/2$ pulse (i.e., 90_x° pulse) is represented as X, and a π_x pulse as $2X$. Phase shifts of 180° are represented by a bar over the X or Y. Hence, a 180_{-y}° pulse (i.e., a 180° pulse applied along the $-y$-axis) will be described as $2\overline{Y}$ (since it is 180° phase-shifted from the 180° pulse applied along the $+y$-axis). A composite pulse sequence known as GROPE-16 represents the following cluster of pulses: $3\overline{X}4XY3\overline{Y}4YX$ (i.e., $270_{-x}^\circ–360_x^\circ–270_{-y}^\circ–360_y^\circ–90_x^\circ$). It compensates for up to 20% error in pulse width and an offset error of $0.5B_1$ (Shaka and Freeman, 1983).

Composite pulses have also been used in overcoming problems due to sample overheating during broadband decoupling experiments. A widely used pulse sequence is Waltz-16 (Shaka et al., 1983), which may be repre-

24 PONDĚLÍ
MONDAY
MONTAG

83 - 282

Gabriel

25 ÚTERÝ
TUESDAY
DIENSTAG

84 - 281

Marián

26 STŘEDA
WEDNESDAY
MITTWOCH

85 - 280

Emanuel

27 ČTVRTEK
THURSDAY
DONNERSTAG

86 - 279

Dita

PÁTEK
FRIDAY
FREITAG
21
80 - 285
Radek

SOBOTA
SATURDAY
SAMSTAG
22
81 - 284
Leona

NEDĚLE
SUNDAY
SONNTAG
23
82 - 283
Ivona

BŘEZEN
MARCH/MÄRZ

P	Ú	S	Č	P	S	N
					1	2
3	4	5	6	7	8	9
10	11	12	13	14	15	16
17	18	19	20	21	22	23
24	25	26	27	28	29	30
31						

TÝDEN
WEEK WOCHE **12**

sented as $A \, \overline{A} \, \overline{A} \, A$, where $A = \overline{3} \, 4 \, \overline{2} \, 3 \, \overline{1} \, 2 \, \overline{4} \, 2 \, \overline{3}$. The subsequently modified GARP I has a larger bandwidth (Shaka *et al.*, 1985). MLEV-16, comprising the composite 180° pulse *ABBA BBAA BAAB AABB* (where $A = 90^\circ_{-y}-180^\circ_{x}-90^\circ_{-y}$ and $B = 90^\circ_{y}-180^\circ_{-x}-90^\circ_{x}$), was originally developed for more efficient heteronuclear decoupling experiments (Levitt *et al.*, 1982); later it found wide application in Hartmann Hahn echo spectroscopy (*HOHAHA*) (Bax and Davis, 1985; Davis and Bax, 1985). A modified version of MLEV-16 is MLEV-17, which contains another 180°_{x} uncompensated pulse at the end of the MLEV-16 sequence (Bax and Davis, 1985; Davis and Bax, 1985).

✦ *PROBLEM 1.36*

What are the advantages of using composite pulses instead of a unified pulse?

───✦

1.3.14 Phase Cycling

Most NMR experiments use combinations of two or more of the four pulse phases shown in Fig. 1.42. The phase of the pulse is represented by the subscript after the pulse angle. Thus, a 90°_{-y} pulse would be a pulse in which the pulse angle is 90° and $-y$ is its phase (i.e., it is applied in the direction $-y$ to $+y$), so it will cause the magnetization to rotate about the y-axis in the xz-plane.

Phase cycling is widely employed in NMR spectroscopy to suppress artifact signals due to field inhomogeneities or imperfect pulse settings. We have already discussed how phase cycling can be used to suppress image peaks in quadrature detection. As mentioned earlier, there are two separate sections of the computer memory, designated A and B, which digitize the signals received. Let us assume that the signals of 0° phase go into section A while signals with 90° phase go into section B. To eliminate imbalances between the two receiver channels, the signals are cycled so both channels contribute equally to the data. Because each receiver channel is receiving signals of only one phase type (0° or 90°), the switching of the receiver channels is carried out simultaneously with the changing of the phases (*phase cycling*), so absorptive and dispersive signals are recorded on successive scans.

A simple, two-step phase cycling scheme may therefore be employed: The signals of 0° phase and 90° phase pass through signal channels (1) and (2) to sections A and B, respectively, of the computer memory during

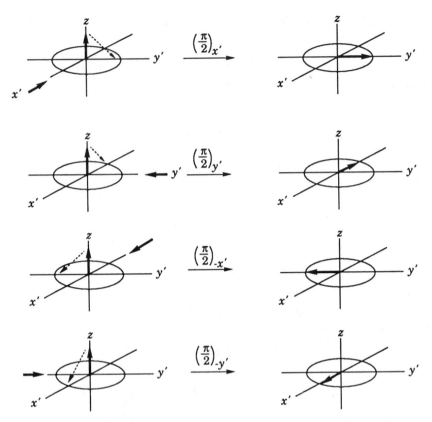

Figure 1.42   A 90° pulse brings the equilibrium magnetization to the $x'y'$-plane. Its orientation in the $x'y'$-plane depends on the direction of the pulse. Applying the pulse along one axis causes the magnetization to rotate in a plane defined by the *other* two axes.

the first cycle; in the second cycle the receiver channels are switched so that both channels contribute equally to the signals (Fig. 1.43).

To suppress other interference effects, the phase of the transmitter pulse is also shifted by 180° and the signals subtracted from sections A and B, leading to the CYCLOPS phase cycling scheme shown in Table 1.4, in which the two different receiver channels differing in phase by 90° are designated as 1 and 2 and the four different receiver pulses (90°_x, 90°_y, 90°_{-x}, and 90°_{-y}) are called x, y, $-x$, and $-y$, respectively.

✦ PROBLEM 1.37

What is phase cycling, and why is it used in NMR spectroscopy?

─── ✦

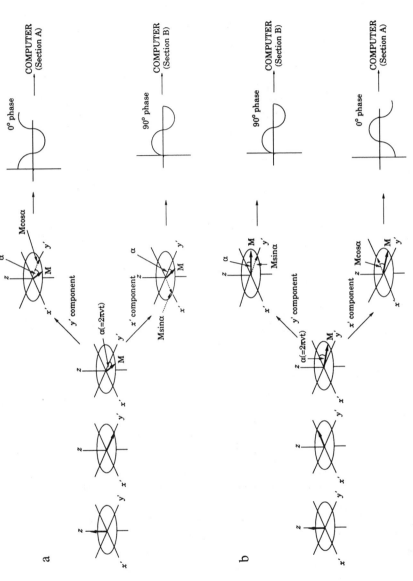

Figure 1.43 The first two steps of the CYCLOPS phase cycling scheme. Any imbalance in receiver channels is removed by switching them so they contribute equally to the regions A and B of the computer memory.

Table 1.4

CYCLOPS Phase Cycling*

Scan	Pulse	Sign A	Sign B	Receiver phase code	Receiver mode	Pulse phase	A	B
1	90°_x	cost ωt	sin ωt	R0	x	x	+1	+2
2	90°_y	sin ωt	−cos ωt	R1	y	y	−2	+1
3	90°_{-x}	−cos ωt	−sin ωt	R2	$-x$	$-x$	−1	−2
4	90°_{-y}	−sin ωt	cos ωt	R3	$-y$	$-y$	+2	−1

*The two memory blocks in the computer are designated A and B; the two receiver channels differing in phase by 90° are shown as 1 and 2.

✦ *PROBLEM 1.38*

What is the effect of phase cycling on the appearance of NMR signals?

—— ✦

1.3.15 Phase Cycling and Coherence Transfer Pathways

Coherence may be considered as a generalized description of transverse magnetization, and it corresponds to a transmission between two energy levels. The magnitude of coherence in each spin system is governed by the coherence level. The difference in magnetic quantum number m_z of two energy levels connected by the same coherence represents the *coherence order, p*. The path describing the progress of a coherence order in a pulse sequence is called a *coherence transfer pathway*. It is possible to change the coherence level by applying a pulse, while in the time interval delay between the pulses the coherence level does not change.

The transitions between energy levels in an *AX* spin system are shown in Fig. 1.44. There are four single-quantum transitions (these are the "normal" transitions A_1, A_2, X_1, and X_2 in which changes in quantum number of 1 occur), one double-quantum transition W_2 between the $\alpha\alpha$ and $\beta\beta$ states involving a change in quantum number of 2, and a zero-quantum transition W_0 between the $\alpha\beta$ and $\beta\alpha$ states in which no change in quantum number occurs. The double-quantum and zero-quantum transitions are not allowed as excitation processes under the quantum mechanical selection rules, but their involvement may be considered in relaxation processes.

Transverse magnetization represents a particular type of coherence involving a change in quantum number p of ± 1. Each coherence σ_{rs} is equal to the difference in magnetic quantum numbers of the nuclei r and s, i.e., the coherence order is $M_r - M_s$, and pulses cause transitions to occur

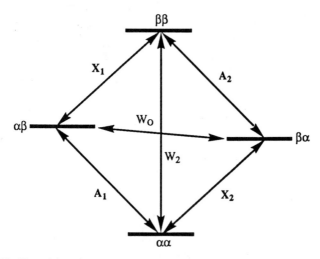

Figure 1.44 Transitions between various energy levels of an *AX* spin system. A_1 and A_2 represent the single-quantum relaxations of nucleus A, while X_1 and X_2 represent the single-quantum relaxations of nucleus X. W_2 and W_0 are double- and zero-quantum transitions, respectively.

between the different coherence orders. At thermal equilibrium, the change in quantum number p is zero, and it is at this point that the coherence transfer pathway is begun by the application of a pulse. For signals to be detected by the receiver, the coherence transfer must end with single quantum coherence (i.e., with $p = -1$).

When a 90° pulse is applied to the sample at equilibrium, the *longitudinal* (*z*) magnetization vanishes (i.e., the population difference between the α and β states decreases to zero) and transverse magnetization is created in the *x′y′*-plane. A phase coherence is now said to exist between the α and β states of the nucleus, since they precess coherently with the same phase (a property conveyed to them by the pulse). The coherence that now exists, termed *single-quantum coherence*, causes a precessing net magnetization of the nucleus, which can be detected in the form of a signal. It is possible to transfer this coherence by applying additional pulses to other states.

Suppose the first pulse resulted in the creation of a phase coherence across the A_1 transition between the αα and αβ states (Fig. 1.44). It is possible to transfer this phase information from the αβ state to the ββ state by applying a selective π pulse across the X_1 transition. The two successive pulses would therefore transfer the phase of the αα state to the ββ state, with the two states now becoming phase coherent with one another.

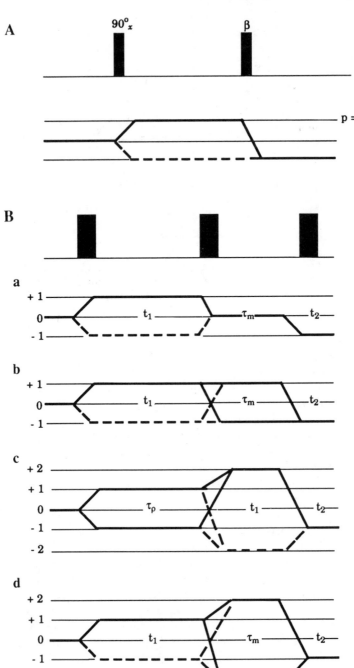

Since the $\alpha\alpha$ and $\beta\beta$ states are separated by a quantum number difference of 2, *double-quantum coherence* is said to have been created. Similarly, in more complex spin systems, coherence can be created between states differing by 2, 3, 4, or higher quantum numbers. The quantum mechanical selection rules stipulate that such *multiple-quantum coherences* do not give rise to detectable magnetization, detectable signals resulting only from single-quantum transitions. Double-, triple-, and other multiple-quantum transitions must therefore be converted into single-quantum coherence before detection. Many different coherence transfer pathways can be used, and by selecting appropriate phase cycling procedures, a particular coherence transfer pathway can be detected exclusively. Some typical coherence transfer pathways in common 2D NMR experiments are shown in Fig. 1.45.

When designing phase cycling procedure, it is necessary to tailor them to the coherence pathways by which the signals reach the receiver. The application of the first pulse causes the z-magnetization to be transferred to some coherence level. The coherence level may then be systematically altered by subsequent applications of one or more pulses, with the sequence of coherence levels by which the signals reach the receiver representing the *coherence pathway*. The phase cycling procedures adopted determine which signals are to be coadded and which signals canceled, according to the coherence pathway chosen. Coherence levels may be negative or positive. For instance, in a nucleus of spin $\frac{1}{2}$, applying a pulse can create a single-quantum coherence of level $+1$ and another coherence of level -1, corresponding to the clockwise and counterclockwise rotating magnetization vectors obtained from the x and y Bloch magnetizations. If the phase of a pulse is shifted, then the behavior of a coherence subjected to such a pulse will be governed by the change in coherence level. For instance, if a pulse is shifted by angle θ, and the change in coherence level is Δ_m, then the effect on the coherence is obtained by multiplying $e^{i\Delta m}$ by the coherence. It is possible thereby to derive a simple rule that would predict which signals

---

Figure 1.45 Coherence transfer pathways in 2D NMR experiments. (A) Pathways in homonuclear 2D correlation spectroscopy. The first 90° pulse excites single-quantum coherence of order $p = \pm 1$. The second mixing pulse of angle β converts the coherence into detectable magnetization ($p = -1$). (B:a) Coherence transfer pathways in NOESY/2D exchange spectroscopy; (B:b) relayed COSY; (B:c) double-quantum spectroscopy; (B:d) 2D COSY with double-quantum filter ($\tau_m = 0$). The pathways shown in (B:a,b, and d) involve a fixed mixing interval (τ_m). (Reprinted from G. Bodenhausen *et al., J. Magn. Resonance,* **58,** 370, copyright © 1984, Rights and Permission Department, Academic Press Inc., 6277 Sea Harbor Drive, Orlando, Florida 32887.)

following a certain coherence pathway will survive a particular phase cycling procedure and which signals would be eliminated (Bain, 1980).

Before the application of a pulse, only equilibrium magnetization exists, directed toward the z-axis corresponding to the zero coherence level for all coherence pathways. When the pulse is applied, two coherence levels, $+1$ and -1, are created during the evolution period that evolve into $Me^{i\omega t}$ and $Me^{-i\omega t}$, respectively, where ω is the absolute Larmor frequency of the nucleus and t is the evolution time. Before detection, the signal is mixed with the carrier frequency W_0 so it comes within the audiofrequency range to be picked up by the receiver. In quadrature detection, the carrier frequency may be considered to be circularly polarized, and mathematically the mixing of the carrier frequency is equivalent to multiplication of each of the two magnetization components by $e^{-i\omega_0 t}$. As a consequence, the magnetization component corresponding to the coherence level of $+1$ is detected, since it is the only signal $Me^{i(\omega-\omega_0)t}$, in the audiofrequency range. Apparently, all coherence pathways will therefore start at zero coherence levels and end at $+1$ coherence levels; since the quadrature receiver is sensitive only to the $+1$ polarization, only the single-quantum coherence is detected.

The signals reaching the receiver from each coherence pathway will, in general, depend on the *pulse phase factor* (and hence on the phase) as well as on the receiver phase factor (i.e., the data routing). To analyze the phase cycling sequence, it is therefore necessary to calculate the change in the coherence level and the phase factor for each pulse in the cycle, including the receiver phase. By multiplying all the phase factors together and coadding the results over the cycle, it is possible to check whether the result is zero or not. If the results add up to zero, then that pathway does not contribute to the signal; if the answer is not zero, then that pathway *may* lead to a signal.

✦ *PROBLEM 1.39*

Summarize the different events during a pulsed NMR experiment, in the form of a flow diagram.

─── ✦

SOLUTIONS TO PROBLEMS

✧ *1.1*

By absorption of continuous energy from the radiofrequency source, transitions of the nuclei occur to higher energy state, β; by relaxation processes, they revert to the lower energy state, α, and an equilibrium is established with a slight Boltzmann excess in the lower α state. This slight excess of population in the lower state will be disturbed on continuous irradiation, and a state of *saturation* will be reached, since the nuclei will not be able to dissipate the extra energy via relaxation processes.

✧ *1.2*

The population difference between two energy levels α and β is directly proportional to the energy difference ΔE,

$$\frac{N_\beta}{N_\alpha} = \exp\frac{-\Delta E}{kt}$$

The energy difference ΔE depends on the strength of the external magnetic field B_0 and the magnetogyric ratio γ. A stronger applied magnetic field B_0 will cause a correspondingly larger separation (ΔE) between the two energy levels and result in a larger difference in the populations of the two levels. Similarly, the energy difference ΔE also depends on the magnetogyric ratio of the nuclear species under observation:

$$\Delta E = \frac{h\gamma B_0}{2\pi.}$$

Since the magnetogyric ratio determines the sensitivity of a nuclear species to the external magnetic field, it has a profound effect on the strength of the NMR signals. For instance, ^1H nuclei will have a Larmor frequency of 300 MHz at an external magnetic field of 7.046 T (tesla), while ^{13}C nuclei will resonate with a Larmor frequency of only 75.435 MHz in the same magnetic field, since the γ for ^{13}C is about a quarter of the γ for ^1H. The signal strength is determined by γ^3, so a ^{13}C signal will be about 64 times weaker than an ^1H signal $[(1/4)^3 = 1/64]$. In practice, a ^{13}C signal is over 6000 times weaker, because it occurs in only 1.1% natural abundance.

✧ *1.3*

No. Since the magnetogyric ratio of ^{13}C is roughly one-fourth that of ^1H, the population difference between the two states (α and β) of ^{13}C nuclei will therefore be about 64 times $[(1/4)^3 = 1/64]$ less than that of the ^1H nuclei.

✧ *1.4*

The frequency with which a nucleus precesses around the applied magnetic field is called its Larmor frequency or Larmor precessional frequency. The Larmor frequency depends on the strength of the applied magnetic field B_0 and on the magnetogyric ratio of the nucleus. When the radiofrequency B_1 is applied in a direction perpendicular to the external magnetic field B_0, and when the value of B_1 matches exactly the Larmor frequency, absorption of energy occurs causing the nucleus to "flip" to a higher-energy orientation. This represents the process of excitation. Through a relaxation process, the nuclei can then relax back to the lower energy level. The energy released during the relaxation process is recorded as an FID (free induction decay), which is then converted into NMR signals through a mathematical operation called Fourier transformation.

✧ *1.5*

The magnetogyric ratio of a nuclear spin represents its response toward the external magnetic field. Nuclear species with larger magnetogyric ratios have larger differences in energy levels ΔE in comparison to nuclear species of smaller magnetogyric ratios, when placed under an external magnetic field of the same strength B_0:

$$\Delta E = \frac{h\gamma B_0}{2\pi}.$$

✧ *1.6*

A higher-MHz NMR spectrometer is always a better choice, since the sensitivity of the experiment is proportional to the frequency of measurement. Moreover, with highly concentrated solutions, the presence of some solid particles can cause an increase in T_1 (FID will be short) and line broadening of the NMR signals will result. Therefore, an optimum concentration (say, 25–50 millimolar solution) is recommended. Of course, ^{1}H-NMR spectra can be readily measured at much lower concentrations, though higher concentrations are necessary for recording ^{13}C-NMR spectra.

✧ *1.7*

The magnetic field strength B_0 has a direct relationship with the Larmor frequencies of nuclei: The stronger the magnetic field, the greater the difference, in hertz (*not* in ppm), between magnetically nonequivalent nuclei having differing chemical shifts. Moreover, the population excess of the lower energy level over the upper energy level increases with increasing magnetic field, B_0, leading to a corresponding increase in the sensitivity of the NMR experiment. The magnitude of the coupling constants, however, remains unaffected by the magnetic field strength.

❖ *1.8*

The probe selection should be based on the actual requirements of a laboratory.

(i) Small-diameter probes are generally suitable when the total availability of the sample is a limiting factor. For natural products or biochemical studies, a $^{13}C/^1H$ probe of 5-mm size is probably the best choice, since the major requirement here is to analyze small quantities of organic samples for the proton and carbon spectra. A 2.5-mm microprobe is also now available for use with special sample tubes having a diameter of 2.5 mm, and it is highly recommended for small samples. An inverse probe is highly desirable for $^1H/^{13}C$ inverse-shift correlation experiments (e.g., HMQC, HMBC). Its use requires certain hardware modifications, if these are not already incorporated.

(ii) For heteronuclear studies, where different types of nuclei are investigated routinely, the broad-band multinuclear probe is an excellent choice, since it can be tuned over a wide frequency range for various elements.

(iii) In organic-synthesis laboratories, where sample quantity is not a limiting factor, the larger probes provide a significant saving in time. A $^{13}C/^1H$ probe of 10–15-mm size is more appropriate.

(iv) A laboratory where nitrogen is the main nucleus to be analyzed, with occasional $^{13}C/^1H$ analysis, a broad-band probe specific for nitrogen should be acquired.

❖ *1.9*

An ideal probe should have the following properties.

(i) It should be made up of a material with a low magnetic susceptibility so it does not distort the static magnetic field B_0 and adversely affect the line shape and resolution of the NMR signals.

(ii) It should have a high field (B_1) homogeneity so it can receive and transmit radiofrequency signals uniformly from all parts of the sample solution.

(iii) It should fulfill the needs of a maximum number of users.

❖ *1.10*

The probe contains the electronics designed to detect the tiny NMR signal. The central component of the probe is a wire that receives the Rf pulse from the transmitter and dissipates it into the sample. It also receives the signal from the sample and transfers it to the receiver circuit. It is therefore necessary to tune the probe wire impedance to match it with those of the transmitter and the receiver. The optimum sensitivity of the NMR experiment will be realized only when this adjustment is made correctly. The probe tuning also minimizes the variable off-resonance effects, and it is therefore essential for the proper reproducibility of the pulse width.

✧ *1.11*

Poor shimming would lead to poor line shape and resolution, as illustrated here:

(a) With proper shimming

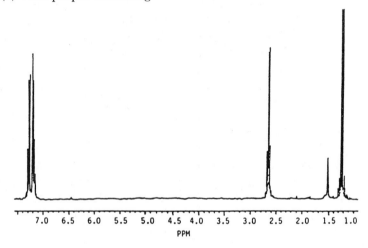

(b) With poor shimming.

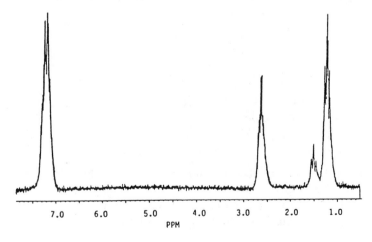

✧ *1.12*

The deuterium lock prevents changes in the static field (B_0) and radio-frequency (B_1) by maintaining a constant ratio between the two. It therefore ensures long-term stability of the magnetic field. If the ^{2}H lock is not applied, a drastic deterioration in the shape of the NMR lines is expected, due to magnetic and radiofrequency inhomogeneities.

(a) With deuterium lock

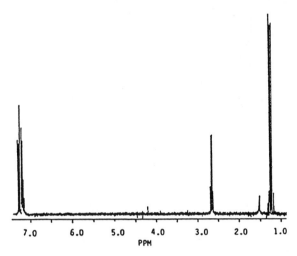

(b) Without deuterium lock.

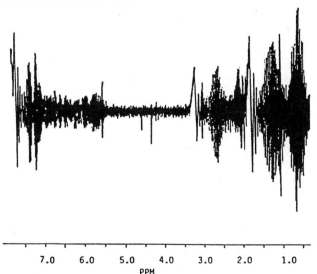

✧ *1.13*

No. The signals from the lock transmitter would obliterate the signals we wish to observe.

✧ *1.14*

A long pulse results in a narrow NMR signal.

✧ *1.15*

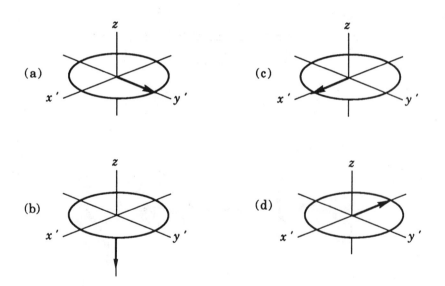

✧ *1.16*

"Hard" and "soft" pulses depend on the pulse width: The shorter the width of the pulse, the "harder," or the more powerful, it will be. Similarly, a long width pulse is "soft" and less powerful. For example, a pulse that requires 25-KHz B_0 for a duration of 20 μs will be "hard," while the same pulse of 20-ms pulse width will be considered "soft."

✧ *1.17*

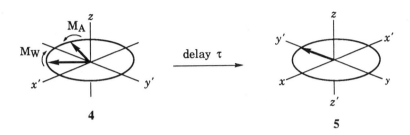

✧ *1.18*
We can use the angular frequency of TMS as the reference frequency of the rotating frame. Deducting this from the Larmor frequency of the signal will leave only the *differential* frequency (or, in other words, the chemical shift) associated with the magnetization vector of the signal.

✧ *1.19*
The amplitude represents a circular motion of the magnetization vector along the *xy*-plane, which slowly relaxes back toward the *z*-axis; the Larmor frequency of the nucleus is inherent to it and would therefore remain unchanged throughout the FID.

✧ *1.20*
The "*tail*" of the FID contains very little information; rather, most of the relevant information is in the initial large-volume portion of the FID envelope. The loss of the "*tail*" of the FID should not, therefore, significantly affect the quality of the data. Another manipulation to compensate for the lost information is exponential multiplication, in which the FID is multiplied by a negative exponential factor.

✧ *1.21*
In pulse NMR we measure in the time domain; i.e., the variation of signal amplitude with time (FID) is recorded. These time-domain data are then subjected to Fourier transformation to convert them into the frequency domain.

✧ *1.22*

If we do not subtract the reference frequency, we will have to process a very large amount of data. For example, on a 500-MHz NMR spectrometer, the frequencies to be processed would be 500,000,000–500,003,600 Hz for protons (since protons normally resonate within 0–12 ppm, i.e., 0–3600 Hz).

✧ *1.23*

It is better to maintain the same number of data points and reduce the spectral width as far as possible. In the alternative case, the improved digital resolution will be at the cost of sensitivity, since it will produce a corresponding increase in acquisition time, AT. Either a greater time period would then be required or a lesser number of scans would be accumulated in the same time period, with a corresponding deterioration in the signal-to-noise ratio.

✧ *1.24*

The folded peaks are easy to identify, since they show different phases than the "normal" signals. On shifting the spectral window to one side, all the "normal" signals will shift in the same direction, and by the same value, as the spectral window. In contrast, the folded signals will move either in the opposite direction or by a different value in the same direction, so their *relative* disposition to other signals in the spectrum will change.

✧ *1.25*

If the gain is too high, then the largest signal of solvent in the spectrum will cause ADC overflow, resulting in severe distortions.

✧ *1.26*

The image peaks resulting from quadrature detection can be easily distinguished from small genuine peaks since they show different phases and move with changes in the reference frequency.

✧ *1.27*

The ratio of the height of an NMR signal to the noise is called the S/N (signal-to-noise) ratio. The simplest way to improve S/N is through signal averaging. Recording a large number of spectra and combining them together will increase S/N. Since the noise contribution is random whereas the signals occur in exactly the same place each time, the signals will build up over a number of scans. This means that over n scans of the experiment, the signal will increase n times, while the noise amplitude will increase by $\sqrt{n}$, to give an overall increase in S/N of $\sqrt{n}$.

✧ *1.28*

Following are examples of spectra recorded with and without apodization.

(a) Spectrum recorded with apodization.

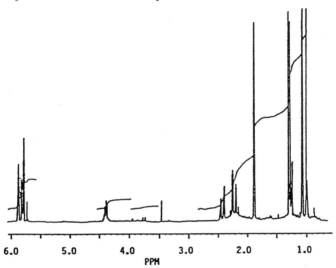

(b) Spectrum recorded without apodization.

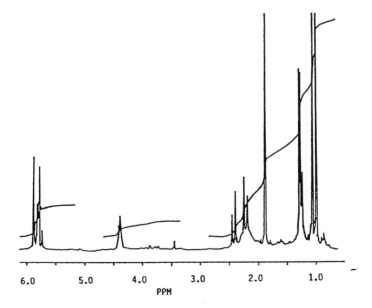

✧ *1.29*

Apodization (exponential multiplication) is used to improve the signal-to-noise ratio, and it does not affect the chemical shifts of the NMR signals.

✧ *1.30*

The simplest definition of *sensitivity* is the signal-to-noise ratio. One criterion for judging the sensitivity of an NMR spectrometer or an NMR experiment is to measure the height of a peak under standard conditions and to compare it with the noise level in the same spectrum. *Resolution* is the extent to which the line shape deviates from an ideal Lorentzian line. Resolution is generally determined by measuring the width of a signal at half-height, in hertz.

✧ *1.31*

(a) Spectrum before resolution enhancement.

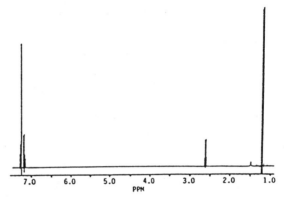

(b) Spectrum after resolution enhancement.

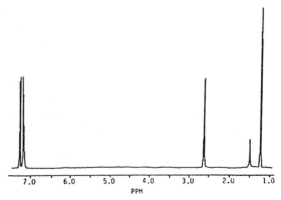

❖ *1.32*

Greatly enhanced sensitivity with very short measuring time is the major advantage of PFT (pulse Fourier transform) experiments. In the CW (continuous wave) experiment, the radiofrequency sweep excites nuclei of different Larmor frequencies, one by one. For example, 500 s may be required for excitation over a 1-KHz range, while in a PFT experiment a single pulse can simultaneously excite the nuclei over 1-KHz range in only 250 μs. The PFT experiment therefore requires much less time than the CW NMR experiment, due to the short time required for acquisition of FID signals. Short-lived unstable molecules can only be studied by PFT NMR.

❖ *1.33*

If the radiofrequency power is too high, the normal relaxation processes will not be able to compete with the sudden excitation (or perturbation), and thermal equilibrium will not be achieved. The population difference (Boltzmann distribution excess) between the energy levels (α and β) will decrease to zero, and the intensity of the absorption signal will also therefore become zero.

❖ *1.34*

Following are some common methods for sensitivity enhancement:

(i) *Sample concentration:* When the sample concentration in a given volume of solvent increases, the number of NMR active nuclei also increases.

(ii) *Temperature:* Slight lowering of the sample temperature increases the energy difference between energy levels and therefore increases the population difference based on the Boltzmann distribution equation.

(iii) *Magnetic field strength:* The population difference (Boltzmann excess) increases with increasing magnetic field.

(iv) *Rf power of B_1:* The signal strength increases with the increase in Rf power. Care must be taken, however, to avoid population saturation.

(v) *Number of scans:* In the pulse Fourier Transform NMR experiment, the signal-to-noise ratio increases as the square root of the number of accumulated scans.

(vi) Polarization transfer and nOe effects also contribute to sensitivity enhancement.

(vii) *Relaxation rate:* Slow-relaxing nuclei cannot attain thermal equilibrium within the pulse interval, which results in lower sensitivity. One way to cope with this problem is to introduce a sufficiently large delay time between the pulses. Addition of small quantities of relaxation reagents can also result in a better signal-to-noise ratio, but this may also produce a change in chemical shifts due to complexation.

❖ *1.35*

The pulse duration controls the extent to which the magnetization vectors are bent. A misalignment of the pulse would lead to various artifact
signals. Including 180° spin-echo pulses can, to some extent, compensate
for setting pulses incorrectly. But in certain experiments (e.g., inverse
NMR experiments), it is extremely important for the success of the experiment that the proper pulse angles be determined and employed.
(a) ¹H-NMR spectrum with proper alignment of pulse width.

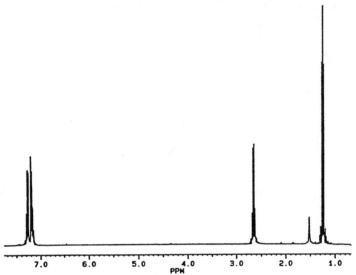

(b) ¹H-NMR spectrum with poorly set pulse width.

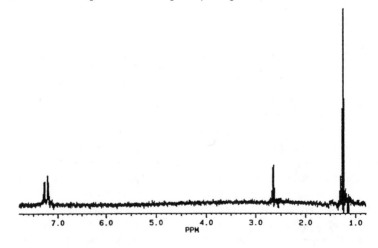

✧ *1.36*

Composite pulses reduce the error introduced due to the delay between the start of the pulse and when it reaches full power. They have also been used to overcome problems of sample overheating during broadband decoupling and in experiments in which pulses have to be applied for long durations.

✧ *1.37*

Phase cycling is widely employed in multipulse NMR experiments. It is also required in quadrature detection. Phase cycling is used to prevent the introduction of constant voltage generated by the electronics into the signal of the sample, to suppress artifact peaks, to correct pulse imperfections, and to select particular responses in 2D or multiple-quantum spectra.

✧ *1.38*

1. ^{1}H-NMR spectrum with phase cycling.

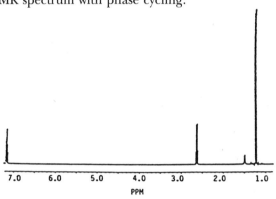

2. ^{1}H-NMR spectrum without phase cycling.

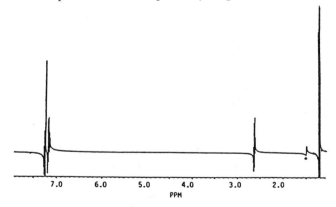

✧ *1.39*

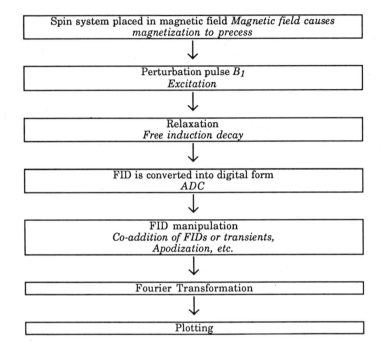

Spin system placed in magnetic field *Magnetic field causes magnetization to precess*

Perturbation pulse B_1
Excitation

Relaxation
Free induction decay

FID is converted into digital form
ADC

FID manipulation
Co-addition of FIDs or transients, Apodization, etc.

Fourier Transformation

Plotting

REFERENCES

Bain, A. D. (1980). *J. Magn. Reson.* **37**(2), 209–216.

Bax, A. (1983). *J. Magn. Reson.* **52**, 76.

Bax, A., and Davis, D. G. (1985). *J. Magn. Reson.* 63, 207; *idem. ibid.* **65**, 355.

Bodenhausen *et al.* (1984). *J. Magn. Reson.* **58**, 370.

Croasmun, W. R., and Carlson, M. K. *Two-dimensional NMR spectroscopy applications for chemists and biochemists.* VCH, New York, p. 13.

Davis, D. G., and Bax, A. (1985). *J. Amer. Chem. Soc.* **107**, 7197.

Derome, E. (1987). *Modern NMR techniques for chemistry.* Pergamon, Oxford, pp. 78–79.

Ernst, R. R. (1966). *Adv. Magn. Reson.* **2**, 1–135.

Harris, R. K. (1989). *Nuclear magnetic resonance spectroscopy.* Longman Scientific & Technical, Essex, England, p. 5.

Homans, S. W. (1990). *A dictionary of concepts in NMR.* Oxford University Press, Oxford, pp. 127–129.

Jeener, J. (1971). Paper presented at Amper Summer School, Basko Polje, Yugoslavia.

Kessler, H., Mronga, S., and Gemmecker, G. (1991). *Magn. Reson. Chem.* **29**, 527–557.

Lawn, D. B., and Jones, A. J. (1982). *Aust. J. Chem.* **35**, 1717.

Levitt, M., Freeman, R., and Frenkiel, T. A. (1982). *J. Magn. Reson.* **47**, 313.

Lindon, J. C., and Ferrige, A. G. (1980). *Prog. NMR Spectroscopy* **14**, 27–66.

Marco, A. De, and Wuethrich, K. (1976). *J. Magn. Reson.* **24**, 201.

Nielsen, N. C., Bildsöe, H., Jakobsen, H. J., and Sörensen, O. W. (1986). *J. Magn. Reson.* **66,** 456.

Shaka, A. J., and Freeman, R. (1983). *J. Magn. Reson.* **55**(3), 487.

Shaka, A. J., Keeler, J., and Freeman, R. (1983). *J. Magn. Reson.* **53**(2), 313.

Shaka, A. J., Barker, P. B., and Freeman, R. (1985). *J. Magn. Reson.* **64**(3), 547.

Thomas, D. M., Bendall, M. R., Pegg, D. T., Doddrell, D. M., and Field, J. (1981). *J. Magn. Reson.* **42,** 298.

Wagner, G., Wüthrich, K., and Tschesche, H. (1978). *Eur. J. Biochem.* **86,** 67.

Wesener, J. R., and Gunther, H. (1985). *J. Magn. Reson.* **62,** 158.

CHAPTER 2

Spin-Echo and Polarization Transfer

2.1 SPIN-ECHO FORMATION IN HOMONUCLEAR AND HETERONUCLEAR SYSTEMS

Artifact signals generated due to field inhomogeneities or errors in setting pulse widths may be suppressed by *spin-echo* production. Let us consider a heteronuclear AX spin system, in which nucleus A is a proton and nucleus X is a carbon. If the behavior of nucleus X is examined, then its magnetization will be affected by nucleus A in two different ways, depending on whether nucleus A is in the lower energy (α) state (i.e., oriented with the applied magnetic field) or in the higher energy (β) state (oriented against the applied field). The magnetization of nucleus X can be considered to be made up of two components, $\mathbf{M}_{XA\alpha}$ and $\mathbf{M}_{XA\beta}$ (where $\mathbf{M}_{XA\alpha}$ is the magnetization vector of nucleus X when coupled to the lower energy (α) state of the neighboring nucleus A, and $\mathbf{M}_{XA\beta}$ is the magnetization vector of nucleus X when coupled to the higher energy (β) state of the nucleus A). For convenience, we will designate $\mathbf{M}_{XA\alpha}$ as $\mathbf{M}_1$ and $\mathbf{M}_{XA\beta}$ as $\mathbf{M}_2$ in the following discussions.

2.1.1 Spin-Echo Production

The basic pulse sequence for the production of a spin-echo is illustrated in Fig. 2.1. The behavior of ^{13}C vectors in a heteronuclear CH sys-

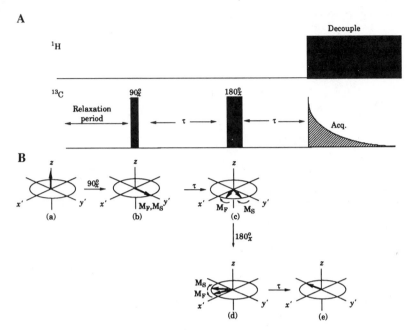

Figure 2.1 (A) Spin-echo pulse sequence. (B) Effect of spin-echo pulse sequence on ^{13}C magnetization vectors; (e) represents the coalescing of the two component vectors by the 180°_x refocusing pulse.

tem is shown. The first 90°_x pulse on the ^{13}C nuclei serves to bend their z-magnetization to the y'-axis. During the subsequent delay period τ, suppose that two component vectors $\mathbf{M}_F$ and $\mathbf{M}_S$ are generated, with $\mathbf{M}_F$ precessing a little faster than $\mathbf{M}_S$. The 180°_x pulse on the ^{13}C-nuclei results in the $\mathbf{M}_F$ and $\mathbf{M}_S$ vectors flipping across the x'-axis and adopting *mirror image* positions in the $x'y'$-plane (Fig. 2.1c,d). During the subsequent delay period τ, which is kept identical to the first τ delay period, the two magnetization vectors travel the remaining distance and become focused on the y'-axis, producing a *spin-echo*. The advantage of the 180° *refocusing pulse* is that it compensates for any errors due to field inhomogeneities. For instance, if we assume that the vector $\mathbf{M}_F$ was closer to the x'-axis and the vector $\mathbf{M}_S$ farther from it before the application of the 180°_x pulse, then the 180°_x pulse would cause a *mirror image jump* across the x'-axis, and vector $\mathbf{M}_F$ would adopt a position a little farther from the $-y'$-axis than vector $\mathbf{M}_S$. Similarly, $\mathbf{M}_S$ would come to lie a little nearer the $-y'$-axis due to the 180°_x pulse. The faster vector $\mathbf{M}_F$ has now to cover a greater distance than the slower vector $\mathbf{M}_S$ in the second τ period, so the original errors in the positioning of the vectors $\mathbf{M}_F$ and $\mathbf{M}_S$ are compensated by the "equal and opposite error" caused by the refocusing 180°_x pulse. As a result, refocusing occurs precisely along the $-y'$-axis at the end of the second τ period.

The 180°_x pulse applied to nucleus X under observation causes a change of its *position* but does not affect the *direction* of rotation of its magnetization vectors. This is because the direction of rotation of the vectors of nucleus X depends on the spin states of nucleus A to which nucleus X is coupled, and it is only when the spin states of nucleus A are interchanged (for instance, by irradiation of nucleus A with a 180° pulse) that the directions of rotation of the vectors of nucleus X undergo a reversal. Since refocusing at the end of the second delay period occurs along the $-y'$-axis, a *negative* signal is produced, *irrespective of the precession frequency of the nucleus*. The pulse sequence used to remove field inhomogeneities is:

$$90^\circ_x-\tau-180^\circ_x-\tau-\text{echo}$$
$$90^\circ_x-\tau-180^\circ_{-x}-\tau-\text{echo}.$$

The phase alteration of the 180° pulse and coaddition of the resulting FIDs serves to cancel the pulse imperfections, thereby producing accurate spin-echoes (Fig. 2.2).

The spin-echo experiment therefore leads to the refocusing not only of the individual nuclear resonances but also of the field inhomogeneity components lying in front or behind those resonances, a maximum negative amplitude being observed at time 2τ after the initial 90° pulse. The frequency of rotation of each signal in the rotating frame will depend on its chemical shift; and after the vector has been flipped by the 180° pulse, it

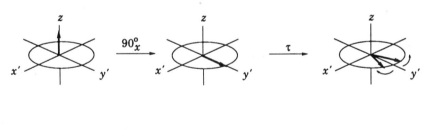

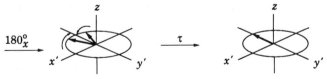

Figure 2.2 Effect of 180°_x pulse on phase imperfections resulting from magnetic field inhomogeneities. Spin-echo generated by 180°_x refocusing pulse removes the effects of magnetic field inhomogeneities.

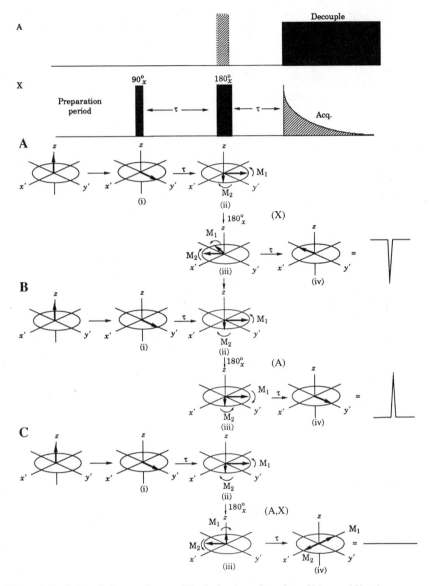

Figure 2.3 Spin-echo experiment. The behavior of nucleus X in an AX spin system is shown. (A) Application of the second 180°_x pulse to nucleus X in the AX heteronuclear system results in a spin-flip of the two X vectors across the x'-axis. But the direction of rotation of the two X vectors does not change, and the two vectors therefore refocus along the $-y'$ axis. The spin-echo at the end of the τ period along the $-y'$ axis results in a negative signal. (B) When the 180°_x pulse is applied to nucleus A in the AX heteronuclear system, the spin-flip of the X vectors

will continue to rotate in the same direction and at the same frequency till it decays due to transverse relaxation effects. If we acquire data at the point at which each echo is produced, Fourier transformation would lead to a spectrum. By varying the delay time τ, we can collect a series of spectra from which the transverse relaxation rate R_2 constants can be measured by plotting the intensity of each signal as a function of time, provided that each signal is a singlet. In the case of multiplets, a more complex pulse sequence is used to cancel the multiplet effects and then to measure the transverse relaxation delay.

✦ *PROBLEM 2.1*

The spin-echo is used to suppress the production of spurious signals due to field inhomogeneities or to eliminate errors in the setting of pulse widths. It is also possible to use the spin-echo to follow the decay of transverse magnetization and to determine the transverse relaxation time (T_2). How might we do this in practice?

── ✦

2.1.2 Spin-Echo Production in a Heteronuclear AX Spin System

In a heteronuclear AX spin system in which nucleus X is being observed, three different cases can be considered, depending on whether the 180° pulse is applied to nucleus A only, nucleus X only, or simultaneously to both nuclei A and X (Fig. 2.3).

In the first case, the 180°_x pulse is applied to nucleus X, causing the two vectors of nucleus X to flip across the x'-axis. But no change occurs in the *direction* of their rotation during the second delay period, so at the end of this period the vectors are refocused along the $-y'$-axis, producing a *negative* signal.

In the second case, the 180°_x pulse is applied only to nucleus A, causing an exchange of spin labels of the A spin states to occur, so the direction of rotation of the X magnetization vectors is reversed during the second

──

across the x'-axis does not occur; only their direction of rotation changes due to relabeling of the A spin states. Spin-echo formation therefore takes place along the y'-axis, producing a positive signal. (C) Simultaneous application of the 180°_x pulse to both nuclei A and X will not only cause a spin-flip but also exchange the direction of their rotation so that at the end of the 2τ delay period, they lie along the x'-axis.

delay period, resulting in a refocusing of the X vectors to occur along the y'-axis at the end of the second delay period, thereby producing a positive signal.

In the third case, the 180°_x pulses are applied *simultaneously* to both nuclei A and X, so the direction of rotation of the X vectors changes, and a flip-over to mirror image positions across the x'-axis also occurs. This results in the two vectors aligning themselves along the x'-axis at the end of the second delay period. Since the detector is assumed to be aligned to detect signal components along the y'-axis, no signal is produced. This third case is also applicable to *homonuclear spin systems,* in which the 180°_x pulse will be nonselective, affecting both the A and X nuclei.

Hence it is clear that if the two delay periods before and after the 180°_x pulses are kept identical, then refocusing will occur only when a *selective* 180°_x pulse is applied. This can happen only in a heteronuclear spin system, since a 180°_x pulse applied at the Larmor frequency of protons, for instance, will not cause a spin flip of the ^{13}C magnetization vectors.

✦ PROBLEM 2.2

What is the effect of applying a spin-echo on chemical shifts?

─── ✦

✦ PROBLEM 2.3

Describe the effects of a 180°_x refocusing pulse in each of the following situations:

(i) Single line dephased due to field inhomogeneity.

(ii) A doublet resulting from heteronuclear coupling, assuming that vectors of nucleus A are being observed and that the 180°_x pulse is applied selectively on nucleus A.

─── ✦

✦ PROBLEM 2.4

Will the vectors of a doublet be refocused at time 2τ in a spin-echo experiment?

─── ✦

✦ PROBLEM 2.5

M_α and M_β are magnetization components of the X vector of a heteronuclear AX spin system, shown here at a certain delay after the application of a 90°_x pulse. Draw the vector positions and their direction of rotation after each of the following radiofrequency pulses:

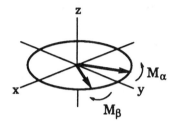

(i) 180°_x pulse applied to nucleus X
(ii) 180°_x pulse applied to nucleus A
(iii) 180°_x pulses applied to nuclei A and X

✦

2.1.3 Attached Proton Test (APT), Gated Spin-Echo (GASPE), and Spin-Echo Fourier Transform (SEFT)

When a 90°_x pulse is applied to the ^{13}C nuclei, then during the subsequent delay interval τ after the pulse, the magnetization vectors of the ^{13}C nuclei of CH_3, CH_2, CH, and quaternary carbons do not rotate synchronously with one another but rotate with characteristically different angular velocities. If the value of the delay time is kept at $1/J$ seconds, then CH_2 and quaternary carbons give signals with positive amplitudes while CH_3 and CH carbons give signals with negative amplitudes. This provides a method, variously known as APT (Attached Proton Test), GASPE (Gated Spin-Echo), or SEFT (Spin-Echo Fourier Transform) for distinguishing between them.

The pulse sequence used in the APT experiment is shown in Fig. 2.4; the movement of the magnetization vectors at various delay intervals τ after the 90°_x ^{13}C pulse is shown in Fig. 2.5. The initial 90°_x ^{13}C pulse bends the magnetization vectors of the CH_3, CH_2, CH, and quaternary carbons so they come to lie together along the y'-axis in the $x'y'$-plane. During the subsequent delay interval, these vectors separate from each other and rotate in the $x'y'$-plane at their characteristic Larmor frequencies. The CH_3, CH_2, and CH carbon magnetization vectors also become split into 4-, 3-, or 2-vector components, respectively, due to coupling with their attached protons. The detector, which by convention is regarded as located along the y'-axis, detects only the sum of these individual vector components, with the signal amplitude increasing to a maximum

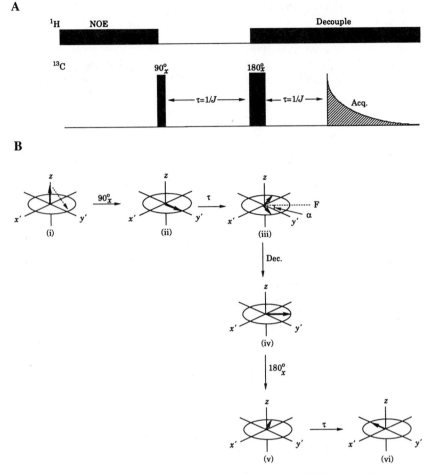

Figure 2.4 (A) Pulse sequence for the gated spin-echo (GASPE) or attached proton test (APT) experiment. (B) Effect of the pulse sequence on the ^{13}C magnetization vectors of a CH group.

value when the vector sum approaches the y'-axis and decreasing as they move away from it toward the $-y'$-axis. Since this cosinusoidal modulation in signal strength with delay time τ depends on the positions (or the respective angular velocities) of the split magnetization components, it is termed *J-modulation*.

The decoupler is off during the first $1/J$ delay period, so it is during this period that the effect of *J*-splitting comes into play and the coupling information is provided. This information is contained in the phase and magnitude of the signal, which in turn are dependent on the positions,

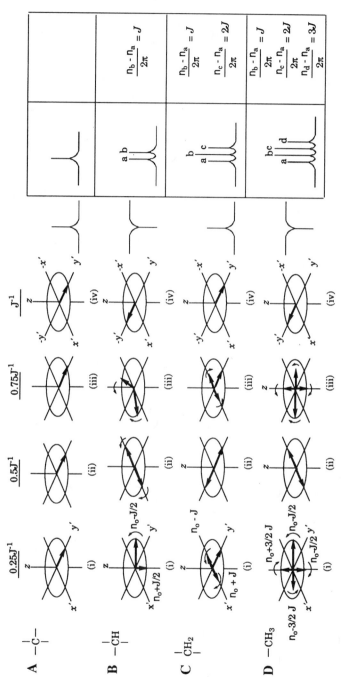

Figure 2.5 Evolution of magnetization vectors of C, CH, CH$_2$, and CH$_3$ carbons in the $x'y'$-plane after the 90$^\circ_x$ pulse. The magnetization vectors of C, CH, CH$_2$, and CH$_3$ carbons evolve to different extents during the delays set at $\tau = 0.25J^{-1}$, $0.5J^{-1}$, $0.75J^{-1}$, and J^{-1}. At J^{-1}, the quaternary and CH$_2$ carbons appear along the $+y'$-axis, giving positive signals, while the CH$_3$ and CH carbons appear along the $-y'$-axis, giving negative signals. This forms the basis of GASPE (or APT) experiments.

angular velocities, and number of individual split magnetization components. The decoupler is switched on at the beginning of the second delay interval that causes the split magnetization vectors to collapse into resultant singlets. The 180°_x pulse causes these vectors to adopt mirror image positions across the x'-axis (Fig. 2.4). At the end of the second delay interval, the CH_2 and quaternary carbons appear with positive phases, while the CH_3 and CH carbons appear with negative phases, thus allowing the CH_3 and CH carbons to be distinguished from the CH_2 and quaternary carbons (Fig. 2.5).

The dependance of signal phases and intensities on delay time τ in the APT experiment is shown in Fig. 2.6. If J_{CH} is assumed to be 125 Hz and the delay time is accordingly set at $1/J = \frac{1}{125} = 8$ ms, then, as is apparent from Fig. 2.6, the quaternary and CH_2 carbons appear with maximum positive amplitudes while the CH_3 and CH carbons afford maximum negative amplitudes [see vertical line at (a)]. If the delay time is adjusted to 6 ms [see vertical line at (b)], then quaternary carbons still appear with similar positive amplitudes, the CH_2 carbons have weaker positive amplitudes, and the CH_3 and CH carbons have weak negative amplitudes. The

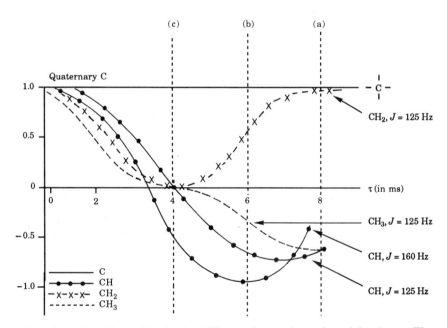

Figure 2.6 Signal intensities in the APT experiment depend on delay time τ. The signal intensities of CH_3, CH_2, and CH carbons are shown as various curves as a function of τ.

signal magnitudes at the end of the 2τ period are given by the following equations:

For quaternary carbons (C): $\quad M_{2\tau} = M_o e^{-2\tau/T_2}$

For methine carbons (CH): $\quad M_{2\tau} = M_o e^{-2\tau/T_2} \cos \pi\tau J$

For methylene carbons (CH$_2$): $\quad M_{2\tau} = \frac{1}{2}\tau\, M_o e^{-2\tau/T_2}(1 + \cos 2\pi\tau J)$

For methyl carbons (CH$_3$): $\quad M_{2\tau} = \frac{3}{4}M_o e^{-2\tau/T_2}(\cos \pi\tau J + \frac{1}{3}\cos 3\pi\tau J)$.

One disadvantage of the APT experiment is that it does not readily allow us to distinguish between carbon signals with the same phases, i.e., between CH$_3$ and CH carbons or between CH$_2$ and quaternary carbons, although the chemical shifts may provide some discriminatory information. The signal strengths also provide some useful information, since CH$_3$ carbons tend to be more intense than CH carbons, and the CH$_2$ carbons are usually more intense than quaternary carbons due to the greater nuclear Overhauser enhancements on account of the attached protons.

In interpreting APT spectra, we have to be careful about erroneous phasing of the signals that may occur if the actual J_{CH} values differ significantly from those set for the experiment. For instance, J_{CH} may be about 125 Hz for alkyl groups, so the optimum $1/J$ will be 8 ms, but J_{CH} in alkenes may be about 160 Hz, so $1/J$ for such functional groups should be set at 6.25 ms (Fig. 2.6). Such alkyl CH carbons will therefore afford the most intense negative signals at $2\tau = 16$ ms, while in alkene CH carbons the most intense negative signals will appear at $2\tau = 12.5$ ms. If 2τ is set at 16 ms, then the alkene CH carbons could appear with weak negative intensities; if the J values differ significantly, we may even obtain signals with the "wrong" phasing, leading to misinterpretation of the results.

With τ set at $\frac{1}{2}J$, the quaternary carbons generally appear with greater intensity than the other carbons, which will be of near-zero intensities, thereby allowing them to be distinguished, particularly from the CH$_2$ carbons, as compared to the "normal" APT spectrum, in which both CH$_2$ and quaternary carbons appear with positive amplitudes. A *difference* APT spectrum, in which an APT spectrum recorded with τ set at $\frac{2}{5}J$ is subtracted from another APT spectrum recorded with τ set at $\frac{3}{5}J$, can provide useful information. The methyl carbons will then appear with reduced intensities in the difference spectrum as compared to the methine carbons, allowing us to distinguish between them.

The errors in the APT spectra may arise due to (a) variation in the T_2 relaxation times of different carbons, (b) wide variation in J_{CH} values, and (c) modulations caused by long range $^{13}C-^{1}H$ couplings. A procedure known as ESCORT (Error Self Compensation Research by Tao Scrambling) has been developed to yield cleaner subspectra with reduced J "cross-talk" (Madsen *et al.*, 1986). This involves replacing the normal APT spectral editing procedure, which is based on using an average $^1J_{CH}$ coupling constant, by a linear combination of experiments that is partly independent of the error in setting J values.

The variation in the T_2 relaxation times of the individual carbons can be minimized by keeping the time between excitation and detection constant (Radeglia and Porzel, 1984).

✦ PROBLEM 2.6

How can you distinguish between C, CH, CH$_2$, and CH$_3$ carbons in the APT experiment?

─── ✦

✦ PROBLEM 2.7

The GASPE (or APT) spectrum of ethyl acrylate is shown here. Assign the signals to the various carbons.

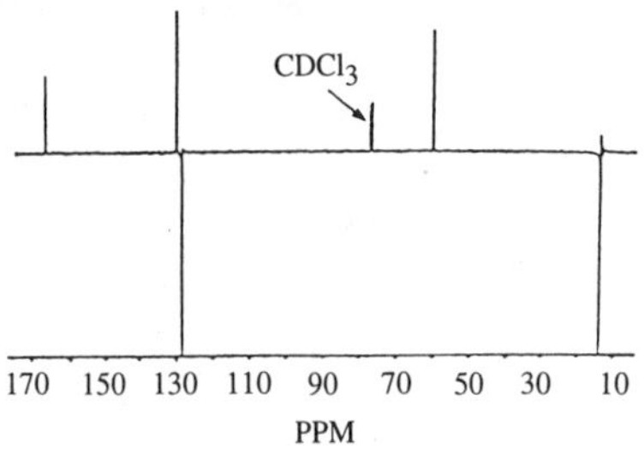

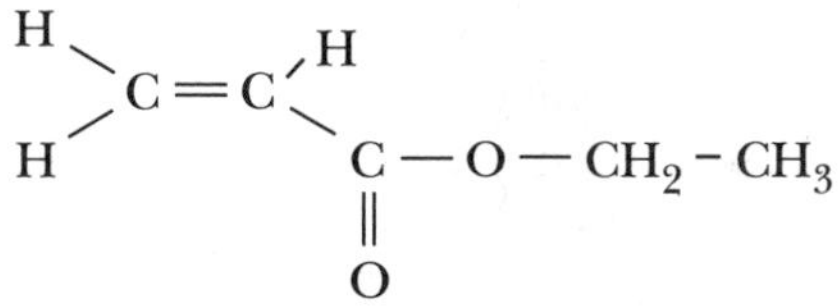

Ethyl acrylate

─── ✦

2.1.4 Nonobservable Magetization by 90° Pulses: Mixing Spin States

In the preceding discussion we have been concerned with the generation and detection of single-quantum coherence (i.e., $\Delta M = +1$) magnetization. This magnetization corresponds to the creation of a vector in the $x'y'$-plane, which is detected. There are many methods to generate magnetization with transitions other than $\Delta M = +1$. Such *multiple-* or *zero-quantum coherences* are "invisible," since they do not induce any fluctuations in a receiver directly, and they do not follow the selection rules. However, through the application of a suitable pulse or a set of pulses, these coherences can be converted into single-quantum transitions before being detected. Such magnetizations can be generated simply by using two 90° pulses separated by a time period. The two 90° pulses together serve to prepare the spin system; and after this preparation period, the coherences are allowed to evolve in the time period τ, before some of them are converted into a single-quantum coherence by a third 90° pulse before detection. The second 90° pulse thus serves to mix the orders of coherences, converting some of the single-quantum coherence generated by the first 90° pulse into other orders of coherence, whereas the third 90° pulse does the reverse—converts nonsingle-quantum coherence into single-quantum coherence.

✦ *PROBLEM 2.8*

What is the difference between single-quantum coherence and zero- or multiple-quantum coherences?

── ✦

✦ *PROBLEM 2.9*

Can the simple vector presentation be used to display the effects of a 90° pulse on zero- or multiple-quantum magnetizations?

── ✦

2.2 CROSS-POLARIZATION

The strength of an NMR signal depends on the gyromagnetic ratio γ of a nucleus, which in turn determines the population difference ΔE between the upper and lower energy states. In a given applied magnetic field, B_0: $\Delta E = \gamma h\, B_0/2\pi$ where h is Planck's constant. Certain nuclei, such as ^{13}C, occur in low natural abundance and have unfavorable magnetogyric ratios, leading to poor sensitivity. Thus the magnetogyric ratio γ of ^{13}C is about a quarter that of ^{1}H, and since the signal obtainable from a nucleus is proportional to γ^3 of that nucleus, the lower γ of ^{13}C results in weakening of its signal intensity by a factor of $(\frac{1}{4})^3 \cong 64$. Moreover, since ^{13}C has a natural abundance of 1.1%, this leads to a further hundredfold reduction in the sensitivity of the signal, so that ^{13}C signals are some 6000-fold weaker than ^{1}H signals. Some increase in sensitivity is achievable through nuclear Overhauser enhancement. Alternatively, we can transfer magnetization from ^{1}H to ^{13}C nuclei to intensify the ^{13}C signals, a process known as *cross-polarization*. Such population-transfer or polarization-transfer procedures are now used extensively in many NMR experiments, such as INEPT and DEPT.

Some of the most important 2D experiments involve chemical shift correlations between either the same type of nuclei (e.g., $^1H/^1H$ homonuclear shift correlation) or between nuclei of different types (e.g., $^1H/^{13}C$ heteronuclear shift correlation). Such experiments depend on the modulation of the nucleus under observation by the chemical shift frequency of other nuclei. Thus, if ^1H nuclei are being observed and they are being modulated by the chemical shifts of other ^1H nuclei in the molecule, then homonuclear shift correlation spectra are obtained. In contrast, if ^{13}C nuclei are being modulated by ^1H chemical shift frequencies, then heteronuclear shift correlation spectra result. One way to accomplish such modulation is by transfer of polarization from one nucleus to the other nucleus. Thus the magnitude and sign of the polarization of one nucleus are modulated at its chemical shift frequency, and its polarization transferred to another nucleus, before being recorded in the form of a 2D spectrum. Such *polarization transfer* between nuclei can be accomplished by the simultaneous application of a pair of 90° pulses on the two nuclei involved in the exchange of polarization, provided the spin components are properly aligned.

Let us consider a simple heteronuclear $^1H/^{13}C$ spin system in which the ^1H polarization is transferred to ^{13}C nuclei. The first 90°_x ^1H pulse bends the z-magnetization to the $+y'$-axis. During the subsequent time interval t_1 the ^1H magnetization is split into counterrotating components that, at the end of $t_1 = \frac{1}{2}J$ period, are aligned along the x'-axis of the rotating frame. The Rf field is then phase shifted in order to allow a 90° pulse to be applied along the y'-axis. This pulse causes the two magnetization component vectors lying along the x'-axis to rotate in the $x'z$-plane so they come to lie along the z-axis. The two vectors α_H and β_H are now directed in opposite directions along the z-axis (see Fig. 2.7, IIe), whereas in the equilibrium state (see Fig. 2.7, IIa) they were both pointing toward the z-axis. What we have therefore succeeded in doing is to invert the population of *one* of the two ^1H spin states. Since the ^{13}C nuclei, which are coupled to ^1H nuclei, have their spin states determined by the spin labels of the ^1H spins, the inversion (or relabeling) of one of the ^1H spins causes a corresponding relabeling of one of the ^{13}C spins (shown as $^\alpha C$ and $^\beta C$ in Fig. 2.7, IIf, with $^\beta C$ having undergone the inversion in the illustration). A 90°_x pulse (along the x'-axis) rotates these two ^{13}C vectors so they come to lie along the y'-axis (IIg). This antiphase magnetization can then be detected immediately, or it can be allowed to evolve for a certain time before detection (IIi). Or if decoupling is applied to the ^1H nuclei during detection, it can be made to focus into a single component before detection.

Clearly the extent to which the ^1H nuclei line up along the x'-axis (IId) will depend on the duration of t_1, which in turn will determine the extent of the antiphase z-magnetization created by the second 90° ^1H pulse. The

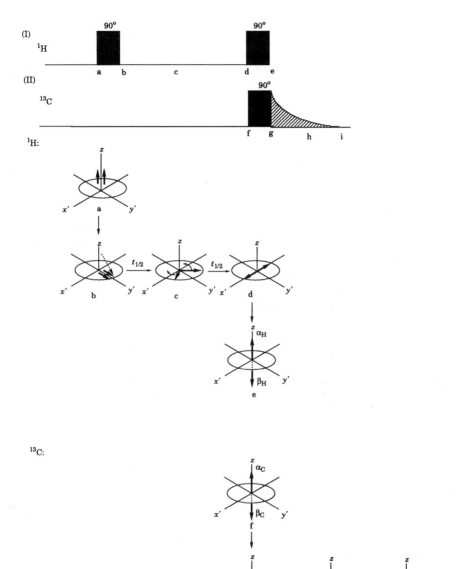

Figure 2.7 Pulse sequence for a heteronuclear AX spin system representing polarization transfer from ^{1}H to ^{13}C nuclei.

extent of divergence of the two magnetization components of A between the two 90° ^{1}H pulses depends on J_{HC}, while the positions of the *resultant* of these two vectors will depend on the extent of precession due to the chemical shift of ^{1}H. Hence, the positioning of the component vectors (and consequently the extent of antiphase z-magnetization created by the second 90° pulse) will depend both on the magnitude of the coupling constant J_{CH} as well as on the chemical shift δ_H. The antiphase polarization of A along the x'-axis (in IId) before the application of the second 90° pulse is modulated by both the chemical shift and the coupling constant frequencies, and it is this information that is transferred to the ^{13}C nuclei by polarization transfer before detection.

The experiment just discussed represents a heteronuclear spin system. In a homonuclear case, the separate 90° ^{13}C pulse necessary in the heteronuclear system is not required, since the second 90° ^{1}H pulse affects the coupled partner ^{1}H nucleus as well. The nucleus detected therefore has its two transitions antiphase with respect to each other, corresponding to the states represented in Fig. 2.7, IIg, IIh, etc. at detection.

The ^{1}H/^{13}C energy level diagram in Fig. 2.8 will help to clarify how polarization transfer, i.e., inversion of one of the ^{1}H spin states, intensifies the two antiphase ^{13}C lines. The coupled ^{1}H/^{13}C nuclei can be represented by four energy levels, corresponding to (a) ^{1}H: α, ^{13}C: α, (b) ^{1}H: α, ^{13}C: β, (c) ^{1}H: β, ^{13}C: α, and (d) ^{1}H: β, ^{13}C: β (i.e., $\alpha\alpha$, $\alpha\beta$, $\beta\alpha$, and $\beta\beta$) states. In the lowest energy ($\alpha\alpha$) orientation, both ^{1}H and ^{13}C nuclei are aligned with the applied magnetic field; in the highest energy ($\beta\beta$) orientation, both are aligned against the applied magnetic field. In the two other energy levels ($\alpha\beta$ and $\beta\alpha$), the two nuclei are not parallel to each other. The population of the lowest energy ($\alpha\alpha$) state is $(\frac{1}{2})\gamma_H + (\frac{1}{2})\gamma_C$, while in the highest energy ($\beta\beta$) state the population is $-(\frac{1}{2})\gamma_H - (\frac{1}{2})\gamma_C$. In

---

Figure 2.8 (a) Energy levels, populations, and single-quantum transitions for a CH system. (b) (i) Populations at thermal equilibrium (Boltzmann distribution) responsible for the normal ^{13}C intensities, for example, of the doublet of CHCl$_3$. The amplitudes and intensities corresponding to this population difference is shown in (c) (i). (b) (ii) Population inversions of protons through the ^{1}H$_1$ transition results in an increased negative difference (-3) between levels 3 and 4, and an increased positive difference ($+5$) between levels 1 and 2. The corresponding amplitudes and intensities of CH doublet are shown in (c) (ii). (b) (iii) The alternative exchange of proton populations achieved through ^{1}H$_2$ transition (b) (i) gives rise to a reverse situation in (b) (iii) to that seen in (b) (ii). The population differences are: level 3–level 4 = $5 - 0 = +5$, and level 1–level 2 = $1 - 4 = -3$. Fig. (c) (iii) shows corresponding amplitudes and intensities of the CH doublet.

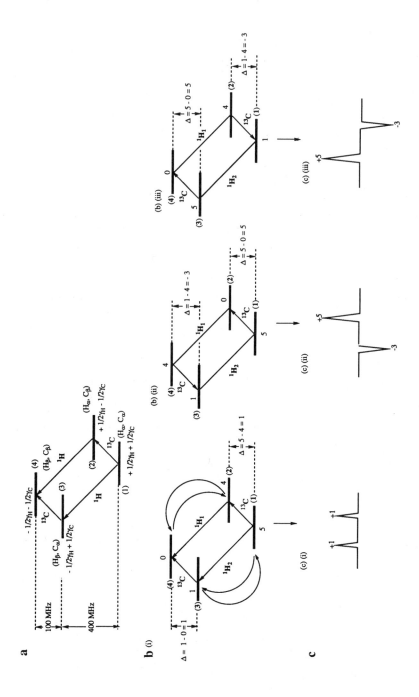

the two intermediate $\alpha\beta$ and $\beta\alpha$ states, the respective populations will be $(\frac{1}{2})\gamma_H - (\frac{1}{2})\gamma_C$ and $-(\frac{1}{2})\gamma_H + (\frac{1}{2})\gamma_C$. This is shown in Fig. 2.8a.

Suppose, for clarity, we add a common factor $[(\frac{1}{2})\gamma_H + (\frac{1}{2})\gamma_C]$ to all four energy levels so the energy difference between them does not change. The values of these energy levels will then become γ, γ_C, γ_H, and $\gamma_H + \gamma_C$. Since γ_H is about four times γ_C, let us also assume that $\gamma_H = 4$ and $\gamma_C = 1$. The populations of the four energy levels will then be 0, 1, 4, and 5. The population difference between the two ^{13}C spin states before the application of the ^1H polarization transfer pulse corresponds to the lower energy state minus the upper energy state, i.e. $1 - 0 = 1$ or $5 - 4 = 1$.

Applying the polarization transfer pulse inverts the population of one of the two ^1H states, producing a corresponding change in the populations of the ^{13}C states. This occurs because the two ^{13}C transitions share a common energy level with this proton transition. Thus, if the populations 0 and 4 of one of the ^1H transitions are exchanged, then the ^{13}C population differences become $5 - 0 = 5$ and $1 - 4 = 3$. The original ^{13}C population differences were $5 - 4 = 1$ and $1 - 0 = 1$, so we see that an intensification of the ^{13}C signal amplitudes has occurred. The *net* population difference was $(5 - 4) + (1 - 0) = 2$; after the polarization transfer pulse the *net* population difference $(5 - 3)$ is still 2. There is therefore no net transfer of magnetization, the intensification of the positive contribution being exactly balanced by a corresponding change in the negative contribution. However, since the original ^{13}C line intensities *before* the polarization transfer pulse were $+1$ and $+1$ and *after* the pulse they are $+5$ and -3, we see a significant increase in the intensity of the individual antiphase ^{13}C signals has occurred (Fig. 2.8c, ii). Figure 2.8c, iii, shows that the alternative exchange of proton populations (5 with 1) through the other ^1H transition leads to the ^{13}C population differences of $5 - 0 = 5$ and $1 - 4 = -3$, *i.e.*, the first peak of the ^{13}C doublet has an intensity of $+5$, and the second peak has an intensity of -3. The spectra therefore show an enhancement of the order of $\gamma_H/\gamma_C = \sim 4$: The two ^{13}C transitions would therefore have asymmetric intensities with an intensification of the individual peaks by factors of approximately $+5$ and -3.

Population transfer experiments may be selective or nonselective. Selective population transfer experiments have found only limited use for signal multiplicity assignments (Sörensen *et al.*, 1974) or for determining signs of coupling constants (Chalmers *et al.*, 1974; Pachler and Wessels, 1973), since this is better done by employing distortionless enhancement by polarization transfer (DEPT) or Correlated Spectroscopy (COSY) experiments. However, nonselective population transfer experiments, such as INEPT or DEPT (presented later) have found wide application.

✦ PROBLEM 2.10

How does magnetization transfer from ^{1}H to a coupled ^{13}C nucleus (polarization transfer or population transfer) affect the signal intensity of the ^{13}C nucleus?

── ✦

✦ PROBLEM 2.11

What is the difference between nonselective and selective polarization transfer?

── ✦

2.2.1 INEPT

The polarization transfer experiment just described involved inversion of one of the two vector components so that it pointed towards the $-z$-axis, in contrast to the equilibrium situation, in which both components were pointing towards the $+z$-axis. It was this inversion of the ^{1}H spins that caused a corresponding change in the population difference of the coupled ^{13}C nuclei, which was detected. INEPT (Insensitive Nuclei Enhancement by Population Transfer) is a nonselective polarization transfer experiment (Knoth, 1979; Avent and Freeman, 1980; Bolton, 1980; Burum and Ernst, 1980; Doddrell *et al.*, 1980; Morris, 1980a, 1980b; Pegg *et al.*, 1981a,b), in which polarization is transferred from all protons to all nuclei having the appropriate ^{1}H/^{13}C coupling.

The pulse sequence employed is basically similar to that already described, except that (a) a 180° ^{1}H pulse is applied in the middle of the evolution period to refocus the effects due to proton chemical shift evolution, and (b) a 180° ^{13}C pulse is applied simultaneously with the 180° ^{1}H pulse, which serves to invert the carbon spin labels so the ^{13}C vectors continue to diverge from one another. The basic INEPT pulse sequence applied to CH doublets is shown in Fig. 2.9. Application of the 90° ^{1}H pulse aligns the vectors along the y'-axis. During the subsequent $\tau = \frac{1}{4}J$ period, the two vectors of the ^{1}H nucleus fan out in the $x'y'$-plane. Applying the 180°_{x} ^{1}H pulse causes them to jump across the x'-axis and adopt mirror image positions in the $x'y'$-plane. The simultaneous application of the 180°_{x} ^{13}C pulse causes an interchange of the spin labels of the two vectors, resulting in an exchange of their "identities" and accompanied by a corresponding change in the direction of rotation, so that in the subsequent $\tau = \frac{1}{4}J$ period they move away from each other, thereby aligning them along the x'-axis at the end of the evolution

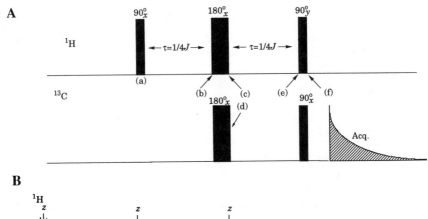

A

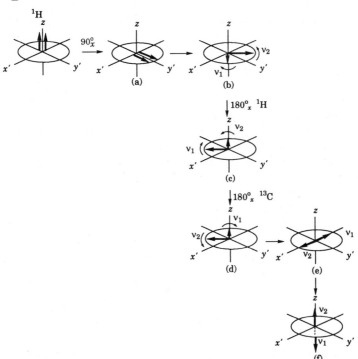

B

Figure 2.9 Pulse sequence for the INEPT experiment. (B) Effect of pulses on ^1H magnetization. Application of the pulse sequence shown results in population inversion of one of the two proton vectors of the CH doublet and therefore causes an intensification of the corresponding ^{13}C lines.

period. As in the previous polarization transfer experiment, a 90°_{y} ^{13}C pulse results in their rotating in the $x'z$-plane so they point in *opposite* directions along the z-axis. Thus, while at equilibrium both component vectors of the ^{1}H doublet pointed towards the $+z$-axis, by applying the INEPT sequence one of these two vector components becomes inverted so it now points towards the $-z$-axis. This $-z$-magnetization created in one of the ^{1}H vectors causes a corresponding change in the population difference of the ^{13}C vectors. This is read by applying a 90°_{x} ^{13}C pulse that inverts the z-magnetization (*longitudinal* magnetization) of the ^{13}C nuclei to *transverse* magnetization in the $x'y'$-plane. The ^{13}C-NMR spectrum now shows signals that are enhanced by a factor $K = \gamma_{H}/\gamma_{C}$ due to the greater population differences between the upper and lower energy states of the ^{13}C nuclei.

✦ *PROBLEM 2.12*

In what respect is the INEPT experiment superior to the APT or GASPE experiment?

── ✦

✦ *PROBLEM 2.13*

Describe the fundamental properties of an INEPT experiment.

── ✦

2.2.1.1 SIGNAL INTENSIFICATION WITH INEPT

In the case of a CH group, the two lines of the ^{13}C doublet have intensities of $+5$ and -3 (see previous discussion), and since they have different phases, they will partially cancel each other if they were to become focused. A suitable delay is therefore introduced to bring them into phase, the time delay differing for CH_3, CH_2, and CH groups. Once both lines are in phase, then each ^{13}C line exhibits a fourfold increase in intensity. This is significantly superior to the threefold enhancement in sensitivity obtainable by nuclear Overhauser enhancement when recording proton-decoupled ^{13}C-NMR spectra. In practice, a fivefold to sixfold enhancement of sensitivity of INEPT spectra can be obtained, since the repetition rate of the entire INEPT sequence is governed by the relatively shorter *proton* spin–lattice relaxation times, whereas in proton-decoupled ^{13}C spectra the repetition rate of successive scans is determined by the longer ^{13}C spin-lattice relaxation times.

In polarization transfer experiments such as INEPT, the recorded multiplets do not have the standard binomial intensity distributions of

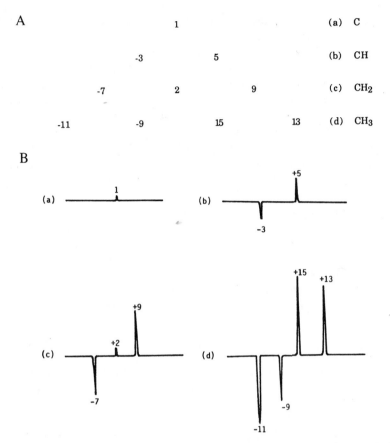

Figure 2.10   (A) Relative intensities in an INEPT spectrum presented as a variation of the Pascal triangle. (B) Signal phase and amplitudes of (a) quaternary, (b) CH (c) CH_2, and (d) CH_3 carbons in a normal INEPT experiment.

1:1 for doublets, 1:2:1 for triplets, and 1:3:3:1 for quartets, etc. Instead, they exhibit intensities—as indicated in the Pascal triangle in Fig. 2.10—of −3:5 for doublets, −7:2:9 for triplets, − 11, −9, 15, and 13 for quartets, etc.

The sensitivity enhancement in the INEPT experiment is particularly marked when nuclei of low magnetogyric ratios are being detected. A comparison of the signal intensities obtained by polarization transfer against those obtained by full NOE for various nuclei is presented in Table 2.1.

Table 2.1

A Comparison of Signal Intensities
Obtained by Polarization Transfer against
Those Obtained by Full nOe from Protons
to the Heteronucleus*

Nucleus	Polarisation transfer	Maximum nOe
^{13}C	3.98	2.99
^{15}N	9.87	-3.94
^{31}P	2.47	2.24
^{29}Si	5.03	-1.52
^{57}Fe	30.95	16.48
^{103}Rh	31.78	-14.98

* The intensities given are relative to those observed without nOe.

2.2.1.2 SPECTRAL EDITING WITH INEPT

INEPT can be used for determining the multiplicities of ^{13}C nuclei coupled to protons. This is done by deleting the 180° ^1H and ^{13}C refocusing pulses and by including a delay Δ the value of which is adjusted to give a particular angle θ, before data acquisition. The CH_3, CH_2, and CH carbon intensities depend on the delay Δ (i.e., on the angle θ), as shown in Table 2.2.

✦ *PROBLEM 2.14*

Based on the equation $K = \gamma_H/\gamma_C$, we would expect about a threefold enhancement in the signal intensities of proton-bearing ^{13}C nuclei in the broad-band decoupled INEPT experiment. In practice, the intensification is more than threefold. Why?

─── ✦

The precession of the ^{13}C multiplet components during the delay Δ by angle θ (where $\theta = \pi J \Delta$) is related to the signal intensities of CH, CH_2, and CH_3 carbons as follows:

CH: $I \alpha \sin \theta$
CH_2: $I \alpha \sin 2\theta$
CH_3: $I \alpha (\frac{3}{4}) (\sin \theta + \sin 3\theta)$.

Table 2.2

Relative Signal Intensities

$\theta*$	CH_3	CH_2	CH
$\pi/4$	$3/2\sqrt{2}$	1	$1/\sqrt{2}$
$\pi/2$	0	0	1
$3\pi/4$	$3/2\sqrt{2}$	-1	$1/\sqrt{2}$

$\theta* = \Delta.$

Since in the rotating frame the doublet vectors of the CH carbons precess at a precessional frequency of $+J/2$ and $-J/2$, if the delay Δ is set at $\frac{1}{2}J$ (i.e., $\theta = 90°$) then only CH carbons will appear. With Δ set at $\frac{1}{4}J$, all protonated carbons will appear with positive amplitudes; with Δ set at $\frac{3}{4}J$, CH and CH_3 carbons will produce positively phased signals and CH_2 carbons will give negatively phased signals. The quaternary carbons do not appear, since there are no attached protons from which polarization transfer can occur. Comparing the spectra at various Δ values with the broad-band decoupled spectrum yields a procedure for determining carbon multiplicities (*spectral editing*). This is illustrated in Fig. 2.11. To distinguish between the three kinds of carbon atoms (CH_3, CH_2, and CH), we need to record three different spectra with the delays Δ adjusted to make θ equal $\pi/4$, $\pi/2$, and $3\pi/4$. Suitable combinations of these spectra allow us to generate subspectra containing CH, CH_2, or CH_3 carbons.

As mentioned already, the INEPT spectra are typified by the antiphase character of the individual multiplets. The INEPT ^{13}C-NMR spectrum of 1,2-dibromobutane is shown, along with the normal off-resonance ^{13}C-NMR spectrum, in Fig. 2.12. Doublets show one peak with positive phase and the other with negative phase. Triplets show the outer two peaks with positive and negative amplitudes and the central peak with a weak positive amplitude. Quartets have the first two peaks with positive amplitudes and the remaining two peaks with negative amplitudes.

Since the natural magnetization has been removed by phase alternation, we cannot apply ^{1}H decoupling during data acquisition, because there is no *net* magnetization contributing to any of the multiplets. Decoupling would therefore result in the mutual cancellation of the antiphase peaks within each multiplet when they collapse together, leading to the disappearance of the signals. It is therefore necessary to collect coupled spectra with

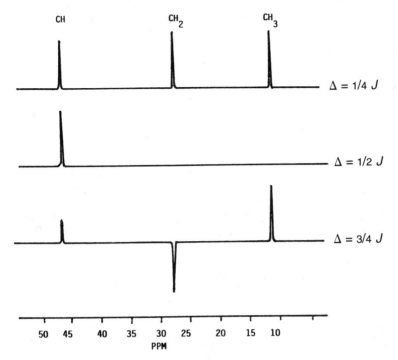

Figure 2.11 The behavior of CH, CH₂, and CH₃ groups during spectral editing.

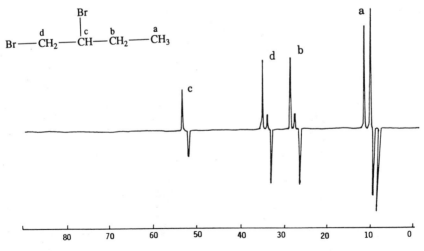

Figure 2.12 INEPT spectrum of 1,2-dibromobutane.

antiphase peaks within multiplets; and if decoupled spectra are required, we need to record a refocused INEPT experiment or a variant of it.

✦ *PROBLEM 2.15*

How can spectral editing in INEPT be used to generate separate spectra for CH, CH_2, and CH_3 groups, and how is it superior to the traditional off-resonance decoupled spectra?

_____ ✦

2.2.2 Refocussed INEPT

The INEPT experiment can be modified to allow the antiphase magnetization to be precessed for a further time period so that it comes into phase before data acquisition. The pulse sequence for the refocused INEPT experiment (Pegg *et al.*, 1981b) is shown in Fig. 2.13. Another delay, D_3, is introduced and 180° pulses applied at the center of this delay simultaneously to both the ^1H and the ^{13}C nuclei. Decoupling during data acquisition allows the carbons to be recorded as singlets. The value of D_3 is adjusted to enable the desired type of carbon atoms to be recorded. Thus, with D_3 set at $\frac{1}{4}J$, the CH carbons are recorded; at $\frac{1}{8}J$, the CH_2 carbons are recorded; and at $\frac{1}{6}J$, all protonated carbons are recorded. With D_3 at $\frac{3}{8}J$, the CH and CH_3 carbons appear out of phase from the CH_2 carbons.

The additional delay causes a decrease in signal strength due to loss of magetization from transverse relaxation. Moreover, severe phase distortions, particularly for CH_3 carbons, can produce anomalous results. A further modification of INEPT known as INEPT$^+$ incorporates an additional

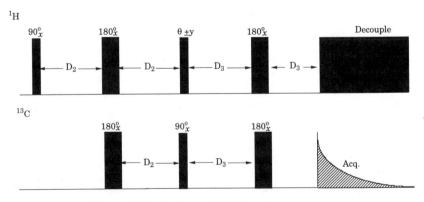

Figure 2.13 Refocused INEPT pulse sequence.

90° purging-pulse on the ^{1}H nuclei before detection (Sörensen and Ernst, 1983). This serves to remove the unwanted phase and multiplet anomalies. Many other variants of the INEPT experiment are known (Bodenhausen and Ruben, 1980; Eich *et al.*, 1982; King and Wright, 1983; Wagner, 1983; Bax and Drobny, 1985; Neuhaus *et al.*, 1985), but their discussion is beyond the scope of this book.

✦ *PROBLEM 2.16*

What are the characteristic differences between basic INEPT and refocused INEPT experiments?

─── ✦

2.2.3 DEPT

The most widely used method for determining multiplicities of carbon atoms is DEPT (Distortionless Enhancement by Polarization Transfer). This has generally replaced the classical method of recording off-resonance ^{13}C spectra with reduced CH couplings from which the multiplicity could be read directly.

The DEPT experiment (Doddrell *et al.*, 1982) involves a similar polarization transfer as the INEPT experiment, except it has the advantage that all the ^{13}C signals are in phase at the start of acquisition so there is no need for an extra refocusing delay as in the refocused INEPT experiment. Coupled DEPT spectra, if recorded, would therefore retain the familiar phasing and multiplet structures (1 : 1 for doublets, 1 : 2 : 1 for triplets, etc.). Moreover, DEPT experiments do not require as accurate a setting of delays between pulses as do INEPT experiments.

The pulse sequence used in DEPT is shown in Fig. 2.14 with reference to a CH group. A 90_x° ^{1}H pulse bends the equilibrium magnetization from the z-axis to the y'-axis. During the subsequent evolution period, the transverse ^{1}H magnetization is modulated by the CH coupling so it splits into two counterrotating vectors; at the end of the first $\frac{1}{2}J$ delay, the vectors point along the $+x'$- and $-x'$-axis. Next, a 180_x° ^{1}H pulse is applied to remove any field inhomogeneities; simultaneously, a 90_x° ^{13}C pulse serves to bend the ^{13}C magnetization from its equilibrium position along the z-axis to the y'-axis. Since there is no z-magnetization for either the ^{1}H or the ^{13}C nuclei, they are effectively decoupled. Therefore during the subsequent $\frac{1}{2}J$ delay period, they remain stationary at their original posi-

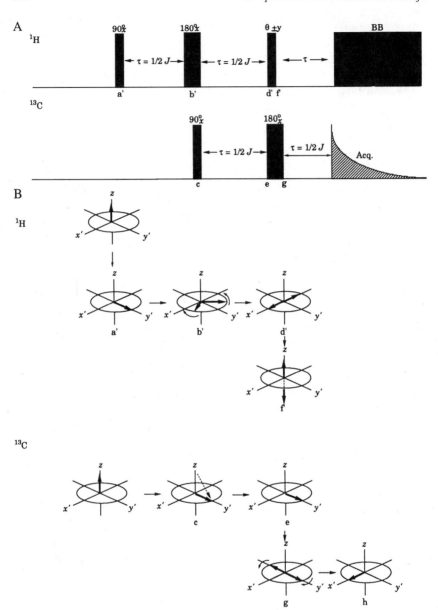

Figure 2.14 (A) Pulse sequence for the DEPT experiment. (B) Effect of the pulse sequence on ^{1}H and ^{13}C magnetization vectors. ^{13}C magnetization can be recorded either as multiplets or, if broad-band decoupling is applied during the acquisition period, as singlets.

tions, with the delay serving only to remove the field inhomogeneity effects by the preceding 180°_x ^{1}H pulse. At the end of the second $\frac{1}{2}J$ delay period, a proton pulse of angle θ is applied. The value of the angle θ is adjusted according to the type of spectrum desired. Thus, if we wish to record only CH carbons, then the value of θ is kept at $90°$ so there is maximum polarization for CH carbons and negligible x'-magnetization for CH_3 and CH_2 carbons.

With the 90°_x pulse we have rotated one of the two ^{1}H spin vectors to the $+z$-axis (i.e., to its original equilibrium position) and the other to the $-z$-axis, i.e., inverted its population. As in the APT and INEPT experiments, this inversion of population of one of the two spin vectors enhances the population difference between the two ^{13}C spin states with which the ^{1}H is coupled (see section 2.2 discussion on polarization transfer for details on this phenomenon) resulting in an intensification of the ^{13}C signal. Simultaneously, a $180°$ pulse is applied to the ^{13}C nuclei to remove any field inhomogeneities during the third $\frac{1}{2}J$ delay period. It is during this last delay that the ^{13}C signals are modulated by ^{13}C–^{1}H coupling, since z-magnetization now exists for the ^{1}H nuclei. The ^{13}C vectors rotate in the $x'y'$-plane, and at the end of the third $\frac{1}{2}J$ delay, the FID is acquired with or without ^{1}H decoupling. With θ set at $90°$, only CH carbons are recorded. With θ at $135°$, the CH and CH_3 carbons are modulated differently from CH_2 carbons so that CH_3 and CH carbons give positively phased signals and CH_2 carbons give negatively phased signals. The two spectra on compar-

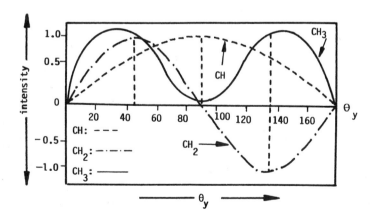

Figure 2.15 In DEPT experiments, signal intensities of CH_3, CH_2, and CH carbons depend on the angle θ_y of the last polarization pulse. For instance, at $\theta_y = 90°$, only CH carbons can be seen, while at $135°$, CH_3 and CH carbons will appear with one phase and CH_2 carbons will appear with the opposite phase.

ison with the broad-band decoupled spectrum suffice to identify CH_3, CH_2, CH, and quaternary carbons. With θ at 45°, all protonated carbons (CH_3, CH_2, and CH) appear with positive phases. The dependence of signal intensities of CH_3, CH_2, and CH carbons in the DEPT experiment on the angle θ_y of the last polarization pulse is shown in Fig. 2.15. The type of DEPT spectra obtained depends on the angle θ_y of the polarization pulse, and DEPT spectra are less dependent than INEPT spectra on the delay times between pulses, an error of ±20% in the setting of these delays giving acceptable DEPT spectra.

Many modifications of DEPT, spectra such as DEPT[+], DEPT[++] (Sörensen and Ernst, 1983), DEPT GL (Sörensen et al., 1983), modified DEPT (MODEPT) (Sörensen et al., 1984), universal polarization transfer (UPT) (Bendall et al., 1983), and phase oscillations to maximize editing (POMMIE) (Bendall and Pegg, 1983; Bulsing et al., 1984) have been reported.

✦ PROBLEM 2.17

DEPT (θ = 135° and 90°) and broad-band decoupled ^{13}C-NMR spectra of ethyl acrylate are shown. Assign the signals to various carbons of the molecule.

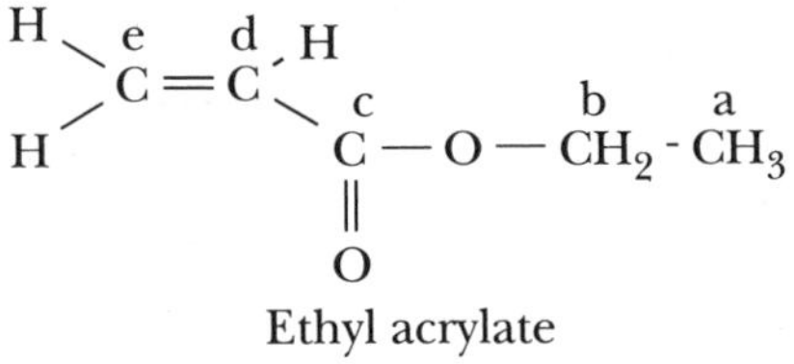

Ethyl acrylate

DEPT (θ = 135°)

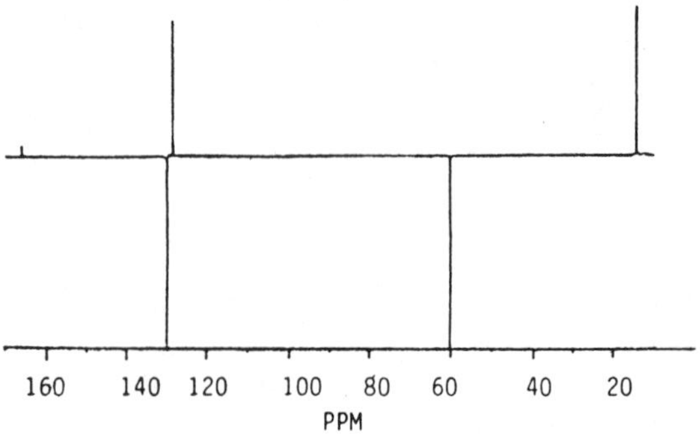

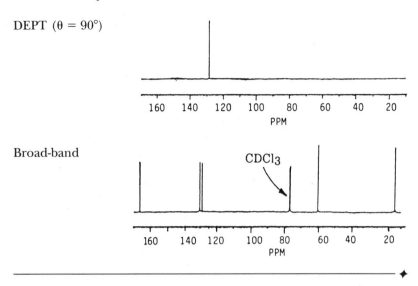

◆

◆ PROBLEM 2.18

What are the differences between INEPT and DEPT experiments? Why is DEPT considered a *distortionless* experiment?

◆

◆ PROBLEM 2.19

Why do quaternary carbons not appear in polarization transfer experiments?

◆

2.3 POLARIZATION TRANSFER IN REVERSE

In the preceding experiments, polarization was being transferred from a nucleus with a high magnetogyric ratio, such as ^1H, to a nucleus with a low magnetogyric ratio, such as ^{13}C. It is possible to transfer the polarization from a nucleus with low magnetogyric ratio, e.g., ^{13}C, to the one with a higher magnetogyric ratio, e.g., ^1H, before detection of the transferred magnetization. Such a *reverse transfer of polarization* has advantages of both selectivity and sensitivity.

Consider the selective polarization transfer in the normal (forward) sense, i.e., from a ^1H nucleus to a ^{13}C nucleus to which ^1H is attached. To carry out this transfer from a given ^1H nucleus, we have to identify and irradiate it selectively. This may not always be possible, particularly in larger molecules such as lipids or proteins, due to extensive overlap, or because the ^1H signals may be obscured by a large water signal. In the corresponding ^{13}C-NMR spectrum, however, it is easier to identify and irradiate individual carbons selectively because of the greater dispersion of the signals, particularly if enrichment of the particular carbon has been carried out. This is especially useful in studies of metabolic pathways. For instance, if we feed a ^{13}C-labeled drug to a living system and we wish to study its metabolism, we can use the ^{13}C label as a handle for detecting its attached protons selectively. The ^{13}C label helps us identify the carbon unambiguously because of its more intense signal as compared to other unenriched carbons in the same molecule, and polarization transfer from such a carbon to the corresponding protons would therefore allow us to follow readily the fate of such protons in the corresponding metabolite.

The inverse INEPT (Bodenhausen and Ruben, 1980) and inverse DEPT (Brooks *et al.*, 1984) experiments utilize such an approach. In the inverse INEPT experiment, successive 90° pulses are applied to the ^{13}C nucleus, followed by a ^1H read pulse. Protons not coupled to the ^{13}C nucleus are suppressed by presaturation of the entire ^1H-NMR spectrum before the polarization transfer, so only those signals will be detected that are generated by polarization transfer from the ^{13}C nucleus.

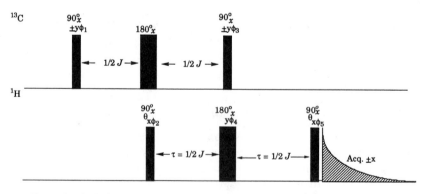

Figure 2.16 Pulse sequence for the inverse (reverse) DEPT experiment.

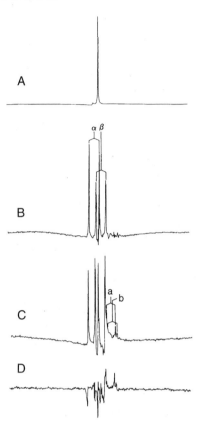

Figure 2.17 Application of the reverse DEPT pulse sequence to monitor ^{13}C-labeled glucose by mouse liver-cell extract. (A) Normal FT spectrum. (B) Reverse DEPT spectrum showing the α- and β-anomeric proton resonances. (C) Two different CH$_2$ proton resonances, a and b, appear after 1.5 h of metabolism. (D) Edited ^{1}H spectrum confirming that the CH$_2$ resonances arise from metabolic products. (Reprinted from *J. Magn. Resonance* **56**, Brooks *et al.*, 521, copyright © 1984, Academic Press.)

Transfer of polarization from ^{13}C nuclei to ^{1}H nuclei and their subsequent detection leads to a 16-fold increase in sensitivity because the ^{13}C magnetization is being measured indirectly through detecting it *via* the nucleus with the higher magnetogyric ratio (i.e., ^{1}H). Irradiation of the protons between the scans causes a further threefold increase in the population of the ^{13}C nuclei due to nOe, so an overall 50-fold increase in sensitivity is achievable in contrast to direct ^{13}C measurements. However, because of the longer relaxation times of the ^{13}C nuclei, the scan rate has to be reduced; and since ^{1}H signals are split, a 15-fold to 20-fold increase in sensitivity may actually be achieved.

The pulse sequence used in the reverse DEPT experiment is shown in Fig. 2.16. Presaturation of the protons removes all ^{1}H magnetization and pumps up the ^{13}C population difference due to nOe. Broad-band decoupling of the ^{13}C nuclei may be carried out. The final spectrum obtained is a one-dimensional ^{1}H-NMR plot that contains only the ^{1}H signals to which polarization has been transferred—for instance, from the enriched ^{13}C nucleus.

Applying the reverse DEPT pulse sequence to monitor ^{13}C-labeled glucose by mouse liver-cell extract is shown in Fig. 2.17. The α- and β-anomeric proton resonances are shown in the starting material; these are transformed to CH_2 proton resonances in the metabolite.

Many subspectral editing techniques alternative to DEPT, such as SEMUT (Subspectral Editing using a Multiple Quantum Trap) (Bildsöe *et al.*, 1983) and SEMUT GL, have been developed that utilize the fact that the transfer of magnetization to unobservable multiple-quantum coherence for CH_3, CH_2, and CH spin systems is dependent on the last flip angle θ. However, these experiments have not been widely used.

✦ PROBLEM 2.20

What are the principles governing reverse polarization transfer experiments? Why are reverse experiments more sensitive than "normal" experiments?

———————————————————————————————— ✦

✦ PROBLEM 2.21

^{13}C-NMR Chemical Shift Assignments by DEPT

The broad-band decoupled and DEPT spectra of podophyllotoxin, $C_{22}H_{22}O_8$, isolated from *Podophyllum hexandrum*, are shown. Although the molecule contains 22 carbons, only 18 signals are visible in the broad-band ^{13}C-NMR spectrum. This would obviously mean that some of these signals represent more than one carbon. Pairs of C-10/C-14, C-11/C-13, and 11-OCH_3/13-OCH_3 are magnetically equivalent while C-6 and C-7 could show accidental overlap. With this information, assign chemical shifts to various carbon atoms in the molecule.

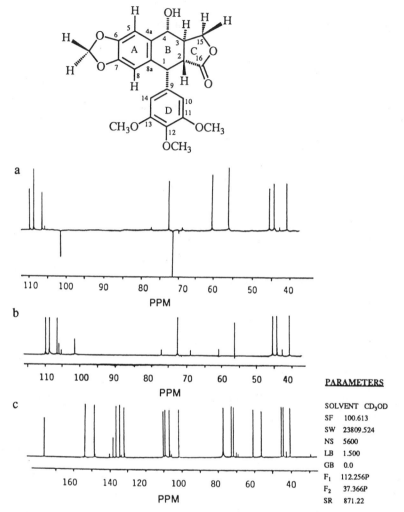

(a) DEPT (θ = 135°), (b) DEPT (θ = 90°), and (c) broad-band decoupled ^{13}C-NMR spectra of podophyllotoxin.

✦ **PROBLEM 2.22**

The DEPT and broad-band decoupled ^{13}C-NMR spectra of vasicinone, $C_{11}H_{10}N_2O_2$, isolated from a plant *Adhatoda vasica,* are shown. Spec-

trum (a) displays CH signals with positive phase and CH₂ signals
with negative phase; (b) shows only CH signals. Signals at $\delta\sim160$
represent two quaternary carbons in broad-band decoupled spectrum
(c). Assign the ¹³C-NMR chemical shifts to the various carbons in
the molecule.

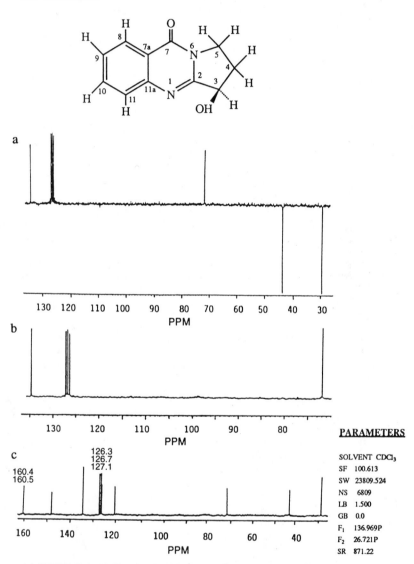

(a) DEPT ($\theta = 135°$), (b) DEPT ($\theta = 90°$) and (c) broad-band decoupled ¹³C-
NMR spectra of vasicinone.

✦ PROBLEM 2.23

^{13}C-NMR Chemical Shift Assignments by DEPT

The DEPT and broad-band decoupled ^{13}C-NMR spectra of thermopsine, $C_{15}H_{20}N_2O$, are shown. Assign resonances to the various carbons.

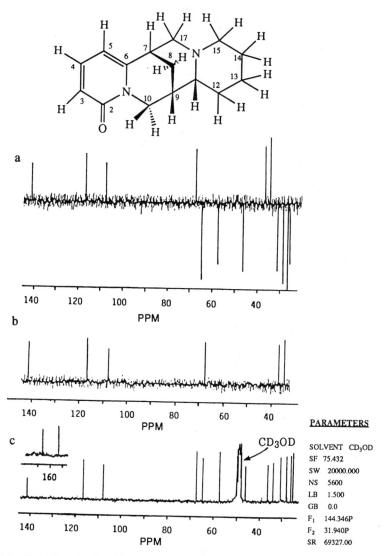

(a) DEPT ($\theta = 135°$), (b) DEPT ($\theta = 90°$), and (c) broad-band decoupled ^{13}C-NMR spectra of thermopsine.

✦ *PROBLEM 2.24*

¹³C-NMR Chemical Shift Assignments by DEPT

The DEPT and broad-band decoupled ¹³C-NMR spectra of malabarolide-A₁, $C_{18}H_{24}O_7$, a furanoid sesquiterpene isolated from *Tinospora malabarica* are presented. The signals for the C-9 and C-10 carbons resonating at δ 39.2 and 39.0, respectively, are embedded in the solvent peak. The peak at δ~72.0 is a cluster of two closely resonating signals at δ72.0 (CH) and 72.5 (CH). Similarly signal at δ~35.0 contain resonance at δ35.0 (CH₂) and 35.5 (CH₂). Assign the signals to the respective carbons.

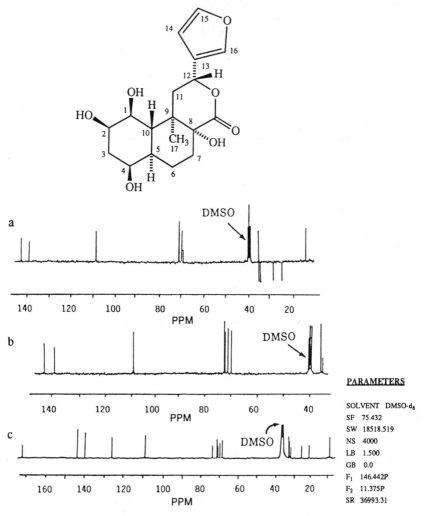

(a) DEPT (θ = 135°), (b) DEPT (θ = 90°), and (c) broad-band decoupled ¹³C-NMR spectra of malabarolide-A₁.

PARAMETERS

SOLVENT DMSO-d₆
SF   75.432
SW   18518.519
NS   4000
LB   1.500
GB   0.0
F₁   146.442P
F₂   11.375P
SR   36993.31

✦ PROBLEM 2.25

^{13}C-NMR Chemical Shift Assignments by GASPE

The GASPE spectrum of podophyllotoxin is shown. The signals at δ 56.0, 108.6, and 152.0 each represent two carbons in identical magnetic environments, while the signal at δ 147.6 also represents two carbons that accidentally appear at the same chemical shift. Assign chemical shift values to various protonated and quaternary carbons in the structure.

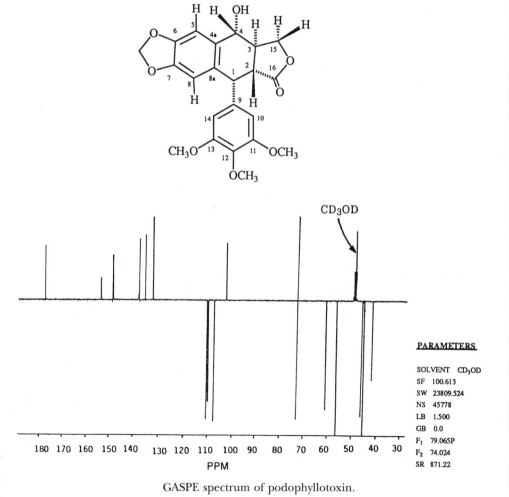

GASPE spectrum of podophyllotoxin.

PARAMETERS

SOLVENT   CD₃OD
SF   100.613
SW   23809.524
NS   45778
LB   1.500
GB   0.0
F₁   79.065P
F₂   74.024
SR   871.22

✦ PROBLEM 2.26

^{13}C-NMR Chemical Shift Assignments by GASPE

The GASPE spectrum of vasicinone is shown. The peak at ~δ 126.5 is a cluster of three peaks at δ 126.3 and 126.7 representing methine carbons. Similarly, the signal at δ ~160 on the positive phase of the spectrum represents two close singlets at δ 160.4 and 160.5. Predict the chemical shift values of various protonated and quaternary carbons in the structure.

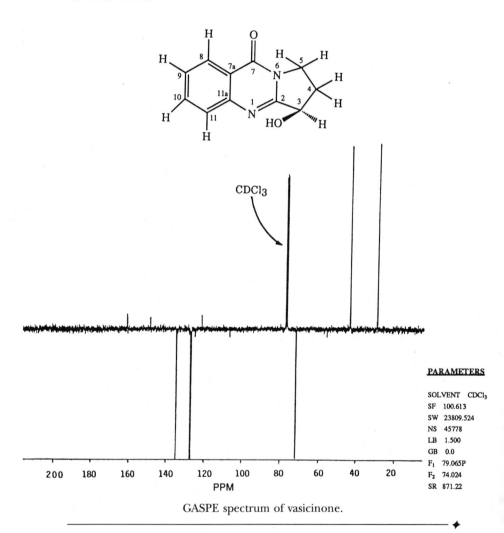

GASPE spectrum of vasicinone.

PARAMETERS

SOLVENT CDCl₃
SF   100.613
SW   23809.524
NS   45778
LB   1.500
GB   0.0
F₁   79.065P
F₂   74.024
SR   871.22

SOLUTIONS TO PROBLEMS

✧ *2.1*

The spin-echo is an elegant method for the measurement of transverse relaxation time T_2. In practice, this is done by repeating the spin-echo experiment many times, with different delay intervals (2τ, 4τ, 6τ, . . . etc.). The following sequence is used:

$$90^\circ_x - \tau - 180^\circ_x - \tau(\text{1st echo}) - \tau - 180^\circ_x - \tau_x(\text{2nd echo}) \ldots .$$

The intensities of the echoes (signal intensities) are than plotted as a function of delays. This yields a straight line whose gradient is $-1/T_2$. The decay in the intensities of echoes (i.e., of the succession of signals which follow the FID), is determined solely by T_2.

✧ *2.2*

Nuclei resonating at different chemical shifts will also experience similar refocusing effects. This is illustrated by the accompanying diagram of a two-vector system (acetone and water), the nuclei of which have different chemical shifts but are refocused together by the spin-echo pulse ($\mathbf{M}_A$ = magnetization vector of acetone methyl protons, $\mathbf{M}_W$ = magnetization vector of water protons).

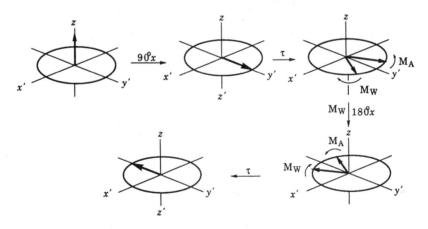

✧ *2.3*

(i) The 180°_x pulse will refocus the magnetization vectors, dephased due to field inhomogeneities and other instrumental errors. Following is the vector representation of this refocusing effect:

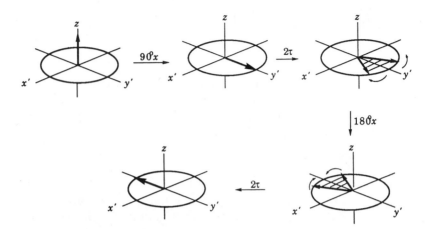

(ii) Assuming that we are observing the magnetization vector of nucleus A, which is coupled with nucleus X in an AX heteronuclear spin system, on application of a $90°_x$ pulse, the magnetization vector of A will come to lie on the y'-axis. Coupling with the two spin states of nucleus X will cause the A vector to split into two components. One of the A vectors (associated with the α state of nucleus X) will rotate at a frequency $\delta + (J_{AX}/2)$, while the other A vector (associated with the β state of X) will rotate at a frequency $\delta - (J_{AX}/2)$. Applying a $180°_x$ selective pulse on A at the end of the first τ period will rotate these vectors onto their mirror image positions. During the second τ period, they will continue to rotate towards the $-y'$-axis so, at the end of this delay, they will realign together along the $-y'$-axis. This can be readily visualized by the following vector diagram:

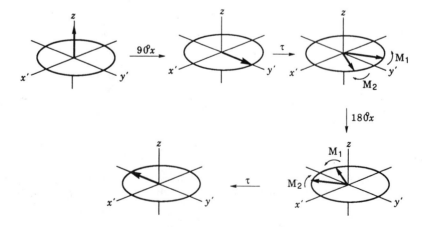

✧ 2.4

No. Since the direction of precession of the vectors is reversed during the second half of the sequence, they continue to diverge from each other. The effect of a spin-echo experiment on spin-spin coupling in a first-order homonuclear spin system is shown in the following vector representation:

The initial 90°_x pulse will bring vector **A** of a homonuclear AX system to lie along the $+y'$-axis. After a delay τ, vector **A** will be split into $\mathbf{A}_\alpha$ and $\mathbf{A}_\beta$ due to its coupling with nucleus X. Labels $\mathbf{A}_\alpha$ and $\mathbf{A}_\beta$ represent the effects of α and β spin states of the coupled nucleus X. The two components $\mathbf{A}_\alpha$ and $\mathbf{A}_\beta$ will move away from each other, creating a relative phase difference of $2\pi J\tau$ by the end of the first delay. Applying a 180°_x pulse to nuclei A and X at this point will have two distinct effects: First it will rotate $\mathbf{A}_\alpha$ and $\mathbf{A}_\beta$ onto mirror image positions. Second, it will also flip the spin labels of the coupled nucleus, i.e., interchange $\mathbf{A}_\alpha$ with $\mathbf{A}_\beta$. Since these are the labels that define the sense of rotation of the individual components, the two component $\mathbf{A}_\alpha$ and $\mathbf{A}_\beta$ will now move in opposite directions to their previous directions of rotation, i.e., away from each other, building up a phase difference of $4\pi J\tau$ at the time of the echo. The inversion of both coupled spins is the key to echo modulation in several *J*-resolved experiments.

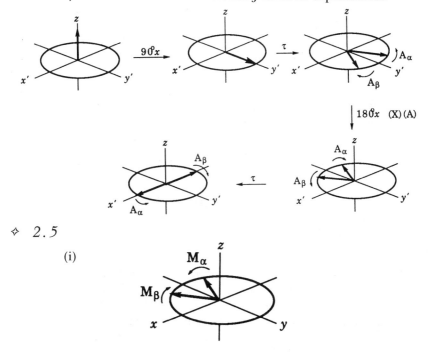

✧ 2.5

(i)

(ii)

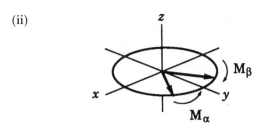

(iii)

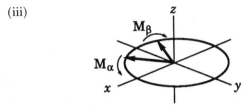

✧ *2.6*

After the 90°_x pulse, the transverse magnetization vectors of ^{13}C nuclei of C, CH, CH_2, and CH_3 do not rotate synchronously with one another but rotate with characteristically different angular velocities during the same delay interval. This results in their appearing with differing (positive or negative) amplitudes. This forms the basis of the APT experiment.

✧ *2.7*

As explained in the text, the signals for CH_3 and CH carbons will appear with one phase, whereas CH_2 and quaternary carbons will appear with opposite phase. In the GASPE spectrum shown, the negative phased signals represent the CH_3 and CH carbons in the molecule, and CH_2 and C carbons appear with positive amplitude. The signal at δ 14.0 is therefore due to the methyl carbon, while the signal at δ 128.6 is assigned to the CH carbon of the olefinic system. The signals with positive amplitude at δ 60.2 and 128.8 are assigned to the CH_2 carbon of the ethyl substituent and the terminal CH_2 carbon of the double bond, respectively. The signal at δ 167.5 with positive amplitude could be assigned to the ester carbonyl carbon.

✧ *2.8*

Single-quantum coherence is the type of magnetization that induces a voltage in a receiver coil (i.e., Rf signal) when oriented in the *xy*-plane. This signal is observable, since it can be amplified and Fourier-transformed into a frequency-domain signal. Zero- or multiple-quantum coherences do not obey the normal selection rules and do not induce any voltage in the receiver coil. These nonobservable magnetizations can only be detected indirectly after conversion to single-quantum coherence.

✧ *2.9*

No. The vector presentation is suitable for depicting single-quantum magnetizations but is not appropriate when considering zero-, double-, and higher-order quantum coherences. Quantum mechanical treatment can be employed when such magnetizations are considered.

✧ *2.10*

The signal intensity of the ^{13}C nucleus depends on the population difference between its two energy states, α and β. At thermal equilibrium, the ^{1}H population difference is about four times higher than that of ^{13}C nuclei, so there are four times as many protons in Boltzmann excess population of the lower energy state as there are ^{13}C nuclei (based on their respective gyromagnetic ratios). Inversion of one of the ^{1}H spin states causes a corresponding intensification of the population of the ^{13}C nuclei, since the ^{13}C transitions share a common energy level with this proton transition. The following diagram explains the phenomenon in a rather simplified fashion.

    (i) Populations at thermal equilibrium: ^{1}H population difference = (1) − (3) or (2) − (4); ^{13}C population difference = (3) − (4) or (1) − (2):

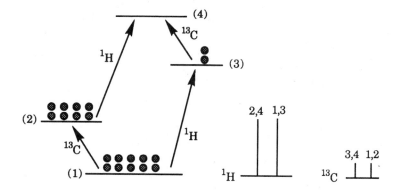

(ii) Populations after inverting one of the ^{1}H transitions:

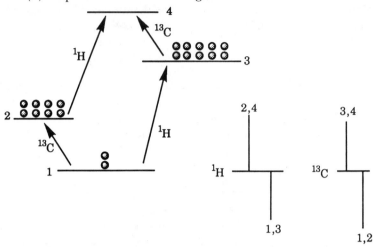

✧ *2.11*

Nonselective polarization transfer, as implied by the term, represents a process that allows simultaneous polarization transfer from *all* protons to *all* X nuclei. In *selective* polarization transfer, however, the population of just *one* nucleus is inverted at any one time. The selective polarization transfer sequence therefore cannot be used to generate a proton-decoupled ^{13}C-spectrum containing *all* sensitivity-enhanced ^{13}C resonances.

✧ *2.12*

The APT pulse sequence provides limited information about the number of hydrogens bonded to the carbons in a molecule, since it does not readily allow us to distinguish between the CH_3 and CH carbons or between CH_2 and quaternary carbons. The INEPT spectrum not only can yield information about the multiplicity of all the carbons, but also affords sensitivity-enhanced ^{13}C signals due to polarization transfer.

✧ *2.13*

The INEPT experiment has the following important characteristics:

(i) It enhances the sensitivity of an insensitive nucleus X, such as ^{13}C or ^{15}N, by a factor of γ_H/γ_X due to population transfer from an attached sensitive nucleus such as ^{1}H, and by a factor of $1 + \gamma_H/\gamma_X$ due to NOE generated by broad-band decoupling.

(ii) It reduces the dependence of signal strength on γ.

(iii) The INEPT experiment, after incorporation of a variable delay Δ, also provides a means for determining the number of protons attached to the ^{13}C nuclei in the molecule (i.e., establishing the multiplicity of the carbon signals). It is thus possible to generate separate subspectra for CH, CH_2, and CH_3 groups.

✧ *2.14*

The nuclear Overhauser effect resulting from the broad-band decoupling during the decoupled INEPT experiment also contributes to the signal enhancement of the ^{13}C lines.

✧ *2.15*

The precession of CH, CH_2, and CH_3 vectors during Δ occurs at different angular velocities. In order to distinguish the three kinds of carbon resonances, it is essential to record three spectra with variable delays Δ. The delay should be adjusted to make θ (where $\theta = \pi J\Delta$) equal $\pi/4$, $\pi/2$, and $3\pi/4$. The $\theta = \pi/4$ spectrum will contain all the resonances, $\theta = \pi/2$ will give only CH signals, and the spectrum obtained with $\theta = 3\pi/4$ will exhibit all CH_2 signals inverted and CH and CH_3 signals on positive phase. Careful examination of these three spectra allows generation of CH, CH_2, and CH_3 subspectra.

✧ *2.16*

The basic INEPT spectrum cannot be recorded with broad-band proton decoupling, since the components of multiplets have antiphase disposition. With an appropriate increase in delay time, the antiphase components of the multiplets appear in phase. In the refocussed INEPT experiment, a suitable refocusing delay is therefore introduced that allows the ^{13}C spin multiplet components to get back into phase. The pulse sequences and the resulting spectra of podophyllotoxin (Problem 2.21) from the two experiments are given below:

(i) INEPT

^{1}H $\qquad\qquad$ $90°_x–\tau–180°_x–\tau–90°_y$
^{13}C $\qquad\qquad$ $180°_x–90°_x–$Acquisition
INEPT $= 135°$

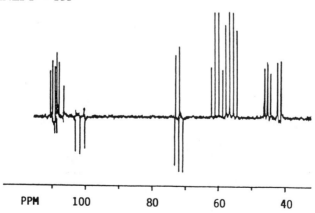

PPM $\qquad$ 100 $\qquad$ 80 $\qquad$ 60 $\qquad$ 40

INEPT = 45°

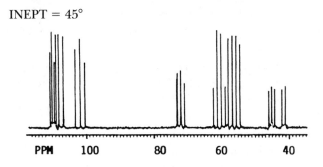

(ii) Refocussed INEPT
^{1}H $90°_x-(¼)J-180°_x-(¼)J-90° ± y-Δ/2-180°-Δ/2-$Decoupling
^{13}C $180°_x-90°_x-180°_x-$Acquisition
INEPT = 135°

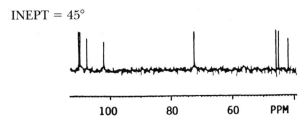

INEPT = 45°

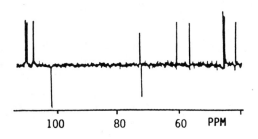

(iii) Broad-band decoupled

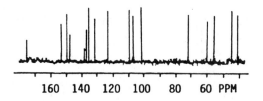

❖ 2.17

The broad-band decoupled ^{13}C-NMR spectrum of ethyl acrylate shows five carbon resonances; the DEPT ($\theta = 135°$) spectrum displays only four signals; i.e., only the protonated carbons appear, since the quaternary carbonyl carbon signal does not appear in the DEPT spectrum. The CH and CH$_3$ carbons appear with positive amplitudes, and the CH$_2$ carbons appear with negative amplitudes. The DEPT ($\theta = 90°$) spectrum displays only the methine carbons. It is therefore possible to distinguish between CH$_3$ carbons from CH carbons. Since the broad-band decoupled ^{13}C spectrum contains all carbons (including quaternary carbons), whereas the DEPT spectra do not show the quaternary carbons, it is possible to differentiate between quaternary carbons from CH, CH$_2$, and CH$_3$ carbons by examining the additional peaks in the broad-band spectrum versus DEPT spectra. The chemical shifts assigned to the various carbons are presented around the structure.

$$
\begin{array}{c}
128.8 \quad 128.6 \diagup \text{H} \\
\text{H}_2\text{C} = \text{C} \diagdown \underset{167.5}{} \quad 60.2 \quad 14.0 \\
\text{C} - \text{O} - \text{CH}_2 - \text{CH}_3 \\
\| \\
\text{O}
\end{array}
$$

❖ 2.18

Both experiments are based on polarization transfer from sensitive nuclei to insensitive nuclei, and therefore the major portions of their pulse sequences are common. The INEPT experiment, without refocusing and decoupling, however, yields spectra with "*distorted*" multiplets. For instance, the two lines of a doublet appear in antiphase with respect one another. Similarly, the central line of a triplet may be too small to be visible, while the outer two lines of the triplet will be antiphase to one another. Introducing a variable refocusing delay Δ and broad-band decoupling in the INEPT sequence can convert this experiment into a more useful one.

In the DEPT experiment, all the signals of the insensitive nuclei are in phase at the start of acquisition, so no refocusing period Δ (with accompanying loss in sensitivity) is required. Since the multiplets appear in-phase, it is called a *distortionless* experiment. Moreover, DEPT spectra depend on the angle θ of the last polarization transfer pulse, and are less dependent on the delay times $\frac{1}{2}J$ between the pulses. An error of $\pm 20\%$ in the estimation of J values still affords acceptable DEPT spectra. This offers a distinct advantage over INEPT, which requires a more accurate setting of delays between the pulses.

✧ **2.19**

In these experiments the signals of insensitive nuclei appear through polarization transfer from the more sensitive nuclei. Since quaternary carbons lack any attached hydrogen atoms, they cannot benefit from the polarization transfer and therefore do not appear in the spectra resulting from these experiments.

✧ **2.20**

Reverse polarization transfer utilizes the fact that protons, due to their greater gyromagnetic ratio, have greater *sensitivity* while carbons or other low-abundant nuclei, because of their greater dispersion, have a higher *selectivity*. By transferring polarization from a less sensitive nucleus (such as ^{13}C) to a more sensitive nucleus (e.g., ^1H) and then recording the carbon nuclei through their effects on proton magnetizations, we transfer the desired selectivity to the resulting spectrum, which has the character of a carbon spectrum. As compared to direct ^{13}C detection, the reverse polarization transfer signals are about 16 times stronger than with "normal" detection.

✧ **2.21**

An examination of the structure shows that the aromatic ring at the C-1 contains three pairs of identical carbons (the two —OCH_3 carbons at δ 56.0, the two carbons bearing the —OCH_3 groups at δ 152.0, and the two carbons *ortho* to the methoxy group at δ 108.6). This accounts for three out of the four "missing" signals. The fourth "missing" carbon could be the one at δ 147.6, due to the accidental overlap of the two oxygen-bearing carbons in ring A to which the methylenedioxy group is attached. Only 11 resonances are visible in the DEPT (θ = 135°) spectrum. This allows the quaternary carbon signals to be distinguished in the broad-band spectrum, which are at δ 131.1, 133.4, 135.5, 137.0, 147.6, 152.0, and 174.0. Quaternary signals between δ 135 and δ 155 generally represent oxygen-bearing aromatic carbons. For instance, the signal at δ 152.0 can be assigned to the methoxy-bearing magnetically equivalent C-11 and C-13, while the signal at δ 137.0 may be assigned to the C-12. The upfield chemical shift of C-12 is due to the electron-donating mesomeric effect of the methoxy groups at C-11 and C-13. The quaternary signal at δ 147.6 may be due to accidental overlap of the oxygen-bearing C-6 and C-7. The signal for the C-16 carbonyl is readily distinguishable at δ 174.0, while the assignments for C-4a, C-8a, and C-9 at δ 133.4, 131.1, and 135.5 could be interchangeable and need additional heteroCOSY experiments [such as Heteronuclear Multiple Bond Connectivity (HMBC), see later] to be distinguished from one another. Two signals appear with negative phase in the DEPT 135° spectrum at δ 71.0 and 101.4, representing the two methylene carbons. The low-field methylene signal at δ 101.4 is characteristic for the methylenedioxy carbon containing two electron-withdrawing oxygen atoms, while the signal at δ 71.0 is assigned to C-15. The methyl carbons of the —OCH$_3$ groups appear at δ 56.0 and 60.7. Differentiation between CH$_3$ and CH carbons is possible from DEPT 90°, which shows only CH signals. The chemical shift assignments to the various carbons of podophyllotoxin are as follows:

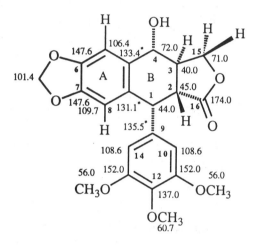

✧ *2.22*

The broad-band ^{13}C-NMR spectrum of vasicinone (spectrum c) shows eleven carbon resonances. The DEPT spectra (θ = 135°, 90°) display signals of protonated carbons only. By comparison with the broad-band decoupled spectrum, it is easy to identify the peaks of the quaternary carbons (δ 120.9, 148.0, 160.5, and 160.4). The DEPT spectrum (θ = 90°) shows signals for methine carbons only (δ 72.2, 126.3, 126.7, 127.1, 134.5). Signals with negative amplitudes in the DEPT 135° spectrum are those of methylene carbons (δ 29.4 and 43.7); signals with positive amplitudes are for methine carbons. Two of the downfield quaternary signals, at δ 160.5 and 160.4, are characteristic of the α,β-unsaturated amide carbonyl carbon (C-7) and iminic carbon (C-2), although in this case the assignments are so close they could be interchangeable. The remaining two quaternary signals, at δ 120.9 and

148.0, may be assigned to C-11a and C-7a, respectively, based on the chemical shift values. The signals for CH_2 carbons at δ 29.4 and 43.7 can be assigned to C-4 and C-5, respectively. The downfield chemical shift of the latter carbon is due to its close proximity to the electron-withdrawing N-6. The most upfield methine signal, at δ 72.2, is readily assigned to the hydroxy-bearing C-3. The remaining CH signals, at δ 126.3, 126.7, 127.1, and 134.5, are due to the aromatic methine carbons. Of these, the chemical shift for C-11 is readily distinguishable (δ 134.5), due to the electron-withdrawing effect of the iminic N-1; the assignments for the remaining three methine carbons (C-8, C-9, and C-10) could be interchangeable and need additional heteroCOSY or HMBC experiments to be distinguished from one another. The chemical shift values are presented around the structure of vasicinone. The chemical shifts printed in italic are for quaternary carbons, and they do not appear in the DEPT spectra.

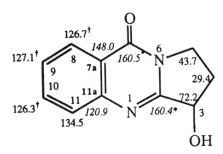

*,†: assigned values are interchangeable

✧ 2.23

The resonances for all 15 carbons of thermposine are visible in the broad-band decoupled ^{13}C-NMR spectrum, whereas the DEPT 135° spectrum (CH_3 absent, CH upright, and CH_2 inverted) exhibits only 13 signals. Six CH carbon resonances appear in the DEPT 90° spectrum. An interpretation of these three spectra therefore indicates the presence of seven CH_2, six CH, and two quaternary carbons. Two quaternary signals appeared at δ 153.6 and 167.7 in the broad-band decoupled spectrum. The latter signal (δ 167.7) is characteristic of an α,β-unsaturated amide carbonyl (C-2 in this case), the former signal (δ 153.6) can be easily assigned to the remaining quaternary carbon, C-6, attached to the amidic nitrogen.

The DEPT spectra (b and c) showed six methine signals, at δ 34.2, 36.6, 67.4, 107.7, 116.7 and 141.2. The last three most downfield signals can be readily assigned to C-5, C-3, and C-4 of the pyridone nucleus. The downfield chemical shift of C-4 is due to its β-disposition to the electron-withdrawing amidic carbonyl. The remaining three methine carbons, resonating at δ 34.2, 36.6 and 67.4, may be assigned to C-9, C-7, and C-11, respectively. Once again the downfield chemical shift of C-11 (δ 67.4) can be justified, based on its attachment to N-16. The methylene signals resonating at δ 25.4, 26.4, 28.3, 30.9, 46.3, 57.4, and 64.6 can be assigned to C-14, C-8, C-12, C-13, C-10, C-15, and C-17 respectively, based on their chemical shifts. For instance, the three downfield methylene signals at δ 46.3, 57.4 and 64.6 can be assigned to the methylenes next to the nitrogen atoms. The amidic nitrogens have less electron-withdrawing effect, which is why C-10 resonates somewhat upfield (δ 46.3) as compared to C-17 (δ 64.6) and C-15 (δ 57.4). C-17 (δ 64.6) is unusually downfield but appears at its characteristic value for such alkaloidal systems, emphasizing the importance of making assignments after comparison with other, closely related structures. Some other assignments cannot be made unambiguously without comparison with other closely related compounds and heteronuclear shift-correlation studies. These assignments are presented around the structure of thermopsine.

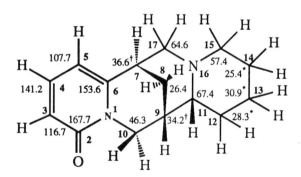

*,† = **assigned** values are interchangeable.

✧ *2.24*

The broad-band decoupled ^{13}C-NMR spectrum of malabarolide-A$_1$ shows signals for all 18 carbon atoms. The DEPT spectrum (θ = 135°) exhibits 14 signals for protonated carbons. It is therefore possible to identify four signals of quaternary carbons, i.e., δ 39.2 (C-9), 74.7 (C-8), 126.0 (C-13), and 171.8 (C-18 carbonyl). The DEPT (θ = 90°)

spectrum displays signals for only the methine carbons. The peaks at
δ 108.9, 139.7, and 143.4 are accordingly assigned to the C-14, C-16,
and C-15 carbons of the furan moiety, respectively. Four downfield
methine carbons, resonating at δ 67.3, 71.1, 72.0 and 72.5, may be
assigned to the C-4, C-12, C-2, and C-1 carbons, since they bear oxygen
functionalities, the assignments to the latter three carbons being interchangeable. The two remaining methine carbons, at δ 35.8 and 39.0,
are due to the C-5 and C-10 carbons, respectively. The methyl signal
(δ 17.0) having a positive phase in the DEPT 135° spectrum is due to
C-17. Four methylene carbons, C-3, C-6, C-7, and C-11, appear as negative phased signals at δ 35.5, 25.6, 29.4, and 35.0, respectively. The
^{13}C-NMR assignments to the various carbons are presented around
the structure.

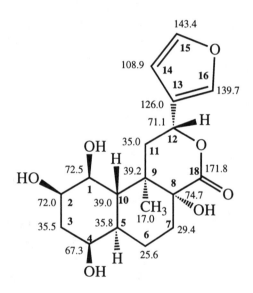

✧ 2.25

The GASPE spectrum of podophyllotoxin shows 17 peaks having positive and negative amplitudes. Eight signals with positive phase (pointing
upwards) are due to quaternary and CH_2 carbons. This is fairly easy to
determine, since the solvent signals (CD_3OD having no H, behaves like
a quaternary carbon) also appear with the same phase. The remaining
nine resonances, having negative phase, are due to CH_3 and CH
carbons.

Let us first consider overlapping signals. Since the signals at δ 56.0
(CH_3), 108.6 (CH), and 152.0 (—C—) each represent two carbons in

identical magnetic environments, they can be assigned to two OCH_3 (substituted on C-11 and C-13), C-10, and C-14 methine carbons and C-11 and C-13 methoxy-bearing quaternary carbons, respectively. The quaternary signal at δ 147.6 represents the overlapping resonances of C-6 and C-7, which accidentally appear at the same chemical shift. The signals with positive phase, representing CH_2 and quaternary carbons, will be considered next. The peaks at δ 101.4 and 71.0 may be readily assigned to the CH_2 of the methylenedioxy group and C-15 (geminal to oxygen), respectively, based on their characteristic chemical shifts. The remaining six signals with this phase represent the nine quaternary carbons. Two of these (δ 152.0 and 147.6) represent two-carbon signals that have already been assigned to C-11/C-13 and C-6/C-7. The downfield quaternary carbon signal, at δ 174.0, is readily assigned to the lactone carbonyl carbon. The remaining four signals (δ 137.0, 135.5, 133.4, and 131.1) could be assigned to C-12, C-9, C-4a, and C-8a, with interchangeable assignments.

Signals with negative phase (pointing downwards) are due to CH_3 and methine carbons. The peaks at δ 56.0 and 60.7 may be assigned to the methoxy carbons, while the remaining signals are due to the methine carbons. As is apparent, most of the assignments are based on chemical shift values, and there is always an element of doubt in the differentiation of the quaternary carbons from the CH_2 carbons and of the CH carbons from the CH_3 carbons. Polarization transfer experiments (DEPT, etc.) are therefore superior to the GASPE or APT experiments, in both sensitivity and multiplicity assignments.

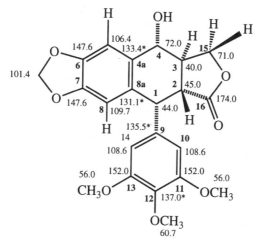

*Assignments are interchangeable

✧ **2.26**

The GASPE (APT) spectrum of vasicinone displays signals for all 11 carbons. The peaks having positive amplitude in the GASPE spectrum are due to the quaternary and CH_2 carbons, while signals with negative amplitude represent CH carbons (the compound does not have any methyl groups). The peaks at δ 29.4 and 43.7 are accordingly assigned to the C-4 and C-5 methylene carbons of ring D. The downfield chemical shift of C-5 (δ 43.7) is due to its proximity to the amidic nitrogen atom. Weak quaternary signals at δ 120.9, 148.0, 160.4, and 160.5 are due to C-11a, C-7a, C-2, and C-7 (carbonyl) quaternary carbons. Due to the very close chemical shifts of the signals at δ 160.4 and 160.5, their unambiguous assignment is not possible. The five CH signals with negative amplitudes, at δ 126.3, 126.7, 127.1, 134.5 and 72.2, can be readily ascribed to the four aromatic CH carbons and to the oxygen-bearing C-3. The chemical shift values are presented on the structure of vasicinone.

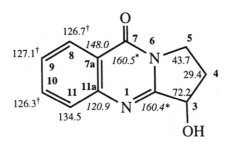

*,† = Assignments are interchangeable

REFERENCES

Avent, A. G., and Freeman, R. (1980). *J. Magn. Reson.* **39**(1), 169–174.
Bax, A., and Drobny, G. (1985). *J. Magn. Reson.* **61**, 306–320.
Bendall, M. R., Pegg, D. T., Tyburn, G. M., and Brevard, C. (1983). *J. Magn. Reson.* **55**(2), 322–328.
Bendall, M. R., and Pegg, D. T. (1983). *J. Magn. Reson.* **52**, 164–168.
Bildsöe, H., Donstrup, S., Jakobsen, H. J., and Sorensen, O. E. (1983). *J. Magn. Reson.* **53**, 154–162.
Bodenhausen, G., and Ruben, D. J. (1980). *Chem. Phys. Lett.* **69**(1), 185–189.
Bolton, P. H. (1980). *J. Magn. Reson.* **41**(2), 287–292.
Brooks, W. M., Irving, M. G., Simpson, S. J., and Doddrell, D. M. (1984). *J. Magn. Reson.* **56**, 521–526.
Bulsing, J. M., Brooks, W. M., Field, J., and Doddrell, D. M. (1984). *J. Magn. Reson.* **56**, 167–173.

Burum, D. P., and Ernst, R. R. (1980). *J. Magn. Reson.* **39**(1), 163–168.

Chalmers, A. A., Pachler, K. G. R., and Wessels, P. L. (1974). *Org. Magn. Reson.* **6**(8), 445–447.

Doddrell, D. M., Pegg, D. T., and Bendall, M. R. (1982). *J. Magn. Reson.* **48**(2), 323–327.

Doddrell, D. M., Pegg, D. T., Bendall, M. R., Brooks, W. M., and Thomas, D. M. (1980). *J. Magn. Reson.* **41**(3), 492–495 (1980).

Eich, G., Bodenhausen, G., and Ernst, R. R. (1982). *J. Am. Chem. Soc.* **104**, 3731–3732.

King, G., and Wright, P. E. (1983). *J. Magn. Reson.* **54**, 328–332.

Knoth, W. H. (1979). *J. Am. Chem. Soc.* **101**(3), 760–762.

Madsen, J. C., Bildsoe, H., and Jacobsen, H. J. (1986). *J. Magn. Reson.* **67**, 243–257.

Morris, G. A. (1980a). *J. Am. Chem. Soc.* **102**(1), 428–429.

Morris, G. A. (1980b). *J. Magn. Reson.* **41**(1), 185–188.

Neuhaus, D., Wagner, G., Vasak, M., Kaegi, J. H. R., and Wuethrich, K. (1985). *Eur. J. Biochem.* **151**(2), 257–273.

Pachler, K. G. R., and Wessels, P. L. (1973). *J. Magn. Reson.* **12**, 337.

Pegg, D. T., Bendall, M. R., and Doddrell, D. M. (1981a). *J. Magn. Reson.* **44**(2), 238–249.

Pegg, D. T., Doddrell, D. M., Brooks, W. M., and Bendall, M. R. (1981b). *J. Magn. Reson.* **44**(1), 32–40.

Radeglia, R., and Porzel, A. (1984). *J. Prakt. Chem.* **326**(3), 524–528.

Sörensen, O. W., Doenstrup, S., Bildsoe, H., and Jakobsen, H. J. (1983). *J. Magn. Reson.* **55**(2), 347–354.

Sörensen, U. B., Bildsoe, H., and Jakobsen, H. J. (1984). *J. Magn. Reson.* **58**, 517–525.

Sörensen, O. W., and Ernst, R. R. (1983). *J. Magn. Reson.* **51**, 477–489.

Sörensen, S., Hansen, R. S., and Jakobsen, H. J. (1974). *J. Magn. Reson.* **14**(2), 243–245.

Wagner, W. (1983). *J. Magn. Reson.* **55**, 151–156.

CHAPTER 3

The

Second

Dimension

In the one-dimensional NMR experiments discussed earlier, the FID was recorded immediately after the pulse, and the only time domain involved (t_2) was the one in which the FID was obtained. If, however, the signal is not recorded immediately after the pulse but a certain time interval (t_1) is allowed to elapse before detection, then during this time interval (the *evolution period*) the nuclei can be made to interact with each other in various ways, depending on the pulse sequences applied. Introduction of this second dimension in NMR spectroscopy, triggered by Jeener's original experiment, has resulted in tremendous advances in NMR spectroscopy and in the development of a multitude of powerful NMR techniques for structure elucidation of complex organic molecules.

Two-dimensional NMR spectroscopy may be defined as a spectral method in which the data are collected in two different time domains: acquisition of the FID (t_2), and a successively incremented delay (t_1). The resulting FID (data matrix) is accordingly subjected to two successive sets of Fourier transformations to furnish a two-dimensional NMR spectrum in the two frequency axes. The time sequence of a typical 2D NMR experiment is given in Fig. 3.1. The major difference between one- and two-dimensional NMR methods is therefore the insertion of an evolution time, t_1, that is systematically incremented within a sequence of pulse cycles. Many experiments are generally performed with variable t_1, which is incremented by a constant Δt_1. The resulting signals (FIDs) from this experiment depend

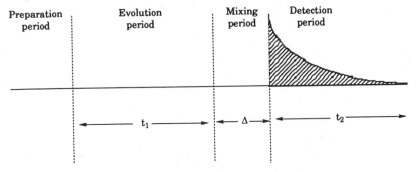

Figure 3.1 The various time periods in a two-dimensional NMR experiment. Nuclei are allowed to approach a state of thermal equilibrium during the preparation period before the first pulse is applied. This pulse disturbs the equilibrium polarization state established during the preparation period, and during the subsequent evolution period t_1 the nuclei may be subjected to the influence of other, neighboring spins. If the amplitudes of the nuclei are modulated by the chemical shifts of the nuclei to which they are coupled, 2D-shift-correlated spectra are obtained. On the other hand, if their amplitudes are modulated by the coupling frequencies, then 2D J-resolved spectra result. The evolution period may be followed by a mixing period Δ, as in Nuclear Overhauser Enhancement Spectroscopy (NOESY) or 2D exchange spectra. The mixing period is followed by the second evolution (detection period) t_2.

on the two time variables (t_1, t_2). Two sets of successive Fourier transformations are therefore required to convert the signals obtained into the two frequency axes, v_2 and v_1.

A convenient way to understand the modern 2D NMR experiment is in terms of magnetization vectors. Figure 3.2 presents a pulse sequence and the corresponding vector diagram of a 2D NMR experiment of a single-line ^{13}C spectrum (e.g., the deuterium decoupled ^{13}C-NMR spectrum of CDCl$_3$).

The preparation period allows the spin system to relax back to its equilibrium state by T_1 and T_2 relaxation processes. The first 90°_x pulse bends the magnetization of the sample (M_z) so it becomes aligned with the y'-axis in the $x'y'$-plane. During the subsequent evolution period (t_2), the magnetization vector will rotate in the $x'y'$-plane. By the end of a certain time interval, t_2, the vector would have moved through an angle α, which is proportional to the evolution time t_1 and the frequency v (angle $\alpha = 2\pi v t_1$ radians).

The vector may be considered to have two components, component **A,** aligned along the y'-axis with magnitude $M \cos (2\pi v t_1)$, and component **B,** aligned along the x'-axis with magnitude $M \sin (2\pi v t_1)$. The second 90°_x

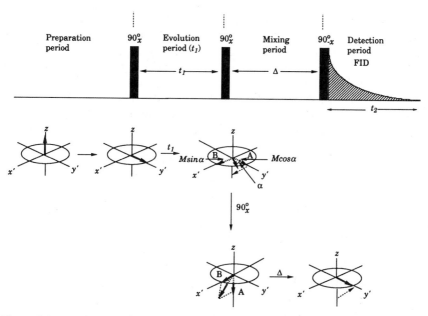

Figure 3.2 An example of a two-dimensional NMR pulse experiment. The time axis may be divided into a *preparation period* (during which the nuclei are allowed to reach an equilibrium state) and an *evolution period* t_1, during which the nuclei interact with each other. In some experiments (e.g., NOESY), this may be followed by a mixing period. Finally, the magnetization is detected during the *detection period* t_2. The experiment is repeated with incremented values of t_1, and a second set of Fourier transformations (after transposition of data) gives the 2D plot. The second 90°_x pulse serves to remove the y' component **A** of the main magnetization (having magnitude $M \cos \alpha$) by bending it away to the $-z$-axis, leaving component **B** along the x'-axis (with magnitude $M \sin \alpha$). After the *mixing period* Δ, the magnetization along the z-axis is returned to the y'-axis by the 90°_{-x} pulse for detection. The signal intensity therefore varies sinusoidally as a function of the angle by which the main vector has diverged from the y'-axis.

pulse at the end of the evolution period rotates component **A** of the magnetization, aligned with the y'-axis, onto the $-z'$-axis, leaving behind component **B,** located along the x'-axis, with magnitude $M \sin (2\pi \upsilon t_1)$ in the $x'y'$-plane. The magnitude of both components **A** and **B** will also be reduced to a certain extent by the inherent relaxation that occurs during the evolution period t_1. Component **B,** which remains in the $x'y'$-plane after the second 90°_x pulse, is detected as transverse magnetization in the correlated spectroscopy (COSY) experiment. If the FID is recorded immediately after the second 90°_x pulse, the detector will "catch" the transverse magnetization at its maximum amplitude. The transverse magnetization would, however, decrease with time as the magnetization vector moves away from the y'-axis in the $x'y'$-plane. It is therefore possible to perform the experiment with discrete changes in the evolution time t_1 and to record many FIDs as a function of t_1. Fourier transformation of each of these FIDs will yield the corresponding "normal" 1D NMR spectrum, with the peak appearing at frequency v_1, except that the amplitude of the peak will oscillate sinusoidally according to sin $(2\pi v t_1)$ in the various spectra thus obtained.

It is therefore apparent that the first series of Fourier transformations of the data (FIDs) with respect to t_2 yield the so called "one-dimensional" spectra, in which the amplitude of the peak oscillates (with frequency v_2) as a function of t_1 (Fig. 3.3):

$$S(t_1, t_2) \xrightarrow{\text{F.T. over } t_2} S(t_1, v_2).$$

If the resulting "one-dimensional" spectra are arranged in rows one behind the other (a process known as *data transposition*), and the peak is viewed along a *column* (i.e., at 90° to the rows), then the sinusoidally oscillating peak gives the appearance of a pseudo-FID (or *interferogram*, Fig. 3.4), with the sinusoidal variation of the peak intensity being evident with respect to the vertical axis (i.e., along t_1). The second Fourier transformation of this interferogram with respect to t_1 then generates a two-dimensional NMR spectrum:

$$S(t_1, v_2) \xrightarrow{\text{F.T. over } t_1} S(v_1, v_2).$$

In the case of a Nuclear Overhauser Enhancement Spectroscopy (NOESY) experiment, a third *mixing period* is inserted after the second 90_x° pulse, and a third 90_x° pulse is applied at the end of the mixing period (Fig. 3.2). The final 90_x° pulse does not affect component **B** lying along the x'-axis (proportional to $M \sin \alpha$), so this component is not converted into detectable magnetization. Component **A** lying along the $-z$-axis is, however, returned to the y'-axis by the third 90_x° pulse for detection. Component **B** along the x'-axis is therefore successively removed at various values of t_1, and component **A** is returned to the y'-axis and detected. The change in signal strength (ignoring loss due to relaxation) depends on the magnitude of component **B** removed, which in turn depends on the sine of angle α [i.e., $M \sin (2\pi v t_1)$]. As different experiments are performed at various values of t_1, the NMR signal is accordingly seen to oscillate sinusoidally as a function of t_1. Data transposition (i.e., arrangement in rows) followed by Fourier transformation of the resulting *columns* of oscillating peaks will give the 2D NOESY spectrum.

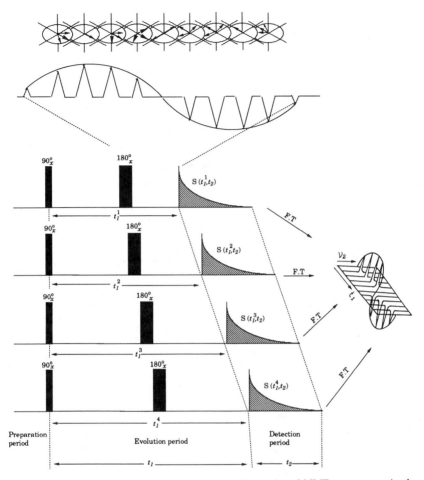

Figure 3.3 The mechanics of obtaining a two-dimensional NMR spectrum. As the t_1 value is varied, the magnetization vectors are "caught" during detection at their various positions on the $x'y'$-plane. The value of the detection time t_2 is kept constant. The first set of Fourier transformations across t_2 is followed by transposition of the data, which aligns the peaks behind one another, and a second set of Fourier transformations across t_1 then affords the 2D plot.

The first Fourier transformation of the FID yields a complex function of frequency with real (cosine) and imaginary (sine) coefficients. Each FID therefore has a real half and an imaginary half, and when subjected to the first Fourier transformation the resulting spectrum will also have real and imaginary data points. When these real and imaginary data points are arranged behind one another, vertical columns result. This transposed data

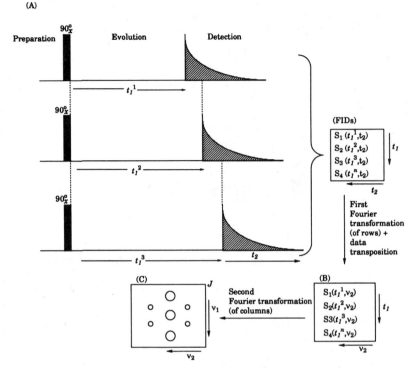

Figure 3.4 Schematic representation of the steps involved in obtaining a two-dimensional NMR spectrum. (A) Many FIDs are recorded with incremented values of the evolution time t_1 and stored. (B) Each of the FIDs is subjected to Fourier transformation to give a corresponding number of spectra. The data are transposed in such a manner that the spectra are arranged behind one another so that each peak is seen to undergo a sinusoidal modulation with t_1. A second series of Fourier transformations is carried out across these columns of peaks to produce the two-dimensional plot shown in (C).

matrix when subject to a second Fourier transformation will furnish a 2D NMR spectrum with both real (absorptive) and imaginary (dispersive) components. The resulting spectrum can be displayed in the absolute-value mode, which contains both these contributions, or in the pure-absorption mode, which contains real contributions only.

It is immaterial which Fourier transformation is carried out first and which second (i.e., whether the function is first transformed with respect to the t_1 or the t_2 variable). Usually, it is more economical to transform individual FIDs (transformation with respect to t_2) and then, after data transposition, to transform the resulting interferograms with respect to t_1 to produce the 2D spectra (or, more precisely, into sections of the 2D NMR spectrum).

There are basically three main types of 2D NMR experiments: *J*-resolved, shift correlation through bonds (e.g., COSY), and shift correlations through space (*e.g.*, NOESY). These spectra may be of homonuclear or heteronuclear type involving interactions between similar nuclei (e.g., protons) or between different nuclear species (e.g., ^1H with ^{13}C).

In 2D *J*-resolved experiments, the precession due to chemical shift frequencies is suppressed by refocusing or reversal in the second half of the evolution period. Precession due to couplings is present in at least one of the two halves of the evolution period. The modulation of the nucleus being observed by the *J*-coupling frequencies of the nuclei with which it is coupled leads to 2D *J*-resolved spectra, which contain chemical shifts of the observed nuclei along one axis (F_2-axis) and their splittings by *J*-coupling along the other axis (F_1-axis).

In shift-correlated spectra (e.g., COSY), the signals of the observed nuclei are modulated by the chemical shift frequencies of the partner nuclei to which they are coupled, and often by *J*-coupling frequencies as well. The result is that after the second Fourier transformation, a 2D plot is obtained in which the unmodulated signals lie along the diagonal axis while cross-peaks lying on either side of the diagonal occur at the chemical shifts of the coupled nuclei, i.e., centered at the coordinates $F_2 = \delta_A$ and $F_1 = \delta_B$, where A and B are the coupled nuclei. In NOESY spectra, the chemical shifts of nuclei lying close to one another in space are correlated.

✦ *PROBLEM 3.1*

What was so exciting about Jeener's original experiment? What is the major difference between 1D and 2D NMR experiments?

─── ✦

✦ *PROBLEM 3.2*

How is it possible to obtain information about the behavior of a spin system during the evolution time t_1?

─── ✦

✦ *PROBLEM 3.3*

When there is only one time variable during a 2D experiment, i.e., t_1, why do we need to process the data through *two* Fourier transformation operations?

─── ✦

✦ *PROBLEM 3.4*

What are the major benefits of 2D NMR spectra?

——————————————————————————————— ✦

✦ *PROBLEM 3.5*

A number of 2D NMR experiments, such as NOESY, have a mixing period incorporated in their pulse sequence. In principle, precession of *xy*-magnetization also occurs during the mixing period. Why do we not need to have a third Fourier transformation to monitor the precession frequencies that occur during the mixing period?

——————————————————————————————— ✦

3.1 DATA ACQUISITION IN 2D NMR

A number of parameters have to be chosen when recording 2D NMR spectra: (a) the pulse sequence to be used, which depends on the experiment required to be conducted, (b) the pulse lengths and the delays in the pulse sequence, (c) the spectral widths SW_1 and SW_2 to be used for F_1 and F_2, (d) the number of data points or time increments that define t_1 and t_2, (e) the number of transients for each value of t_1, (f) the relaxation delay between each set of pulses that allows an equilibrium state to be reached, and (g) the number of preparatory dummy transients (DS) per FID required for the establishment of the steady state for each FID. Table 3.1 summarizes some important acquisition parameters for 2D NMR experiments.

3.1.1 Pulse Sequence

The choice of the pulse sequence to use is of fundamental importance. We must decide carefully what information is required, and choose the right experiment to provide it. Although hundreds of 2D pulse sequences are now available for various experiments, only some have proven themselves to be of general utility. Only such proven techniques should be chosen to solve structural problems.

3.1.2 Setting Pulse Lengths and Delays

Accurate calibration of pulse lengths is essential for the success of most 2D NMR experiments. Wide variations (>20%) in the setting of pulse lengths may significantly reduce sensitivity and may lead to the appearance of artifact signals. In some experiments, such as inverse NMR experiments, accurately set pulse lengths are even more critical for successful outcomes.

Table 3.1
Some Important 2D Acquisition Parameters*

F_1 domain	F_2 domain
$N_1 = TD_1$ = number of FIDs, t_1 domain point pairs	$N_2 = TD_2$ = number of total t_2 domain points (quadrature channels A and B)
SI_1 = total number of data point (size) for F_1 transform; normally, this = $2TD_1$	SI_2 = total number of data point (size) for F_2 transform; normally, this = TD_2
$\pm SW_1$ = frequency domain, in Hz	SW_2 = frequency domain, in Hz (quadrature)
$Hz/PT_1 = 2SW_1/SI_1$ or $4SW_1/SI_1$ (phase sensitive)	$Hz/PT_2 = 2SW_2/SI_2$
$DW_1 = 0.5/SW_1 = t_1$ increment for a single evolution period	$DW_2 = 0.5/SW_2$ (sequential sampling) or $1/SW_2$ (simultaneous sampling)
$DW_1' = DW_{1/2} = t_1$ increment for each half of a split evolution (echo) or for the TPPI method in general	t_2 = increment (dwell time) for quadrature
$t_1 = t_1^o + n(DW_1)$; $n = 0, 1, 2, \ldots$, N_1-1 (evolution period)	$t_2 = t_2^o + n(DW_2)$; $n = 0, 1, 2, \ldots$, N_2-1 (detection period)
$t_1' = t_1/2$	$AQ = t_2^{max}$ (acquisition time)
D_1 = relaxation delay (preparation period)	
t_m, t_d, t_c, etc = fixed mixing periods	
NS = number of transients per FID	
DS = number of preparatory dummy transients per FID	

* Based on the nomenclature and conventions of Bruker software.

✦ PROBLEM 3.6

Why is accurate calibration of pulse widths and delays essential for the success of a 2D NMR experiment?

─── ✦

3.1.3 Spectral Widths (SW₁ and SW₂)

The spectral width SW_2 relates to the frequency domain. With the variation of the evolution period t_1, the intensity and phase of the signals undergo changes that are determined by the behavior of the nuclear spins during the evolution and mixing periods. During the detection period t_2, FIDs are accumulated, each at a discrete value of t_1, and Fourier transformation with respect to t_2 produces a set of spectra each with the frequency domain F_2. Arranging these spectra in *rows*, corresponding to rows of the data (matrix), and Fourier transformation at different points across *columns* of this data matrix creates the second frequency domain F_2. In quadrature detection, the F_2 domain covers the frequency range $SW_2/2$.

Bruker instruments use quadrature detection, with channels A and B being sampled alternately, so the dwell time is given by:

$$DW_2 = \frac{1}{SW_2} \quad \text{and} \quad t_2 = t_2^0 + n(DW_2)$$

where $n = 0, 1, 2, \ldots, N_2 - 1$, t_2^0 is the initial value of t_2, and N_2 is the total number of A and B data points.

In Varian instruments, simultaneous sampling occurs—i.e., points A and B are taken together as pairs—so the dwell time is given by:

$$DW_2 = \frac{1}{SW_2} \quad \text{and} \quad t_2 = t_2^0 + n(DW_2)$$

where $n = 0, 1, 2, \ldots, (N_2/2) - 1$.

In quadrature detection, the transmitter offset frequency is positioned at the center of the F_2 domain (i.e., at $F_2 = 0$; in single-channel detection it is positioned at the left edge). Frequencies to the left (or downfield) of the transmitter offset frequency are positive; those to the right (or upfield) of it are negative.

The spectral width SW_1 associated with the F_1 frequency domain may be defined as $F_1 = SW_1$. The time increment for the t_1 domain, which is the effective dwell time, DW_1 for this period, is related to SW_1 as follows: $DW_1 = (½)SW_1$. The time increments during t_1 are kept equal. In successive FIDs, the time t_1 is incremented systematically: $t_1^0 = t_1 + n(DW_1)$, where $n = 0, 1, 2, 3, \ldots, N_1 - 1$, N_1 is the number of FIDs, t_1^0 is the value of t_1 at the beginning of the experiment, and DW_1 is the time increment for the t_2 domain.

In homonuclear-shift-correlated experiments, the F_1 domain corresponds to the nucleus under observation; in heteronuclear-shift-correlated experiments, F_1 relates to the "unobserved" or decoupled nucleus. It is therefore necessary to set the spectral width SW_1 after considering the 1D spectrum of the nucleus corresponding to the F_1 domain. In 2D *J*-resolved spectra, the value of SW_1 depends on the magnitude of the coupling constants and the type of experiment. In both homonuclear and heteronuclear experiments, the size of the largest multiplet structure, in hertz, determines

SW_1, which in turn is related to the homonuclear or heteronuclear coupling constants. In homonuclear 2D spectra, the transmitter offset frequency is kept at the center of F_1 (i.e., at $F_1 = 0$) and F_2 domains. In heteronuclear-shift-correlated spectra, the decoupler offset frequency is kept at the center ($F_1 = 0$) of the F_1 domain, with the F_1 domain corresponding to the "invisible" or decoupled nucleus.

✦ *PROBLEM 3.7*

Why are we much more likely to have signals outside the spectral width (SW) in an average 2D NMR experiment than in a 1D NMR experiment? Why do spectral widths in a 2D NMR need to be defined very carefully, and what effects will this have on the spectrum?

———————————————————————————————— ✦

3.1.4 Number of Data Points in t_1 and t_2

The effective resolution is determined by the number of data points in each domain, which in turn determines the length of t_1 and t_2. Thus, though digital resolution can be improved by zero-filling (Bartholdi and Ernst, 1973), the basic resolution, which determines the separation of close-lying multiplets and line widths of individual signals, will not be altered by zero-filling.

The effective resolution R for N time-domain data points is given by the reciprocal of the acquisition time, AQ: $R = 1/AQ = 2SW/N$. Since there must be at least two data points to define each signal, the digital resolution DR (in hertz per point, or, in other words, the spacing between adjoining data points) must be $DR = R/2 = 2SW/2N$. The real part of the spectrum is therefore obtained by the transform of $2N$ data points, of which N are time-domain points and another N are zeros. This level of zero-filling allows the resolution contained in the time domain to be recorded. Additional zero-filling will lead only to the appearance of a smoother spectrum, but will not cause any genuine improvement in resolution.

The number of data points to be chosen in the F_1 and F_2 domains is dictated not only by the desired resolution but by other, external considerations, such as the available storage space in the computer, and the time that can be allocated for data acquisition, transformation, and other instrument operations. Clearly, to avoid any unnecessary waste of time, we should choose the *minimum* resolution that would yield the desired information. Thus, if the peaks are separated by at least 1 Hz, then the desired digital resolution should be $R/2 = \frac{1}{2}$ Hz to allow for signal separation in the F_2 domain. The resolution considerations in the F_2 and F_1 domains may be different, depending on what information is required from each domain in the experiment. Since the acquisition time AQ is small as compared to the total length of the pulse sequence with its various relaxation delays, it is preferable to choose N_2 (the number of data points in F_2) based on the final digital resolution required *without zero-filling*. The sensitivity and resolution in the F_2 domain will be improved as long as t_2 is equal to or less than the effective relaxation time T_2.

In the case of the t_1 domain, since it is only the number N of data points that determines the resolution, and not the time involved in the pulse sequence with various delays, it is advisable to acquire only half the theoretical number of FIDs and to obtain the required digital resolution by zero-filling. Thus the resolution R_1 in the F_1 domain will be given by $R_1 = 2SW_1/N_1$; that in the F_2 domain is given by $R_2 = 1/AQ = 2SW_2/N_2$.

✦ **PROBLEM 3.8**

What factors dictate the choice of the number of data points (SI_1 and SI_2) in 2D NMR spectroscopy?

─── ✦

3.1.5 Number of Transients

The number of transients (NS) required varies depending on the experiment to be performed. For most experiments, four transients are required to provide quadrature detection and suppress the axial peaks at $F_1 = 0$. Additional phase cycles may be necessary involving 16 or 32 transients if artifacts arising from imperfect 180° pulses are to be suppressed or if quadrature images appear in the F_2 dimension. The number of transients will therefore be the number of cycles in the phase cycling routine multiplied by the minimum number N_1 of FIDs to be acquired. The signal-to-noise ratio in the 2D plot will be defined by the number of transients (Levitt *et al.*, 1984), provided that the time domains are not significantly greater than the relaxation times T_2 or T_2^*. Normally, only the minimal transients required should be used, unless the sample quantities are very small or the experiment is insensitive. Table 3.2 summarizes some of the data requirements for the common NMR experiments.

✦ **PROBLEM 3.9**

What factors govern the minimum number of transients (NS)?

─── ✦

3.1.6 Dummy Scans and Relaxation Delays between Successive Pulses

To reach a steady state before data acquisition, a certain number of "dummy" scans are usually required. If the relaxation delay between the

Table 3.2

Summary of Data Requirements for 2D NMR experiments

Experiment	¹H Freq. (MHz)	TD₂ = SI₂ (words)	Hz/PT₂	TD₁ = SI1/₂ (words)	Hz/PT1	NS	DS	T_r (s)	Time (h)	FIDs
¹H J-resolved (4.5 ppm, ± 18 Hz)	400	4K	0.872	32	0.436	16	2	1.5	0.58	32
¹H COSY, NOESY (4.5 ppm)	400	1K	3.488	256	3.488	16	2	1.5	2.65	256
						32	2	2	8.0	256
¹H COSY-TPPI (4.5 ppm)	400	1K	3.488	256	3.488	16	2	1.5	3.83	256
¹³C–¹H J-res. (gated dec.) (142 ppm, ± 120 MHz)	400	8K	3.488	256	0.977	16	2	2	3.38	256
¹³C–¹H correlated (94 × 4.5 ppm)	400	2K	9.213	256	2,654	16	2	2	3.0	256
¹³C INADEQUATE (142 ppm)	400	4K	6.975	256	55,804	32	4	3	32.0	256

TD_2, t_2 time domain data points; TD_1, t_1 time domain pairs of data points; SI_2 and SI_1, total data points in F_2 and F_1 domains, respectively; Hz/PT$_2$ and Hz/PT$_1$, digital resolution in F_2 and F_1 (real) domains; NS, number of acquired transients; DS, number of dummy transients; Tr, recycle time.

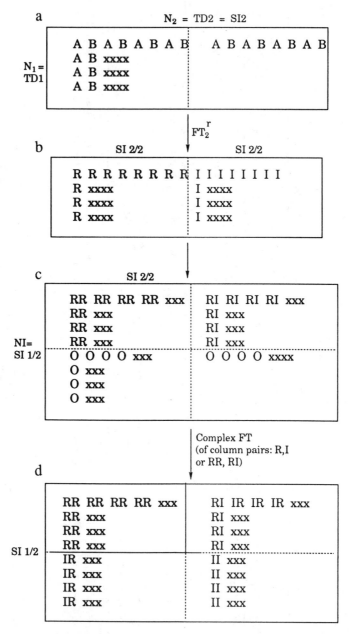

Figure 3.5 Schematic representation of data processing in a 2D experiment (one zero-filling in F_1 and two zero-fillings in F_2). (a) N_1 FIDs composed of N_2 quadrature data points, which are acquired with alternate (sequential) sampling. (b) On a real

repetition time of the pulse sequence is sufficiently long ($5T_1$ of the slowest-relaxing nuclei), then dummy scans may not be necessary. In practice it is usual to have the relaxation delay set at about $2T_1$ or $3T_1$ s and to use two dummy scans if the number of transients (NS) is 4 or 8, or to use four dummy scans if NS is greater than 16.

✦ *PROBLEM 3.10*

What are "dummy" scans, and why are a number of these scans acquired before actual data acquisition?

─── ✦

3.2 DATA PROCESSING IN 2D NMR

At the end of the 2D experiment, we will have acquired a set of N_1 FIDs composed of N_2 quadrature data points, with $N_2/2$ points from channel A and $N_2/2$ points from channel B, acquired with sequential (alternate) sampling. How the data are processed is critical for a successful outcome. The data processing involves (a) dc (direct current) correction (performed automatically by the instrument software), (b) apodization (window multiplication) of the t_2 time-domain data, (c) F_2 Fourier transformation and phase correction, (d) window multiplication of the t_1 domain data and phase correction (unless it is a magnitude or a power-mode spectrum, in which case phase correction is not required), (e) complex Fourier transformation in F_1, (f) coaddition of real and imaginary data (if phase-sensitive representation is required) to give a magnitude (M) or a power-mode (P) spectrum. Additional steps may be tilting, symmetrization, and calculation of projections. A schematic representation of the steps involved is presented in Fig. 3.5.

The first set of Fourier transformations in 2D NMR experiments, such as COSY, produces signals in the F_2 dimension. These signals have real (R) and imaginary (I) components. The second set of Fourier transformations across t_1 gives signals in the F_1 domain, which also have real (R) and imaginary (I) components, leading to four different quadrants arising from

───

Fourier transformation in t_2 (Ft_2'), N_1 spectra are obtained with real (R) and imaginary (I) data points. For detection in the quadrature mode with simultaneous sampling, a complex Fourier transformation is performed, with a phase correction being applied in F_2. (c) A normal phase-sensitive transform: $R \rightarrow RR$ and $I \rightarrow RI$. (d) Complex FT is applied to pairs of columns, which produces four quadrants, of which only the RR quadrant is plotted.

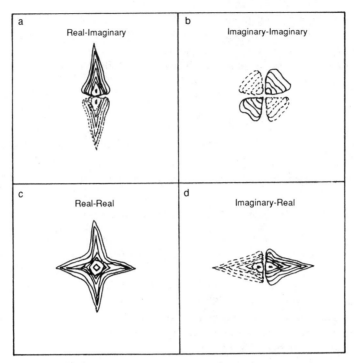

Figure 3.6 The first set of Fourier transformations across t_2 yields signals in ν_2, with absorption and dispersion components corresponding to real and imaginary parts. The second FT across t_1 yields signals in ν_1, with absorption (i.e., real) and dispersion (i.e., imaginary) components; quadrants (a), (b), (c), and (d) represent four different combinations of real and imaginary components and four different line shapes. These line shapes normally are visible in phase-sensitive 2D plots.

the four possible combinations of the real and imaginary components (*RR, RI, IR,* and *II*). The pure absorption-mode signals correspond to the real-real (*RR*) quadrant (Fig. 3.6). Each of these data processing procedures will be considered next, briefly.

3.2.1 DC Correction

Since there is some contribution of the receiver dc to the signal, this needs to be removed. In 1D experiments, the FIDs decay substantially, and the last portion of the FID gives a reasonably good estimate of the dc level.

The computer works out the average dc level automatically, and subtracts it from each of the two quadrature data sets.

In 2D NMR experiments, the FIDs are relatively short and with fewer data points, so dc correction is more difficult to carry out accurately. Phase cycling procedures are recommended whenever required to remove dc offsets before dc correction, which is carried out before the first (F_2) Fourier transform. Since the data points in the F_1 transform arise from frequency-domain spectra, no dc correction is normally required (we expect to see unchanging dc components at $F_1 = 0$).

✦ PROBLEM 3.11

What is dc correction?

─── ✦

✦ PROBLEM 3.12

Why is it not necessary to perform dc correction before the second (F_1) Fourier transformation?

─── ✦

3.2.2 Peak Shapes in 2D NMR Spectra

The apodization functions mentioned earlier have been applied extensively in 1D NMR spectra, and many of them have also proved useful in 2D NMR spectra. Before discussing the apodization functions as employed in 2D NMR spectra we shall consider the kind of peak shapes we are dealing with.

The free induction decay may be considered as a complex function of frequency, having real and imaginary coefficients. There are two coefficients in each point of the frequency spectrum, and these real and imaginary coefficients may be described as the cosine and sine terms, respectively. Each set of coefficients occupies half the memory in the time-domain spectrum. Thus, if memory size is N, then there will be $N/2$ real and $N/2$ imaginary data points in the spectrum. Through the process of phasing, we can display signals in the absorption mode. Such *pure 2D absorption peaks* are characterized by the following property: If we take sections across them parallel to the F_1 or F_2 axes, they produce pure 1D absorption Lorentzian lines (Fig. 3.7a and b). However, they lack cylindrical or elliptical symmetry, and display protruding ridges running down the lines parallel to F_1 and F_2, so they give star-shaped cross-sections (Fig. 3.7c). This distortion can be suppressed by Lorentz-Gaussian transformation.

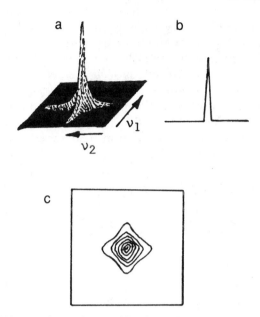

Figure 3.7 Different views of pure 2D absorption peak shapes: (a) a 3D view, (b) a vertical cross-section, i.e., a vertical 2D view, and (c) a horizontal 2D view.

The *pure negative 2D dispersion peak shapes* are shown in Fig. 3.8. The peaks have negative and positive lines, with vanishing signal contributions as they pass through the base line; and they are broadened at the base, so they decay slowly, leading to a poor signal-to-noise ratio. Horizontal cross-sections of such peaks give the appearance of butterfly wings. To improve the signal-to-noise ratio as well as the appearance of 2D spectra, it is desirable to remove the dispersive components.

The *phase-twisted peak shapes* (or mixed absorption-dispersion peak shape) is shown in Fig. 3.9. Such peak shapes arise by the overlapping of the absorptive and dispersive contributions in the peak. The center of the peak contains mainly the absorptive component, while as we move away from the center there is an increasing dispersive component. Such mixed phases in peaks reduce the signal-to-noise ratio; complicated interference effects can arise when such lines lie close to one another. Overlap between positive regions of two different peaks can mutually reinforce the lines (constructive interference), while overlap between positive and negative lobes can mutually cancel the signals in the region of overlap (destructive interference).

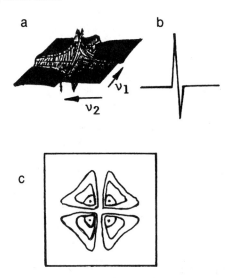

Figure 3.8 Different views of pure negative 2D dispersion peak shapes: (a) a 3D view, (b) a vertical cross-section, and (c) a horizontal cross-section.

Two-dimensional spectra are often recorded in the absolute-value mode. The absolute value A is the square root of the sum of the squares of the real (R) and imaginary (I) coefficients:

$$A = \sqrt{R^2 + I^2}. \tag{1}$$

Although this eliminates negative contributions, since the imaginary part of the spectrum is also incorporated in the absolute-value mode, it produces broad dispersive components. This leads to the broadening of the base of the peaks ("tailing"), so lines recorded in the absolute-value mode are usually broader and show more tailing than those recorded in the pure absorption mode.

✦ *PROBLEM 3.13*

What are the common peak shapes, and why it is necessary to know the peak shapes before applying apodization?

─── ✦

3.2.2.1 APODIZATION IN 2D NMR SPECTRA

Some of the apodization functions described in Section 1.3.11 can also be adopted for use in 2D spectra. The types of functions used will vary according to the experiment and the information wanted.

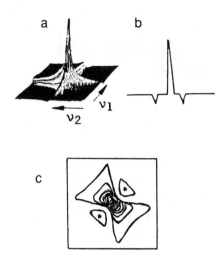

Figure 3.9 Different views of an absorption-dispersion peak: (a) a 3D view, (b) a 2D vertical cross-section, and (c) a 2D horizontal cross-section.

The majority of 2D experiments produce lines with phase-twisted line shapes that cannot be satisfactorily phased. An absolute-value display is then most convenient, since phasing is not required, the lines having positive components only. However, the lines have long dispersive tails, so we need to apply strong weighting functions to suppress such tailing, which would otherwise cause undesirable interference effects between neighboring signals. A price has to be paid for such cosmetic improvement of spectra in terms of sensitivity, since multiplication by weighting functions leads to rejection of some of the signals present in the early part of the FID.

The weighting functions used to improve line shapes for such absolute-value-mode spectra are *sine-bell, sine bell squared, phase-shifted sine-bell, phase-shifted sine-bell squared,* and a *Lorentz-Gauss transformation* function. The effects of various window functions on COSY data (absolute-value mode) are presented in Fig. 3.10. One advantage of multiplying the time domain $S(t_1)$ or $S(t_2)$ by such functions is to enhance the intensities of the cross-peaks relative to the noncorrelation peaks lying on the diagonal.

Another resolution-enhancement procedure used is *convolution difference* (Campbell *et al.,* 1973). This suppresses the ridges from the cross-peaks and weakens the peaks on the diagonal. Alternatively, we can use a shaping function that involves production of *pseudoechoes.* This makes the envelope of the time-domain signal symmetrical about its midpoint, so the dispersion-mode contributions in both halves are equal and opposite in sign (Bax *et al.,* 1979,1981). Fourier transformation of the pseudoecho produces signals

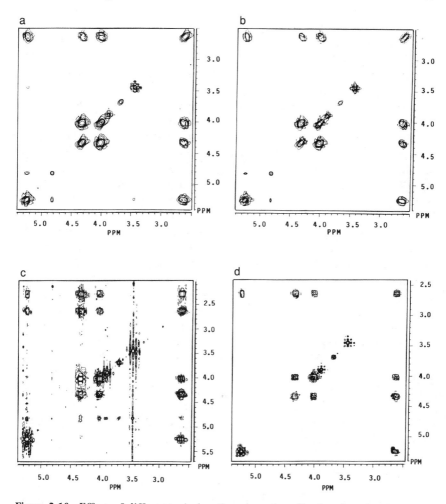

Figure 3.10 Effect of different window functions (apodization functions) on the appearance of COSY plot (magnitude mode). (a) Sine-bell squared and (b) sine-bell. The spectrum is a portion of an unsymmetrized matrix of a ^{1}H-COSY LR experiment (400 MHz in CDCl$_3$ at 303 K) of vasicinone. (c) Shifted sine-bell squared with $\pi/4$. (d) Shifted sine-bell squared with $\pi/8$. (a) and (b) are virtually identical in the case of delayed COSY, whereas sine-bell squared multiplication gives noticeably better suppression of the stronger dispersion-mode components observed when no delay is used. A difference in the effective resolution in the two axes is apparent, with F_2 having better resolution than F_1. The spectrum in (c) has a significant amount of dispersion-mode line shape.

in the pure-absorption mode. A Gaussian shaping function may be employed with the pseudoecho to give it a symmetrical envelope.

The sine-bell functions are attractive because, having only one adjustable parameter, they are simple to use. Moreover, they go to zero at the end of the time domain, which is important when zero-filling to avoid artifacts. Generally, the sine-bell squared and the pseudoecho window functions are the most suitable for eliminating dispersive tails in COSY spectra.

With COSY and 2D *J*-resolved spectra, it is normally necessary to apply weighting functions in both dimensions. Multiplication with a sine-bell squared function is recommended.

Phase-sensitive spectra, such as phase-sensitive COSY, do not show phase-twisted line shapes that arise due to the mixing of the absorptive and dispersive components. The peaks lying on or near the diagonal tend to differ in phase by 90° from the correlation cross-peaks, which appear on either side of the diagonal. This means that if we adjust the phasing of the cross-peaks to the absorption mode, then the diagonal peaks will appear in the dispersive mode, and vice versa. Since it is the cross-peaks that are of interest, we normally adjust the phasing so they appear in the absorptive mode. Improvements in peak shape can be achieved by multiplying with a cosine-bell (sine-bell shifted by 90°) function.

Heteronuclear-shift-correlation spectra, which are usually presented in the absolute-value mode, normally contain long dispersive tails that are suppressed by applying a Gaussian or sine-bell function in the F_1 domain. In the F_2 dimension, the choice of a weighting function is less critical. If a better signal-to-noise ratio is wanted, then an exponential broadening multiplication may be employed. If better resolution is needed, then a resolution-enhancing function can be used.

✦ PROBLEM 3.14

What are the effects of various window functions on the shapes of peaks in a COSY data set?

——————————————————————————————————— ✦

3.2.2.2 F_2 FOURIER TRANSFORMATION

The next step after apodization of the t_2 time-domain data is F_2 Fourier transformation and phase correction. As a result of the Fourier transformations of the t_2 time domain, a number of different spectra are generated. Each spectrum corresponds to the behavior of the nuclear spins during the corresponding evolution period, with one spectrum resulting from each t_1 value. A set of spectra is thus obtained, with the rows of the matrix now containing N real and N imaginary data points. These real and imaginary parts are rearranged so the individual spectra are in rows one behind the other in a matrix constituted by the F_2 and t_1 domains.

3.2.2.3 F_1 Fourier Transformation and Window Multiplication

The matrix obtained after the F_2 Fourier transformation and rearrangement of the data set contains a number of spectra. If we look down the columns of these spectra parallel to t_1, we can see the variation of signal intensities with different evolution periods. Subdivision of the data matrix parallel to t_1 gives columns of data containing both the real and the imaginary parts of each spectrum. An equal number of zeros is now added and the data sets subjected to Fourier transformation along t_1. This Fourier transformation may be either a Redfield transform, if the t_2 data are acquired alternately (as on the Bruker instruments), or a complex Fourier transform, if the t_2 data are collected as simultaneous A and B quadrature pairs (as on the Varian instruments). Window multiplication for t_1 may be with the same function as that employed for t_2 (e.g., in COSY), or it may be with a different function (e.g., in 2D *J*-resolved or heteronuclear-shift-correlation experiments).

✦ PROBLEM 3.15

The two-dimensional data set $S(t_1, t_2)$ requires two Fourier transformation operations. Explain why the time variable t_2 is almost always Fourier transformed before t_1.

                                                           ✦

3.2.2.4 Magnitude-Mode Spectra

If phase-sensitive spectra are not required, then magnitude-mode $P(\omega)$ (or "absolute-mode") spectra may be recorded by combining the real and imaginary data points. These produce only positive signals and do not require phase correction. Since this procedure gives the best signal-to-noise ratio, it has found wide use. In heteronuclear experiments, in which the dynamic range tends to be low, the power-mode spectrum may be preferred, since the S/N ratio is squared and a better line shape is obtained so that wider window functions can be applied.

✦ PROBLEM 3.16

How are magnitude-mode spectra computed from the FID?

                                                           ✦

3.2.2.5 Tilting, Symmetrization, and Projections

Finally, certain other procedures—such as tilting, symmetrization, or plotting of projections—may be required, depending on the type of spectra being recorded. In homonuclear 2D *J*-resolved experiments, *J*-coupling information is present in both t_1 and t_2 domains, so multiplets appear at an angle of 45°. A tilt-correction procedure (Baumann *et al.*, 1981a) is therefore applied mathematically to produce orthogonal *J* and δ axes, making the multiplets more readable. However, this tilt correction leads to a 45° tilt in the F_1 line shape. This can be suppressed by using a window function for F_1, such as multiplication by sine-bell squared, to reduce the tailing of signals in F_1. Symmetrization procedures can also be employed for cosmetic improvement of the spectrum when the peaks occurring in a spectrum are symmetrically arranged—such as the multiplets in a 2D *J*-resolved spectrum, or when the cross-peaks lie symmetrically on either sides of the diagonal in COSY and NOESY spectra. Symmetrization can be carried out by *triangular multiplication,* (Baumann *et al.*, 1981b) in which each pair (*a, b*) of symmetry-related points is replaced by the geometric mean $\sqrt{ab}$. A simpler symmetrization procedure is to replace each pair of symmetry related points by the smaller of the two values *a* and *b*.

We can take slices at various points of the 2D spectrum, along either the F_1 or the F_2 axes or, alternatively, we can record projections. Such subspectra can provide useful information. There are two ways such projections can be produced. The first method involves coaddition through a row or column to create a point in the projection spectrum. In this so-called "sum" mode, weak multiplets in the 2D spectrum (e.g., the 2D *J*-resolved spectrum) will appear much stronger than the noise in the projection spectrum. One disadvantage is a broadening of signals due to the addition of "tails" from neighboring peaks. The other "maximum point" method relies on the selection of the maximum data point from each column or row. This improves line shapes but produces weak signals, and some multiplets may be too weak to be recognizable in the projection spectrum. *In homonuclear experiments, it is therefore advisable to record F_2 projections rather than F_1 projections because of the resulting higher resolution and sensitivity.*

✦ *PROBLEM 3.17*

What could be the possible reasons for noise in 2D NMR spectra, and how can symmetrization be used to improve the quality and readability of the plot?

─── ✦

✦ *PROBLEM 3.18*

What are projection spectra, and how are they different from normal 1D NMR spectra?

✦

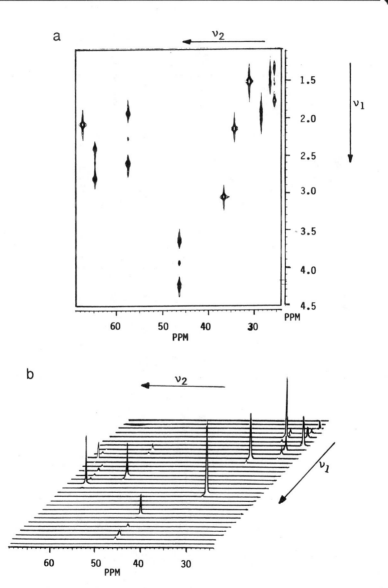

Figure 3.11 (a) Contour plot. (b) Stacked plot.

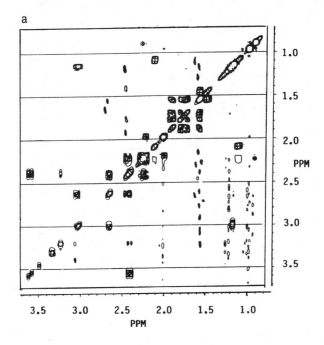

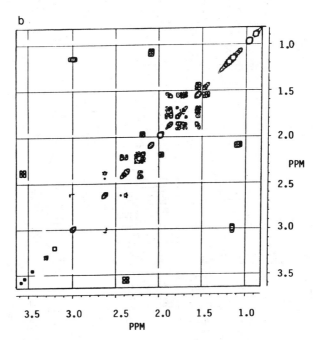

Figure 3.12 Two-dimensional NMR plots recorded at different contour levels. (a) Two-dimensional spectra recorded at low contour level usually have noise lines across the plot. (b) With a higher contour level, many of the noise peaks are eliminated and the peaks become clearer.

3.3 PLOTTING OF 2D SPECTRA

Two-dimensional NMR spectra are normally presented as contour plots (Fig. 3.11a), in which the peaks appear as contours. Although the peaks can be readily visualized by such an "overhead" view, the relative intensities of the signals and the structures of the multiplets are less readily perceived. Such information can be easily obtained by plotting slices (cross-sections) across rows or columns at different points along the F_1 or F_2 axes. Stacked plots (Fig. 3.11b) are pleasing esthetically, since they provide a pseudo-3D representation of the spectrum. But except for providing information about noise and artifacts, they offer no advantage over contour plots. Finally, the projection spectra mentioned in the previous section may also be recorded.

The recording of contour plots involves selecting the optimum contour level. At too low a contour level, noise signals may make the genuine cross-peaks difficult to recognize; if the contour level is too high, then some of the weaker signals may be eliminated along with the noise (Fig. 3.12).

In the discussions that follow, the theory behind the common 2D NMR experiments is presented briefly, the main emphasis being on how newcomers can solve practical problems utilizing each type of experiment.

✦ *PROBLEM 3.19*

How many types of plots are generally used in 2D NMR spectroscopy?

── ✦

SOLUTIONS TO PROBLEMS

✧ *3.1*

Jeener's idea was to introduce an incremented time t_1 into the basic 1D NMR pulse sequence and to record a series of experiments at different values of t_1, thereby adding a second dimension to NMR spectroscopy. Jeener described a novel experiment in which a coupled spin system is excited by a sequence of *two* pulses separated by a variable time interval t_1. During these variable intervals, the spin system is allowed to evolve to different extents. This variable time t_1 is therefore termed the *evolution time*. The insertion of a variable time period between two pulses represents the prime feature distinguishing 2D NMR experiments from 1D NMR experiments.

✧ *3.2*

Information about the behavior of the spin system during the evolution time can be obtained indirectly by observing its influence on a set of FIDs. Since the FIDs obtained portray the positions and status of the magnetization vectors, we can map the behavior of the nuclei during the evolution time from the FIDs.

✧ *3.3*

There are actually two independent time periods involved, t_1 and t_2. The time period t_1 after the application of the first pulse is incremented systematically, and separate FIDs are obtained at each value of t_1. The second time period, t_2, represents the detection period and it is kept constant. The first set of Fourier transformations (of rows) yields frequency-domain spectra, as in the 1D experiment. When these frequency-domain spectra are stacked together (data transposition), a new data matrix, or "pseudo-FID," is obtained, $S(t_1, F_2)$, in which absorption-mode signals are modulated in amplitude as a function of t_1. It is therefore necessary to carry out second Fourier transformation to convert this "pseudo FID" to frequency domain spectra. The second set of Fourier transformations (across columns) on $S(t_1, F_2)$ produces a two-dimensional spectrum $S(F_1, F_2)$. This represents a general procedure for obtaining 2D spectra.

✧ *3.4*

Some important benefits of 2D NMR spectroscopy are:

(i) *J*-Modulated 2D NMR experiments (2D *J*-resolved) provide an excellent method to resolve highly overlapping resonances into readily interpretable and recognizable multiplets.

(ii) Homonuclear-shift-correlation 2D NMR techniques (COSY, NOESY, etc.) offer a way of identifying spin couplings of nuclei. By interpretation of homonuclear 2D correlated spectra, we can obtain information about the spin networks. Similarly, heteronuclear-shift-correlation 2D NMR techniques help to identify one-bond or long-range heteronuclear connectivities.

(iii) Some recently developed 2D NMR techniques, such as Homonuclear Hartmann-Hahn Spectroscopy (HOHAHA) and Heteronuclear Multiple Bond Connectivity (HMBC), are excellent ways to identify nuclei separated by several bonds, thereby allowing chemists to build complex structures from substructural fragments.

(iv) 2D INADEQUATE spectra can allow the direct deduction of the carbon framework of the molecules.

✧ *3.5*

Precession of *xy*-magnetization may also occur during the mixing period. This is normally a constant effect that does not vary from one FID to the next, since the mixing period is a constant time period and changes taking place during it contribute equally in every FID.

✧ *3.6*

Although accurate calibration of pulse lengths and delays is desirable for every type of NMR experiment, it becomes essential in 2D NMR experiments. Errors in pulse length greater than 10–20% usually significantly lower the S/N ratio and increase the amplitude of artifact peaks. When such errors approach ±40–50%, the spectra become severely distorted. This is easily understood, since a 180° pulse with 50% error becomes a 90° pulse and the pulse sequence is transformed into a totally different pulse sequence. Similarly, serious errors in mixing delays, expressed in terms of scalar couplings (i.e., $1/nJ$) are generally catastrophic for the experiment, and they may not only lower the S/N ratio, but may cause complete loss of certain signals. This is illustrated by the following example.

(i) COSY 45° spectrum with accurate pulse lengths

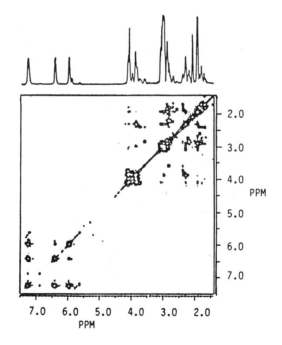

(ii) COSY 45° with poor pulse lengths (45% error) affords cross-peaks with much weaker intensities

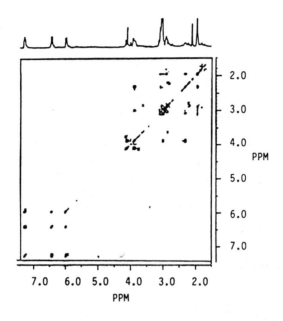

✧ *3.7*

In a 2D NMR experiment it is essential to minimize the size of the data set by cutting down the spectral width so that valuable instrumentation time is not wasted in accumulating noise. However, if the spectral width chosen is too small, then the signals falling outside the spectral width give rise to erroneously positioned peaks, either on the diagonal line or on one of the axes of the 2D spectra. The spectral widths (SW_1 and SW_2) in 2D NMR experiments therefore need to be defined carefully, since they depend not only on the type of the nuclei being detected but also on the kind of experiment. The spectral width SW_2 (for the F_2 domain) is generally defined by the nature of the nucleus being observed during the detection period and the spectral regions to be acquired, while the spectral width SW_1 (for the F_1 domain) depends mostly on the type of experiment. For instance, SW_1 in homonuclear-shift-correlation experiments depends on the observed nucleus, whereas in heteronuclear-shift-correlation experiments it corresponds to the "unobserved" or decoupled nucleus. For J-resolved 2D NMR experiments, SW_1 corresponds to the coupling constants. The complexities in defining SW are the prime reason for peak folding problems encountered in 2D NMR spectroscopy.

❖ *3.8*

The following factors are considered when determining the number of data points, SI, in 2D NMR experiments:

(i) Desired resolution.
(ii) Time available for data acquisition.
(iii) Time required for FT and spectral manipulation.

❖ *3.9*

The requirement of the minimum number of transients (NS) depends on the following factors:

(i) The insensitivity of the 2D experiment; e.g., 32 transients are required in case of the inherently insensitive 2D INADE-QUATE experiment, in contrast to 16 transients for the more sensitive 2D *J*-resolved spectrum.

(ii) To attain the required sensitivity, for example, in order to detect very small couplings, it may become necessary to accumulate more transients.

(iii) Sample quantities: large quantities of sample require a minimum number of transients.

(iv) Appearance of artifacts and quadrature images in the spectrum force more transients to be recorded.

❖ *3.10*

"Dummy" scans are the preparatory scans with the complete time course of the experiment (pulses, evolution, delays, acquisition time). A certain number of these "dummy" scans are generally acquired before each FID in order to attain a stable steady state. Though time-consuming, they are extremely useful for suppressing artifact peaks.

❖ *3.11*

Dc correction is a process by which the contribution of the receiver dc is omitted from the FID. The dc level is generally determined by examining the last (one-fourth) portion of the FID (tail), which is more likely to have the maximum dc contribution of the receiver. The level is then subtracted automatically from each FID of the data set before F_2 Fourier transformation.

❖ *3.12*

Since data points in F_1 transformation are taken directly from the frequency-domain spectra (resulting from F_2 transformations), there is no need for a second dc correction.

❖ *3.13*

There are generally three types of peaks: pure 2D absorption peaks, pure negative 2D dispersion peaks, and phase-twisted absorption-dispersion peaks. Since the prime purpose of apodization is to enhance resolution and optimize sensitivity, it is necessary to know the peak shape on which apodization is planned. For example, absorption-mode lines, which display protruding ridges from top to bottom, can be dealt with by applying Lorentz-Gauss window functions, while phase-twisted absorption-dispersion peaks will need some special apodization operations, such as muliplication by sine-bell or phase-shifted sine-bell functions.

❖ *3.14*

The sine-bell, sine-bell squared, phase-shifted sine-bell, and phase-shifted sine-bell squared window functions are generally used in 2D NMR spectroscopy. Each of these has a different effect on the appearance of the peak shape. For all these functions, a certain price may have to be paid in terms of the signal-to-noise ratio, since they remove the dispersive components of the magnitude spectrum. This is illustrated in the following COSY spectra:

(i) COSY 45° spectrum with sine-bell multiplication

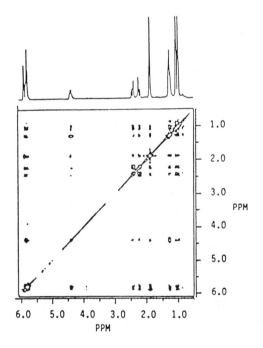

(ii) COSY 45° spectrum with sine-bell squared multiplication

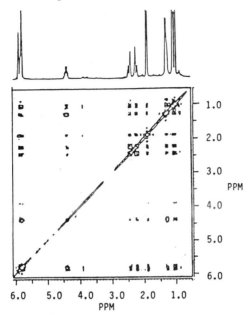

(iii) COSY 45° spectrum with phase-shifted sine-bell multiplication

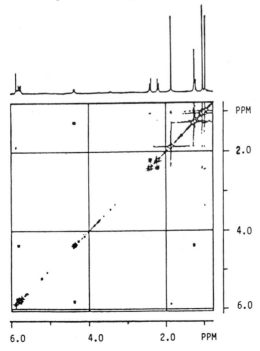

(iv) COSY 45° spectrum with phase-shifted sine-bell squared multi-
      plication

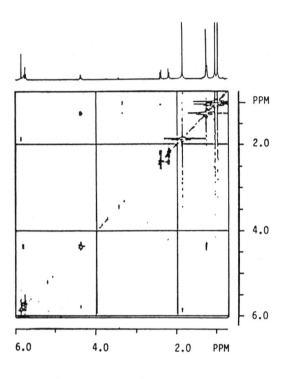

3.15

The second time variable, t_2, is the so called "real-time" variable, representing the time spent in data acquisition, while t_1 is the evolution
interval between the two pulses during the experiment. The effect of
t_1 can only be observed indirectly by noting its influence on t_2. It is
therefore necessary to carry out F_2 transformation first, in order to
generate a series of spectra (rows of the matrix), which is then used
as a pseudo-FID for F_1 transformation in the second step.

✧ *3.16*

The frequency-domain spectrum is computed by Fourier transformation of the FIDs. Real and imaginary components $v(\omega)$ and $i\mu(\omega)$ of
the NMR spectrum are obtained as a result. Magnitude-mode or power-
mode spectra $P(\omega)$ can be computed from the real and imaginary parts
of the spectrum through application of the following equation:

$$P(\omega) = \sqrt{[v(\omega)]^2 + [i\mu(\omega)]^2}.$$

✧ *3.17*

The noise in 2D experiments may arise from a variety of sources, such as unequal increments in evolution time, phase noise in the frequency generation system, quarks in the computer handling of successive FIDs, and quantization errors or digitizer noise. Symmetrization is a cleaning routine that can remove noise from the 2D plot. This process is generally used in 2D spectra that have inherent symmetry, such as COSY, NOESY, and 2D *J*-resolved experiments. In these spectra the genuine signals are either symmetrically arranged about the diagonal (COSY) or symmetrically arranged about the central peak of a multiplet (2D *J*-resolved). The noise is random, and can be easily eliminated by this simple mathematical operation. Only peaks with a symmetry-related partner are retained.

✧ *3.18*

One-dimensional spectra obtained by projecting 2D spectra along a suitable direction often contain information that cannot be obtained directly from a conventional 1D spectrum. They therefore provide chemical shift information of individual multiplets that may overlap with other multiplets in the corresponding 1D spectra. The main difference between the projection spectrum and the 1D spectrum in shift-correlated spectra is that the projection spectrum contains only the signals that are coupled with each other, whereas the 1D ^{1}H-NMR spectrum will display signals for all protons present in the molecule.

✧ *3.19*

There are basically four types of plots:

(a) Projection plots
(b) Contour plots
(c) Stacked plots for pseudo-3D presentation
(d) Slice plots for 1D presentation.

The following are some important presentations of a COSY 45° spectrum:

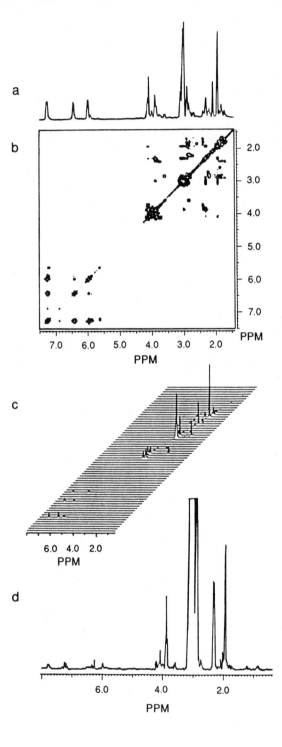

a

b

2.0
3.0
4.0
5.0
6.0
7.0
PPM

7.0   6.0   5.0   4.0   3.0   2.0   PPM
PPM

c

6.0 4.0 2.0
PPM

d

6.0   4.0   2.0
PPM

REFERENCES

Bartholdi, E., and Ernst, R. R. (1973). *J. Magn. Reson.* **11**, 9.

Baumann, R., Kumar, A., Ernst, R. R., and Wüthrich, K. (1981a). *J. Magn. Reson.* **44**, 76.

Baumann, R., Wider, G., Ernst, R. R., and Wüthrich, K. (1981b). *J. Magn. Reson.* **44**, 402.

Bax, A., Mehlkopf, A. F., and Smidt, J. (1979). *J. Magn. Reson.* **35**(3), 373–377.

Bax, A., Freeman, R., and Morris, G. A. (1981). *J. Magn. Reson.* **43**, 333.

Campbell, I. D., Dobson, C. M., Williams, R. J. P., and Xavier, A. V. (1979). *J. Magn. Reson.* **11**(2), 172–181.

Levitt, M. H., Bodenhausen, G., and Ernst, R. R. (1984). *J. Magn. Reson.* **58**, 462.

CHAPTER 4

Nuclear Overhauser Effect

The strength of an NMR signal depends on the difference in population between the ground (α) state and the excited (β) state of a given nucleus. There is a slight excess (Boltzmann excess) of nuclei in the ground state, and it is this excess that is responsible for the NMR signal. Even at 400 MHz, the difference in population between the two states is very small (about 1 in 660,000), and it gives rise to a macroscopic magnetic moment M_z° parallel to the static magnetic field B_0 having a magnitude corresponding to the population difference between the α and β states. This equilibrium difference magnetization M_z° represents the *longitudinal* magnetization, which is directed along the z-axis, i.e., parallel to the applied magnetic field, and it precesses about B_0 at its characteristic Larmor frequency. Application of an Rf pulse bends the direction of the equilibrium magnetization M_z° away from the z-axis and thus creates a transverse component of magnetization M_{xy}, which induces an alternating current within the receiver coil of the NMR spectrometer and is thereby detected as an NMR signal. Clearly, the intensity of this signal will depend on the gyromagnetic ratio of the nucleus under observation, on the population difference between the α and β states of the nucleus that determine the intrinsic strength of the magnetizations, as well as on the angle by which the vector M_z° is bent by the applied Rf pulse. Thus the population difference between the upper and lower states, represented by the longitudinal magnetization, is "sampled" by the Rf pulse, which converts it into a corresponding, proportionately sized transverse magnetization before detection.

The Rf observe pulse generally has very high power, often 100 W or above. It therefore precesses with a very high frequency, typically about 25 kHz, so the time taken to bend the equilibrium magnetization is short, often of the order of 10 ms. Such a pulse would excite the entire proton spectral width uniformly. Broad-band decoupling, on the other hand, does not use such high power, since it has to be applied continuously (for instance, proton noise decoupling during ^{13}C acquisition typically uses 1–10 W with fields of a few kilohertz). In contrast to such high-power Rf irradiation, selective irradiations use much lower power—a milliwatt or less, corresponding to field frequencies of a few tens of hertz. Thus only resonances in a very narrow frequency range (*on resonance spins*) are affected; resonances lying outside this narrow window (*off-resonance spins*) remain unaffected. Moreover, even the on-resonance spins now behave differently than when they were subjected to high Rf fields—under low Rf fields they precess at much lower frequencies (~10 Hz)—so relaxation processes become very important (and indeed result in nOe, as will be seen later).

Moreover, precession under selective irradiation occurs in the longitudinal plane of the rotating frame, instead of rotation in the transverse plane, which occurs during the evolution of the FID. The magnitude of the vector undergoing precession about the axis of irradiation decreases due to relaxation and field inhomogeneity effects.

4.1 nOe AND SELECTIVE POPULATION TRANSFER

In Chapter 2 (Section 2.2) we considered the phenomenon of selective population transfer (SPT). Under conditions of low-power irradiation, the lines of a multiplet may be unevenly saturated. This results in changes in the relative intensities of the multiplet components of the *coupling partner(s)* of the nucleus that has been irradiated. Thus, in an AX system, if nucleus A is irradiated and nucleus X is observed, then the intensity gained by one X line is balanced exactly by the intensity lost by another X line, so the *overall* integral of the X doublet remains unchanged (provided, of course, that there is no nOe enhancement of X). Such SPT distortions must be borne in mind when interpreting nOe difference spectra. And many methods can be used for reducing such distortions involving the use of composite pulses (Shaka *et al.*, 1984), irradiating each line of a multiplet and coadding the results (Neuhaus, 1983), or cycling the decoupler repeatedly around the lines of a target multiplet before each scan during the preirradiation period (Köver, 1984; Williamson and Williams, 1985).

We have previously considered 1D and 2D spectra involving the coupling of nuclei through bonds (*scalar coupling*, or *J-coupling*). Nuclei can also interact with each other directly through space (*dipolar*, or *magnetic, coupling*). This latter form of interaction is responsible for the *nuclear Over-hauser effect* (nOe), which is the change of intensity of the resonance of one nucleus when the transitions of another nucleus that lies close to the first nucleus are perturbed by irradiation. This change in intensity may correspond to an increase, as in small, rapidly tumbling molecules, or to a decrease, as in large, slower-tumbling molecules. It can provide valuable information regarding the relative stereochemistry of the nuclei in a given structure. The perturbation is usually carried out with a weak Rf field, and it eliminates the population difference across the transitions of the saturated nucleus. This triggers a compensatory response from the system through a readjustment of the population difference across transitions of the close-lying nuclei; this response is measured on the NMR spectrometer in the form of the nuclear Overhauser effect. Thus, if nucleus S is being saturated and nucleus I observed, and if the original and final (i.e., after irradiation of nucleus S) intensities of nucleus I are represented by I_0 and I', then nOe may be defined as:

$$\eta_I\{S\} = \frac{(I' - I_0) \times 100}{I_0}, \tag{1}$$

where η_I is the nOe at nucleus I when nucleus S is irradiated.

4.2 RELAXATION

When placed in a strong magnetic field, nuclei such as ^{1}H or ^{13}C, which have a spin quantum number of ½, adopt one of two quantized orientations, a low-energy orientation aligned with the applied field B_0 and a high-energy orientation aligned against the applied field. And they exhibit a characteristic precessional motion at their respective Larmor frequencies. When exposed to electromagnetic irradiation from an Rf oscillator, the nuclei can undergo excitation when the oscillator frequency matches exactly the Larmor frequency of each nucleus. This excitation involves transfer of excess spin population from the lower energy state to the higher energy state. Certain relaxation processes then come into play as the system responds to restore thermal equilibrium. The most important of these is *spin–lattice relaxation*, T_1, in which energy is transferred from the spins to the surrounding lattice as heat energy. For such relaxation to occur, the nuclei must be exposed to local oscillating magnetic fields some of whose frequencies can match exactly their respective precessional frequencies. There can be many sources of such fluctuating magnetic fields; the major source is the magnetic moments of other protons present in the same tumbling molecule. Since the molecule is undergoing various translations, rotations, and internal motions, there is virtually a continuum of energy levels available, and energy exchange can occur readily between the nucleus and the lattice through such *dipole–dipole interactions.*

In addition to dipole–dipole interactions, nuclei may relax by other mechanisms. For example, the presence of paramagnetic materials in solution can cause very efficient relaxation, since magnetic moments of electrons are a thousandfold greater than those of protons. Thus, although such interactions involve much larger internuclear distances, being intermolecular rather than intramolecular, they can still make T_1 so short that no nOe would be observed. It is for this reason that we need to remove dissolved paramagnetic oxygen carefully from solutions of the substance before undertaking nOe measurements.

In nOe we are concerned mainly with *longitudinal relaxation,* i.e., the return of longitudinal magnetization to its original value $\mathbf{M}_z^\circ$, which involves reestablishing the original equilibrium state between the populations of the ground and excited states. This will involve changes in population in the excited and ground states via transitions between the two spin states, so that the original population difference is restored. To return to the original equilibrium state, the *transverse magnetization* generated must also decay to zero, by loss and dephasing of the individual contributions to the transverse magnetization, $\mathbf{M}_{xy}$. But this need not concern us here, since nOe is concerned primarily with longitudinal relaxation, which occurs independent of transverse magnetization.

The number of transitions between upper and lower states depends not only on the populations N_α and N_β of the two states, but also on the efficiency of relaxation, more often called the *transition probability W,* as well as on the temperature of the surrounding lattice. Each NMR transition will be accompanied by a corresponding change in the lattice energy.

✦ *PROBLEM 4.1*

What are the two major uses of the nOe effect in NMR spectroscopy?
───✦

4.3 MECHANISM OF nOe

The *rate* at which dipole–dipole relaxation occurs depends on several factors: (a) the nature of the nucleus, (b) the internuclear distance, r, and (c) the effective correlation time, τ_c, of the vector joining the nuclei (which is inversely proportional to the rate at which the relevant segment of the molecule tumbles in solution). The *magnitude* of the dipole–dipole interaction, however, will depend only on the internuclear distance.

Let us consider two nuclei I and S that are close in space (hence, their dipole–dipole interaction is significant), but that are not scalar-coupled to each other ($J_{I,S} = 0$), with I being the one observed and S being the one irradiated. Let us also assume that both spins form part of a rigid molecule that is tumbling isotropically (i.e., without any preferred axis of rotation). Thus the only major relaxation pathway available is that via dipole–dipole relaxation. Figure 4.1 presents the energy-level diagram of such a system. Since the two spins I and S are not *J*-coupled, each appears as a single resonance line, with the line for nucleus I arising due to transitions I_1 and I_2 and that for nucleus S arising due to transitions S_1 and S_2. Since each nucleus can exist in two different energy states, α and β, there are four possible combinations of these states: $\alpha\alpha$, $\alpha\beta$, $\beta\alpha$, and $\beta\beta$. The $\alpha\alpha$ state corresponds to both nuclei in the lowest energy state, the $\beta\beta$ state represents a combination in which both are in the upper energy state, while in the $\alpha\beta$ and $\beta\alpha$ combinations one of the two nuclei is in the lower energy state while the other is in the upper energy state. A Boltzmann distribution prevails at thermal equilibrium, with the $\alpha\alpha$ state having the highest population, $\alpha\beta$ and $\beta\alpha$ states having some intermediate populations, and the $\beta\beta$ state corresponding to the lowest population. Since the nuclei I and S are

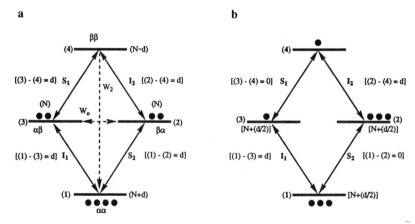

a   **b**

Figure 4.1 (a) Populations at Boltzmann equilibrium before application of a radio-frequency pulse on nucleus S. Nuclei I and S are not coupled. (b) Populations immediately after the application of a pulse on nucleus S. The populations connected by S transitions are readily equalized, while the *difference* of populations connected by the I transitions remains unaffected (i.e., (1)–(3) or (2)–(4) remains unchanged).

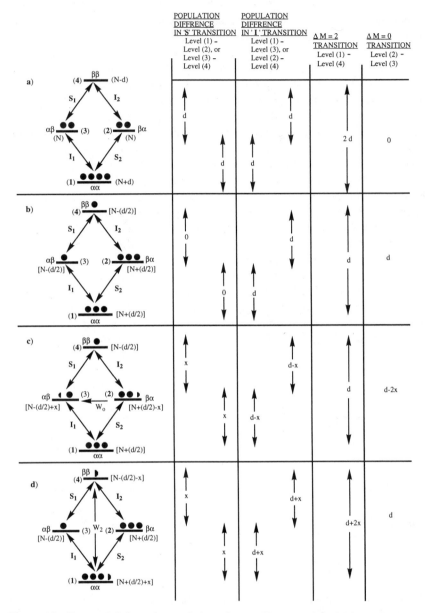

Figure 4.2 Energy levels and populations for an IS system in which nuclei I and S are not directly coupled with each other. This forms the basis of the nuclear Overhauser enhancement effect. Nucleus S is subjected to irradiation, and nucleus I is observed. (a) Population at thermal equilibrium (Boltzmann population).

not J-coupled to one another, the two single-quantum transitions I_1 and I_2 have the same energy, as do the transitions S_1 and S_2.

An interesting feature of the energy diagram is that there are two transitions that involve the *simultaneous* flip of *both* spins. These are the transitions $\alpha\alpha \leftrightarrow \beta\beta$ (a *double-quantum* process, W_2) and $\alpha\beta \leftrightarrow \beta\alpha$ (a *zero-quantum* process, W_0). Indeed, it is these two transitions that are responsible for the nOe effect observed, since it is through them that the saturation of nucleus S affects the intensity of nucleus I. Both W_2 and W_0 are "forbidden" transitions, since they cannot be excited directly by an Rf pulse or lead directly to detectable NMR signals; they are, however, allowed in the context of *relaxation* processes. This highlights an important feature: The selection rules governing excitation processes (i.e., an interaction of spin with an external oscillating field) are different from the selection rules governing relaxation processes (in this case involving energy exchange of the spins with the lattice).

To understand how nOe occurs, we have to consider the following situations: (a) the populations of the nucleus I prevailing at thermal equilibrium before the application of the Rf pulse on nucleus S, (b) populations immediately after the pulse is applied to nucleus S, and (c) populations after the system has had some time to respond, with either W_0 or W_2 being the predominant relaxation pathway.

For simplicity, let us assume that the $\alpha\beta$ and $\beta\alpha$ states have the same population, designated N. The upper energy state ($\beta\beta$) will then have a slightly lower population $(N - d)$, and the lower energy state ($\alpha\alpha$) will have a slightly higher population $(N + d)$, so the energy difference between the lowest and the highest energy states will be $2d$.

When the pulse is applied to nucleus S, the population levels connected by S transitions are rapidly equalized, but there is no *immediate* change in the *difference* in population levels connected by the I transitions. Figure 4.2 shows that on application of the pulse, levels 2 and 4 acquire the same populations as levels 1 and 3, respectively, but the population *difference*

(b) Population immediately after the radiofrequency pulse is applied. (c) Population after the system has had time to respond, with W_0 (zero-quantum transition) being the predominant pathway, as in macromolecules. The population difference between the two energy levels connected by the I transitions is now less $(d - x)$ than this difference (d) at thermal equilibrium. This will lead to a *negative* nOe effect. (d) Same as (c), but with W_2 (double-quantum transition) being the predominant relaxation pathway, as in smaller molecules. The population of the lower energy level (of energy states connected by I transitions) is increased and that of the upper level is decreased. This gives rise to a positive nOe effect.

between the levels connected by the I transitions [i.e., (1) − (3) or (2) − (4)] are unaffected (d in both cases).

Since the equilibrium state has been disturbed, the system tries to restore equilibrium. For this it can use as the predominant relaxation pathways the double-quantum process W_2 (in fast-tumbling, smaller molecules), leading to a positive nOe, or the zero-quantum process W_0 (in slower-tumbling macromolecules), leading to a negative nOe.

Let us first consider the situation in which W_2 is the predominant relaxation pathway. The population states immediately after the pulse are represented in Fig. 4.2b. The system tries to reestablish the equilibrium state by transferring some population x from the $\beta\beta$ state, which had a population of $N − (d/2)$ to the $\alpha\alpha$ state, which had a population of $N + (d/2)$. This transfer of population (x), shown in Fig. 4.2d (represented by a half circle), gives level 4 a somewhat reduced population $[N − (d/2) − x]$ and level 1 a correspondingly increased population $[N − (d/2) + x]$.

If we now examine the result of this population adjustment on the I transitions we find that the upper population of each I transition has decreased by x while the lower level of each I transition has increased by x; the population difference between the upper and lower states (levels 1 and 3) connected by the I_1 transition is

$$[N + (d/2) + x] − [N − (d/2)] = d + x$$

Similarly, the population difference between the states connected by the I_2 transition (levels 2 and 4) is

$$[N + (d/2)] − [N − (d/2) − x] = d + x$$

Hence the population difference between the lower and upper energy states of the two I transitions becomes $d + x$, as compared to the original difference of d at equilibrium. Thus an intensification of the lines for nucleus I will be observed by an amount corresponding to this increased difference x. *This is the positive nuclear Overhauser effect that is encountered in small, rapidly tumbling molecules, in which W_2 is the predominant relaxation pathway.*

In large molecules that tumble slowly, the predominant relaxation pathway is via W_0. This is shown schematically in Fig. 4.2c. A part of the population x is now transferred from the $\beta\alpha$ state to the $\alpha\beta$ state. This causes an *increase* in the population of the upper level of one I transition (I_1, level 3) and a *decrease* in the lower population level of the other I transition (I_2, level 2). As a result, the population *difference* between the lower and upper levels of each I transition is reduced to $d − x$ (i.e., level 1 − level 3, or level 2 − level 4, becomes $d − x$). The reduction in population difference by x as compared to the equilibrium situation (Fig. 4.2a) produces a corresponding *decrease* in the signal intensity of the I nucleus. *Hence, if W_0 is the main relaxation pathway, then saturation of nucleus S will reduce the intensity of the I nucleus; i.e., a negative nuclear Overhauser effect will be observed.* In practice, the relaxation processes involve the single-quantum relaxation pathway W_1 (which causes no change in signal intensity), as well as the double-quantum W_2 and zero-quantum pathways W_0, which respectively enhance or reduce the signal intensities, with W_2 dominating in small, rapidly tumbling molecules and W_0 dominating in large, slowly tumbling molecules.

✦ PROBLEM 4.2

Is it possible to predict the predominant mode of relaxation (zero-quantum or double-quantum) by observing the sign of nOe (negative or positive)?

── ✦

4.4 FACTORS AFFECTING nOe

In addition to the dipole–dipole relaxation processes, which depend on the strength and frequency of the fluctuating magnetic fields around the nuclei, there are other factors that affect nOe: (a) the intrinsic nature of the nuclei I and S, (b) the internuclear distance (r_{IS}) between them, and (c) the rate of tumbling of the relevant segment of the molecule in which the nuclei I and S are present (i.e., the effective *molecular correlation time, τ_c*).

✦ PROBLEM 4.3

Why do we need to involve zero-quantum (W_0) or double-quantum (W_2) processes to explain the origin of the nuclear Overhauser enhancement?

── ✦

The molecular correlation time τ_c represents the time taken for the relevant portion of the molecule containing the nuclei I and S to change from one orientation to another. It is chosen so that it is nearly equal to the *minimum* waiting time between various orientations rather than to the *average* waiting time between the orientations (thus waiting times shorter than τ_c do not occur frequently). The lower limit of the waiting time then corresponds to the upper limit of the frequency range of the fluctuating magnetic fields. Small, rapidly tumbling molecules may have τ_c on the order of 10^{-12} s; large molecules (or small molecules in viscous solutions) may have τ_c that is a thousandfold longer (10^{-8}–10^{-9} s).

In order for relaxation to occur through W_1, the magnetic field fluctuations need to correspond to the Larmor precession frequency of the nuclei, while relaxation via W_2 requires field fluctuations at double the Larmor frequency. To produce such field fluctuations, the tumbling rate should be the reciprocal of the molecular correlation time, i.e., τ_c^{-1}, so most efficient relaxation occurs only when $\nu_0\tau_c$ approaches 1. In very small, rapidly tumbling molecules, such as methanol, the concentration of the fluctuating magnetic fields (*spectral density*) at the Larmor frequency is very low, so the relaxation processes W_1 and W_2 do not occur efficiently and the nuclei of such molecules can accordingly relax very slowly. Such molecules have $\tau_c \ll 1$, and they are said to be in the *extreme narrowing limit*. The line widths are then determined by instrumental factors rather than by other fundamental considerations.

In larger, slowly tumbling molecules (or small molecules in viscous solutions), tumbling occurs very slowly, so fields corresponding to the Larmor precession frequency ν_0 (for relaxation via W_1) or $2\nu_0$ (for relaxation via W_2) cannot be generated sufficiently. Relaxation through the zero-quantum transition W_0 then becomes important, involving a *mutual spin-flip*, i.e., the shifting of energy from one spin to another ($\alpha\beta \leftrightarrow \beta\alpha$). In compounds with molecular weights of over 10,000, the nOe appears with a maximum intensity of -1. W_0 transitions occur between close-lying energy levels, and require only fields of low frequency, which are readily available from slowly tumbling molecules.

While the *rate of change* of dipolar interaction depends on τ_c, its *magnitude* depends only on the internuclear distance and is independent of τ_c. Thus the dipole–dipole relaxation depends on the molecular correlation time τ_c, the internuclear distance r, and the gyromagnetic ratios of the two nuclei, γ_I and γ_S:

$$R_1 = K\gamma_I^2\gamma_S^2 \, (t)^{-6}\tau_c. \tag{2}$$

The *efficiency* of the relaxation process is therefore governed by the r^{-6} relationship, so doubling of the internuclear distance decreases the relaxation by a factor of 64 (since $2^{-6} = \frac{1}{64}$). Moreover, the *strength* of the dipole–dipole interaction depends on the square of the product of the gyromagnetic ratios of the two nuclei, so proton–proton interactions are the main relaxation processes because of the greater gyromagnetic ratio of ^{1}H in comparison to ^{13}C. Deuterium, on the other hand, has a gyromagnetic ratio 6.5 times smaller than that of a proton, so proton–deuterium interaction should be $(6.5)^2$, i.e., about 42 times less than proton–proton interactions, other conditions remaining the same. Carbon atoms relax mainly through attached protons, so quaternary carbons relax much more slowly than protonated carbons, and methyl carbons bearing three hydrogen atoms will normally relax faster than methine carbons that bear only one hydrogen.

4.4.1 Internuclear Distance and nOe

If the intensity of nucleus I before irradiation of the neighboring nucleus S was I_0, and its intensity after irradiation of nucleus S was I_1, than the fractional increase in intensity $\eta_I(S)$ is given by $(I_0 - I_1)/I_0$. If we assume that relaxation of nucleus I occurs only through dipole–dipole relaxation by interaction with nucleus S and that a steady equilibrium state has been reached, then the fractional increase in intensity η_I of nucleus I is given by the equation

$$\eta_I(S) = \frac{(S_0/I_0)(W_2 - W_0)}{2W_1^1 + W_0 + W_2}. \tag{3}$$

If we assume that the extreme narrowing condition exists, then the following simpler expression applies:

$$\eta_I(S) = \frac{\gamma_S}{2\gamma_I} \tag{4}$$

In the case of proton–proton interactions, both nuclei S and I will have the same gyromagnetic ratios, and an implication of the Equation (4) then is that there is an upper limit of 50% on the nOe obtainable, *whatever the distance between nuclei S and I*. This means that the observation of an nOe between two nuclei does not *necessarily* mean they are spatially close to one another, and nOe results must therefore be interpreted with caution. Similarly, as will be seen later, the absence of nOe between two nuclei does not necessarily mean they are far apart. In the case of heteronuclear nOe, since the gyromagnetic ratio of proton (γ_S) is four times the gyromagnetic ratio of carbon (γ_I), γ_S/γ_I can be four times greater than that obtainable in homonuclear nOe.

While the final magnitude of nOe depends, as indicated earlier, on the relaxation pathways W_1, W_2, and W_0, *the initial rate of buildup of nOe (transient nOe) depends only on the rate of cross-relaxation between the nuclei,* and this can provide valuable information about the distance between the nuclei (r). This rate of buildup can be proportional to r^{-6}, where r is the distance between the nuclei. Thus, if the proportionality constant is determined, we can calculate an approximate distance between the two nuclei. The best results are obtained in rigid molecules when the nuclei are less than 3 Å apart. If only direct nOe's are involved in a two-spin

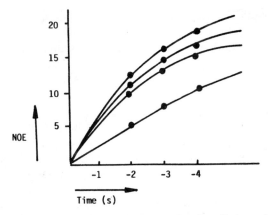

Figure 4.3   nOe buildup with respect to time, after irradiation of bound alanine methyl protons in ristocetin A-tripeptide complex for four different protons. (Reprinted from D. H. Williams *et al., J. Am. Chem. Soc.* **105,** 1332, copyright (1983), with permission from The American Chemical Society, 1155 16th Street, N.W. Washington, D.C. 20036, U.S.A.)

system and no third proton takes part in the relaxation, then a plot of the magnitudes of nOe against time gives an exponential curve (Fig. 4.3). If the size of the nOe at time t is h_t, and if the final steady-state value of nOe is h_∞, then a plot of $h_\infty - h_t$ against t yields a straight line of slope $-k$ (Fig. 4.4). Since $k \propto r^{-6}$, the internuclear distance can be calculated.

4.4.2 Three-Spin System

If there are more than two nuclei exerting relaxation effects on one another, then it is convenient to consider them in *pairs* and to arrive at the overall effect by adding together the effects of various possible pairs. In the case of a three-spin system, we can consider two different situations: (1) the nuclei H_A, H_B, and H_C are arranged in a straight line, and (2) they are in a nonlinear arrangement.

✦ *PROBLEM 4.4*

What is *transient nOe,* and why is it considered to provide a better estimate of the internuclear distance (r) than the normal nOe effect?

———————————————————————————————✦

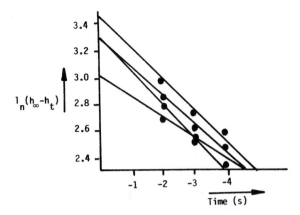

Figure 4.4   Data in Fig. 4.3 when plotted as $(h_\infty - h_t)$ versus time. The slope of the lines represents the internuclear distance r that corresponds to the rate of nOe buildup, which is directly proportional to r^{-6}. (Reprinted from D. H. Williams *et al., J. Am. Chem. Soc.* **105,** 1332, copyright (1983), with permission from The American Chemical Society, 1155 16th Street, N.W. Washington, D.C. 20036, U.S.A.).

✦ *PROBLEM 4.5*

How can you measure the *transient nOe?*

─── ✦

Let us first consider the nuclei H_A, H_B, and H_C as lying in a straight line and equidistant from one another (Fig. 4.5). The central proton, H_B, can relax by interactions with two neighbors, H_A and H_C, while H_A and H_C can relax by interaction with only one neighbor, so H_B can relax twice as quickly as H_A or H_C. If we assume that relaxation can occur only through dipole–dipole relaxation, and that an equilibrium steady state has been

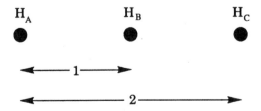

Figure 4.5   Three protons arranged in a straight line, equidistant from each other. Here, H_A is farther from H_C than in upcoming Fig. 4.6.

reached, then the maximum nOe achieved for H_B is 50%, which can occur through relaxation via H_A and H_C. Assuming we are dealing with small molecules causing positive nOes, irradiation of nucleus H_A produces a sizeable positive nOe at the neighboring H_B and a small direct positive nOe at the more distant H_C. Thus H_A and H_C can each potentially enhance H_B by 25%. On the other hand, H_B would be responsible for almost all the relaxation effect on H_A or H_C, since H_A and H_C are twice as far from one another as they are from H_B. Since we already know from the previous discussion that a 64-fold reduction in nOe occurs on doubling the internuclear distance ($r^{-6} = 2^{-6} = \frac{1}{64}$), then the proportion of relaxation effect of H_A on H_C (or H_C on H_A) will be $\frac{1}{64} = 0.8\%$, while the remainder (49.2%) will be due to H_B. Thus while irradiation of H_B can cause 49.2% nOe at H_A or H_C, irradiation of H_A or H_C can cause only 25% nOe at H_B, illustrating that *the nOe between two nuclei will generally differ in the two directions because of the differing relaxation pathways available to the two nuclei.* If the nuclei H_A, H_B, and H_C are arranged linearly but are not equidistant then, obviously, the nucleus nearer to H_B will contribute more to its nOe than the nucleus that is farther from it.

4.4.3 Three-Spin Effects

So far we have considered only direct nOe. However, there is an indirect effect of H_A on H_C (or of H_C on H_A) that should be considered (the so-called *three-spin effect*). Assuming that we are dealing with small molecules causing positive nOe's, irradiation of nucleus H_A causes a sizable positive nOe at the neighboring H_B and a small positive *direct* nOe at the more distant H_C. However, since the irradiation-induced *decrease* in the population difference between the upper and lower states of H_A is causing an *increase* in the intensity (or population difference) of H_B, it is logical to assume that this increase would produce an opposite effect at H_C, i.e., a decrease of its intensity. H_C will therefore experience two opposing effects upon irradiation of H_A, a small positive direct nOe and a larger negative nOe, *so an overall negative nOe will be observed at H_C upon irradiation of H_A.* This effect is most likely when the three nuclei are arranged linearly. Irradiation of H_A will cause alternating +ve and −ve nOe's on other atoms (H_D, H_E, H_F, etc.) in a linear arrangement beyond H_C, though the effect decreases sharply beyond two atoms, so it is not measurable.

✦ PROBLEM 4.6

Following is a linear three-spin system and the observed steady-state nOe's between three nuclei. The distance between nuclei A and B is double that between nuclei A and C ($\gamma_{AB} = 2\gamma_{AC}$). Explain the nOe results, in terms of both relaxation and internuclear distances.

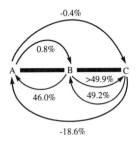

It is possible to distinguish between direct and indirect nOes from their kinetic behavior. The direct nOes grow immediately upon irradiation of the neighboring nucleus, with a first-order rate constant, and their kinetics depend initially only on the internuclear distance r^{-6}; indirect nOes are observable only after a certain time lag. We can thus suppress or enhance the indirect nOe's (e.g., at H_C) by short or long irradiations, respectively, of H_A. A long irradiation time of H_A allows the buildup of indirect negative nOe at H_C, while a short irradiation time of H_A allows only the direct positive nOe effects of H_A on H_C to be recorded.

✦ PROBLEM 4.7

Explain what is meant by *three-spin effects*, or *indirect nOe effects*. When do such indirect effects matter?

4.4.4 Nonlinear Arrangement

If H_A, H_B, and H_C do not lie on the same line (Fig. 4.6), then as H_A comes closer to H_C, the *direct* +ve nOe between H_A and H_C will increase, and a point may come when it totally cancels the larger, *indirect* negative nOe effect exerted by H_A on H_C through H_B. Thus no nOe may be observed at H_C upon irradiation of H_A, *even though H_A and H_C are spatially close!* The absence of nOe between nuclei therefore does not *necessarily* mean they are far from one another.

Molecules with a rigid central core (such as a ring system) and a freely moving side chain may exhibit significant differences in the mobility of the protons in the central ring system as compared to the side chain. These are reflected in their corresponding relaxation rates, with the protons lying

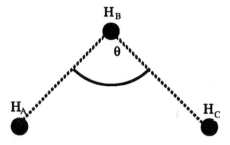

Figure 4.6 A three-proton angular system. H_A, H_B, and H_C are disposed at an angle θ.

on the more mobile side chain relaxing faster than those in the less mobile ring system. These differences in relaxation rates can be used to advantage. For instance, if the protons lying in the two regions have similar chemical shifts so it is difficult to distinguish them from one another, then measurement of relaxation rates may help to identify them.

So far we have been concerned with homonuclear nOe effects. nOe between nuclei of different elements can also be a useful tool for structural investigations. Such heteronuclear nOe effects—for instance, between protons and carbons—can be used with advantage to locate quaternary carbon atoms. Normally, heteronuclear nOe effects are dominated by interactions between protons and *directly bonded* carbon atoms, and they can be recorded as either 1D or 2D nOe spectra.

As stated earlier, since $\eta_I = \gamma_S/2\gamma_I$, and since the gyromagnetic ratio of proton is about fourfold greater than that of carbon, then if ^{13}C is observed and ^1H is irradiated (expressed as $^{13}C\{^1H\}$), at the extreme narrowing limit $\eta_I = 198.8\%$; i.e., the ^{13}C signal appears with a threefold enhancement of intensity due to the nOe effect. This is a very useful feature. For instance, in noise-decoupled ^{13}C spectra in which C–H couplings are removed, the ^{13}C signals appear with enhanced intensities due to nOe effects.

Proton irradiation *before* acquisition of the ^{13}C spectrum results in nOe but no decoupling, whereas proton irradiation *during* ^{13}C data acquisition produces decoupling without nOe. It is therefore possible to separate the two effects by "gating" the decoupler on and off for appropriate time periods (Fig. 4.7). Moreover, since the power needed to induce nOe is much less than that required for decoupling, the power level of the decoupler is reduced during the preirradiation time period and then increased during the acquisition period. This cuts down the heating effects when continuous irradiation with decoupler powers of 2–3 W are employed.

The low intensities of nonprotonated carbons is usually due to their long relaxation times. The addition of a paramagnetic substance such as

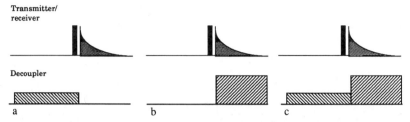

Figure 4.7   Pulse schemes representing separation of decoupling effects from the nOe during X nucleus acquisition. The decoupler is programmed to produce noise-modulated irradiation or composite pulse decoupling at two power levels. Suitable setting of the decoupler may produce either (a) nOe only, (b) proton decoupling only, or (c) both nOe and proton decoupling.

Cr(acac)$_3$ can produce a hundredfold increase in the relaxation rate, with the rate varying linearly with the concentrations of Cr(acac)$_3$, so the nOe is also reduced a hundredfold. Protonated carbons with faster relaxation rates are less affected (about tenfold increase in relaxations rates), and the reduction of nOe is correspondingly less.

An improved procedure for recording heteronuclear nOes is to irradiate individuals lines of the multiplet with a low decoupling power instead of exciting the entire multiplet with high power. This results in greater selectivity of the protons being irradiated and higher sensitivity of the carbon signals (Bigler and Kamber, 1986).

✦ PROBLEM 4.8

What happens in three-spin systems when the A-B-C angles θ are 180° and 78°, respectively, $r_{AB} = r_{BC} = 1$, and spin A is irradiated?

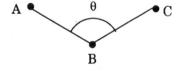

_____ ✦

4.4.5 nOe Difference Spectra

The most widely used nOe experiment is nOe difference spectroscopy. Two different sets of experiments are recorded, one in which certain protons are subjected to irradiation and enhancements are obtained of other nearby protons, and the other without such irradiation so that normal unperturbed spectra are obtained. Subtraction of the perturbed FID from the unperturbed one, followed by Fourier transformation of the difference FID, gives the nOe difference spectrum, which ideally contains only the nOe effects.

The nOe difference spectrum has the advantage that it allows measurements of small nOe effects, even 1% or below. The experiment involves switching on the decoupler to allow the buildup of nOe. It is then switched off, and a $\pi/2$ pulse is applied before acquisition. The nOe is not affected much by the decoupler's being off during acquisition, since the nOes do not disappear instantaneously (the system takes several T_1 seconds to return to its equilibrium state).

+ *PROBLEM 4.9*

Explain the main advantage of nOe difference spectroscopy. Why does it involve a mathematical subtraction of the normal ^{1}H-NMR spectrum from the nOe-enhanced ^{1}H-NMR spectrum?

――― +

4.5 SOME PRACTICAL HINTS

4.5.1 Solvent

This is probably the most important consideration when recording nOe difference spectra. Solvents with sharp intense lock signals, such as d_6-DMSO or d_6-acetone, are preferable to a solvent such as CDCl$_3$, which has a weaker lock signal, or D$_2$O which has a broad lock signal. This is because the lock on FT NMR spectrometers operates by sampling continuously the dispersion-mode deuterium signal of the deuterated solvent. Any field drift would create an error signal, which, because of the dispersion-mode (half positive, half negative) shape of the deuterium signal, would be positive if the drift is on one side and negative if the drift is on the other side. The spectrometer is built to correct this error automatically, but the correction is more precise in solvents with a sharp lock signal (such as d_6-acetone) than in those with weak lock signals (e.g., CDCl$_3$) or broad lock signals (e.g., D$_2$O). D$_2$O is also not very suitable, since its chemical shift is largely temperature dependent. However, if it must be used, as in water-soluble compounds, such as sugars, then addition of 2–3% of d_6-acetone for use as the lock resonance is recommended.

We normally avoid protonated solvents, because the very intense solvent peak will obscure nearby protons, and the dynamic range problem will also reduce the quality of the spectrum. If the solvent has exchangeable protons (such as H_2O or HOD), then saturation transfer processes from solute to solvent or solvent to solute can complicate the spectra.

✦ *PROBLEM 4.10*

Why is degassing of sample solutions in NMR tubes essential before nOe experiments, and why are aqueous solutions (solutions in D_2O) not generally degassed in the NMR tubes?

———————————————————————————————— ✦

✦ *PROBLEM 4.11*

Explain the dependence of the nOe on molecular motion (tumbling).

———————————————————————————————— ✦

4.5.2 Temperature

It is important that constant temperature be maintained throughout the nOe difference experiment. If the instrument is fitted with a constant-temperature device, then it is advisable to adjust it to a few degrees above room temperature so that it maintains a constant temperature accurately.

4.5.3 Sample Purity

Impurities that can lead to sharp decreases in spin-lattice relaxation times, such as paramagnetic metal ions or dissolved oxygen, need to be removed. Paramagnetic ions can be removed by complexation with ethylenediamine tetraacetate (EDTA) by filtering solutions through a chelating resin. Oxygen should be removed through repeatedly freezing the contents of a specially constructed NMR tube by dipping it in liquid nitrogen, evacuating the tube, and then thawing. Such tubes are commercially available, Simply bubbling an inert gas through the tube is not enough for proper degassing of the sample.

The nOe difference spectrum is highly demanding, since even the slightest variation in the spectra recorded with and without preirradiations will show up as artifacts in the difference spectrum (Fig. 4.8). The errors can be random, due to phase instability caused by temperature effects on the Rf circuits, variations in spinner speed, etc. The problem of phase instability is reduced in the latest generation of instruments with digital

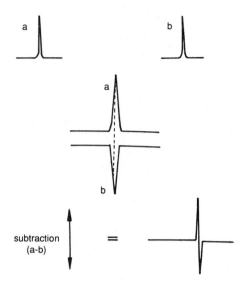

Figure 4.8 A slight variation in the positions of peaks (a) and (b) will not lead to mutual or complete cancellation on subtraction, and a peak of characteristic line shape may result.

frequency synthesizers. If the spinner speed variation is a serious problem, then the difference spectrum can be recorded with the spinner off.

The long-term changes due to variations in temperature, field drift, etc. can be minimized by acquiring alternately the preirradiation data and the control data, at the shortest time intervals possible, and coadding the data later. Typically, a cycle would consist of two to four dummy scans followed by eight or 16 data acquisitions at each preirradiation frequency. If there are only two frequencies in the cycle, then each cycle would take about a minute. In automatic multiple-scan experiments with several irradiation frequencies, each cycle would take a correspondingly longer time, depending on the number of frequencies and the number of data acquisitions at each frequency. The dummy scans are necessary to ensure that the effects of irradiation at the previous frequency in the cycle have disappeared before data acquisition. The need for multiples of four or eight scans arises as a requirement of the CYCLOPS phase cycle.

Avoid having moving metal objects near the magnet when carrying out nOe difference experiments, to prevent random variations in frequency. A small line-broadening (~2 Hz) can also be applied to the spectra before or after subtraction, to reduce subtraction artifacts.

✦ *PROBLEM 4.12*

What precautions are normally taken during sample preparation for the nOe experiment?

── ✦

SOLUTIONS TO PROBLEMS

✧ *4.1*

The nOe experiment is one of the most powerful and widely exploited methods for structure determination. nOe difference (NOED) or the two-dimensional experiment, NOESY, is used extensively for stereochemical assignments. It provides an indirect way to extract information about internuclear distances. The other use of nOe is in signal intensification in certain NMR experiments, such as the broad-band decoupled ^{13}C-NMR experiment.

✧ *4.2*

The positive nOe observed in small molecules in nonviscous solution is mainly due to double-quantum W_2 relaxation, whereas the negative nOe observed for macromolecules in viscous solution is due to the predominance of the zero-quantum W_0 cross-relaxation pathway.

✧ *4.3*

If only single-quantum transitions (I_1, I_2, S_1, and S_2) were active as relaxation pathways, saturating S would not affect the intensity of I; in other words, there will be no nOe at I due to S. This is fairly easy to understand with reference to Fig. 4.2. After saturation of S, the population difference between levels 1 and 3 and that between levels 2 and 4 will be the same as at thermal equilibrium. At this point W_0 or W_2 relaxation processes act as the predominant relaxation pathways to restore somewhat the equilibrium population difference between levels 2 and 3 and between levels 1 and 4 leading to a negative or positive nOe respectively.

✧ *4.4*

Transient nOe represents the rate of nOe buildup. The nOe effect (so-called equilibrium value) itself depends only on the competing balance between various complex relaxation pathways. But the initial rate at which the nOe grows (so-called *transient nOe*) depends only on the rate of cross-relaxation τ_c between the relevant dipolarly coupled nuclei, which in turn depends on their internuclear distance (r).

✧ *4.5*

The rate of growth of nOe (transient nOe) can be measured easily by the following pulse sequence:

$$(180°)\,\text{Sel}—\tau_m—90°—\text{Acquisition}.$$

The procedure involves applying a selective inversion pulse on a selected resonance as the initial perturbation. This is followed by a variable waiting period τ_m before acquisition of a spectrum by application of a 90° pulse. During these variable waiting periods, the nOe will build up on the dipolarly coupled nuclei. A plot of the intensities of the observed nucleus versus the waiting period gives a graph in which the slope represents the transient nOe. The following are transient nOe spectra obtained using the pulse sequence just shown with a variable delay, along with the normal ^{1}H-NMR spectrum.

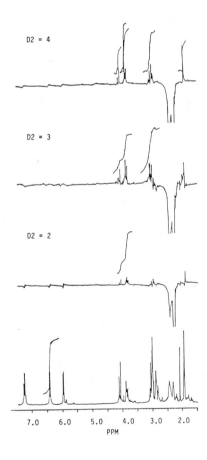

✧ *4.6*

From the figure it is obvious that the central nucleus, B, relaxes almost twice as fast as A and C, because it has two nuclei providing the relaxation pathways. On the other hand, nucleus B contributes to virtually all the relaxation of both A and C. The saturation of B results in about 49.0% nOe of both A and C. Since the internuclear distance between A and B is greater than that between B and C, the nOe effects experienced at B by saturation of A and C are not equal. When C is saturated, the nOe enhancement at B is almost 50% (actually, 49.2%). In contrast, saturation of A gives only a very small nOe at B. This is because of the fact that A contributes very little to the relaxation of B relative to C. The negative enhancements shown in the figure are quite different from the general negative nOe's that arise via a different mechanism. For example, saturation of spin A directly increases the net magnetization of spin B and also, to some extent, of spin C. In other words, decreasing the population difference in A increases the population difference in B and, to a lesser extent, in C by the direct effect. The increased population difference in B, however, will lead to a more pronounced decrease in the population difference in C, thereby canceling the direct positive nOe effect of nucleus A on C leading to an overall negative nOe at C.

✧ *4.7*

Three-spin effects arise when the nonequilibrium population of an enhanced spin itself acts to disturb the equilibrium of other spins nearby. For example, in a three-spin system, saturation of spin A alters the population of spin B from its equilibrium value by cross-relaxation with A. This change in turn disturbs the whole balance of relaxation at B, including its cross-relaxation with C, so that its population disturbance is ultimately transmitted also to C. This is the basic mechanism of indirect nOe, or the three-spin effect.

The magnitude of the indirect nOe effect depends on the geometry of the three-spin system. It is maximum with a linear geometry of the spin system ($\theta = 180°$). When θ decreases, the distance r_{AC} also decreases, and the direct enhancement at C becomes more and more significant. At $\theta = 78°$, the direct and indirect nOe effects become equal. With smaller values of θ, the direct contribution rapidly starts to dominate the nOe at C, and a strong positive enhancement results. This means that indirect nOe effects are to be expected mainly when the spins are close to having a linear geometry.

✧ *4.8*

In a system, when $\theta = 180°$, the distance between spins A and C (r_{AC}) will be a maximum and the nOe between A and C a mimimum. When θ decreases from $180°$, r_{AC} decreases and the direct nOe between A and C increases. As a result of the consequently more effective A-C relaxation, the B-C relaxation process becomes relatively less important, and the indirect negative A-C contribution is correspondingly decreased. When $\theta = 78°$, $r_{AC} = 1.26$, so there will be no net nOe between A and C, even though the A and C spins are very close to each other. This is because the direct (positive) and indirect (negative) nOe effects are equal and opposite.

✧ *4.9*

In the nOe difference spectrum, only the nOe effects of interest remain, while the unaffected signals are removed by subtraction. It does not therefore matter if the nOe responses are small or buried under the unaffected signals, since they show up in the difference spectrum. The main benefit of nOe difference spectroscopy is that it converts the changes in intensity into a form that is more readily recognizable. The difference spectrum is obtained by a process in which a control (normal) spectrum is subtracted from a spectrum acquired with irradiation of a particular signal.

✧ *4.10*

Paramagnetic materials, such as dissolved oxygen and other gases, contribute to the relaxation of nuclei after their initial perturbation. It is therefore necessary to exclude paramagnetic materials, principally oxygen, from the solutions in NMR tubes in order to derive a true picture of internuclear cross-relaxations during the nOe experiment. This is generally achieved by degassing the solutions through freeze-drying. Aqueous solutions are difficult to freeze without the risk of cracking the NMR tubes. They are therefore degassed in a separate flask and then transferred to the NMR tubes.

✧ *4.11*

When the zero-quantum W_0 transition is greater than double-quantum W_2, the nOe enhancements will be negative. Similarly, when W_2 is greater than W_0, the resultant nOe will have a positive sign. The predominance of W_2 and W_0 over one another depends on the molecular motion. It is known that the W_0 transition is maximal when the molecule tumbles at a rate of about 1 KHz, while the W_2 transition is fastest at a tumbling rate of about 800 MHz. On this basis, a rough idea of the sign of nOe can be obtained. For example, small molecules in nonviscous solvents tumble at a faster rate than larger molecules. It is therefore reasonable to expect that for small molecules, W_2 will dominate over W_0, and the sign of the enhancement will therefore be positive.

✧ *4.12*

Following are some important precautions for sample preparation.

(i) Removal of solid particles by filteration.

(ii) Removal of paramagnetic impurities, such as paramagnetic metal ions and molecular oxygen. These metal ions can be removed easily be adding a small amount of ethylenediamine tetraacetate (EDTA). Molecular oxygen can be removed by degassing the sample. Paramagnetic impurities lead to rapid spin-lattice relaxation and therefore reduce the intensities of enhancement.

(iii) The concentration of solute in the sample should not be very large, to avoid aggregation.

REFERENCES

Bigler, P., and Kamber M. (1986). *Magn. Reson. Chem.* **11**(11), 972.

Köver, K. E. (1984). *J. Magn. Reson.* **59,** 485.

Neuhaus, D. (1983). *J. Magn. Reson.* **53,** 109.

Shaka, A. J., Bauer, C., and Freeman, R. J. (1984). *J. Magn. Reson.* **60,** 479.

Williams, D. H., Williamson, M. P., Butcher, D. W. and Hammond, S. J. (1983). *J. Am. Chem. Soc.* **105,** 1332.

Williamson, M. P., and Williams, D. H. (1985). *J. Chem. Soc. Perkin Trans I,* (5), 949.

CHAPTER 5

Important 2D NMR Experiments

5.1 HOMO- AND HETERONUCLEAR J-RESOLVED SPECTROSCOPY

5.1.1 Heteronuclear 2D J-Resolved Spectroscopy

Heteronuclear two-dimensional J-resolved spectra contain the chemical shift information of one nuclear species (e.g., ^{13}C) along one axis, and its coupling information with another type of nucleus (say, ^{1}H) along the other axis. 2D J-resolved spectra are therefore often referred to as J,δ-spectra. The heteronuclear 2D J-resolved spectrum of stricticine, a new alkaloid isolated by one of the authors from *Rhazya stricta*, is shown in Fig. 5.1. On the extreme left is the broadband ^{1}H-decoupled ^{13}C-NMR spectrum, in the center is the 2D J-resolved spectrum recorded as a stacked plot, and on the right is the contour plot, the most common way to present such spectra. The multiplicity of each carbon can be seen clearly in the contour plot.

The mechanics of obtaining a 2D spectrum have already been discussed in the previous chapter. A 1D ^{1}H-coupled ^{13}C-NMR spectrum contains both the chemical shift and coupling information along the same axis. Let us consider what would happen if we could somehow swing each multiplet by 90° *about* its chemical shift so that the multiplet came to lie at right angles to the plane containing the chemical shift information. Thus, if the 1D spectrum was drawn in one plane—say, that defined by the NMR chart paper—then the multiplets would rotate about their respective chemical

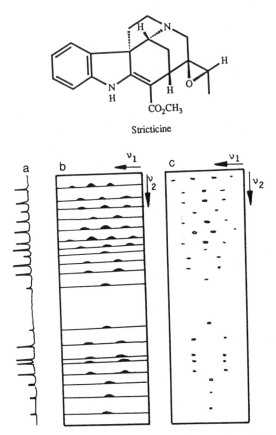

Stricticine

Figure 5.1 (a) Broad-band decoupled ^{13}C-NMR spectrum of the indole alkaloid stricticine; (b) stacked plot presentation of 2D heteronuclear *J*-resolved spectrum of stricticine; (c) 2D heteronuclear *J*-resolved spectrum of stricticine in the form of a contour plot.

shifts so that each multiplet came to lie above and below the plane of the paper, perpendicular to it, crossing the paper at the chemical shift of the nucleus. The chemical shifts would then be defined by an axis lying in one plane (the plane of the paper), and the coupling information would be defined by an axis lying perpendicular to that plane (Fig. 5.2). This is precisely what happens as a result of the 2D *J*-resolved experiment. Since the various multiplets now no longer overlap with one another (because the individual nuclei usually have differing chemical shifts), the multiplicity and *J*-values can easily be read in two separate planes.

    To understand how the chemical shift and multiplicity information is separated, we need to reconsider the spin-echo sequence described earlier.

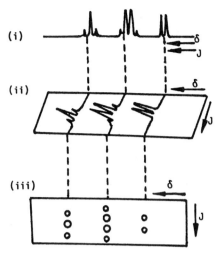

Figure 5.2 Presentation of 2D J-resolved spectra. In the 1D plot (i), both δ and J appeared along the same axis, but in the 2D J-resolved spectrum (ii), the multiplets are rotated by 90° at their respective chemical shifts to generate a 2D plot with the chemical shifts (δ) and coupling constants (J) lying along two different axes. (iii) The 2D J-resolved spectrum as a contour plot.

Let us consider the magnetization vectors of a ^{13}C nucleus of a CH group, for purposes of illustration. A 90°_x ^{13}C pulse will bend the magnetization by 90° to the y'-axis. During the subsequent evolution period, the magnetization vector will be split into two component vectors under the influence of the coupling with the attached protons, and these vectors will oscillate in the $x'y'$-plane.

If, for simplicity, we observe the behavior of these vectors in the rotating frame with respect to the chemical shift frequency of the CH carbon (i.e., if the frame itself is rotating at the frequency of the CH carbon), then they will appear to separate from one another and rotate away in opposite directions in the $x'y'$-plane with an angular velocity of $\pm J/2$ in the rotating frame. As the vectors process away from the y'-axis, the magnitude of their net resultant along the y'-axis will decrease till it becomes zero (when they point along the $+x'$- and $-x'$-axes). As they process away further towards the $-y'$-axis, this resultant will have a growing negative intensity that will reach a maximum when they coalesce along the $-y'$-axis. The two vectors then move back toward the $+y'$-axis, with a corresponding increase in positive signal intensity. This corresponds to the perturbation of the intensity of the ^{13}C signal as a function of the scalar coupling constant, J (Fig. 5.3E, i). A series of spectra are acquired by Fourier transformation of the FIDs at various values of t_1 (Fig. 5.3A). The data are then transposed by the com-

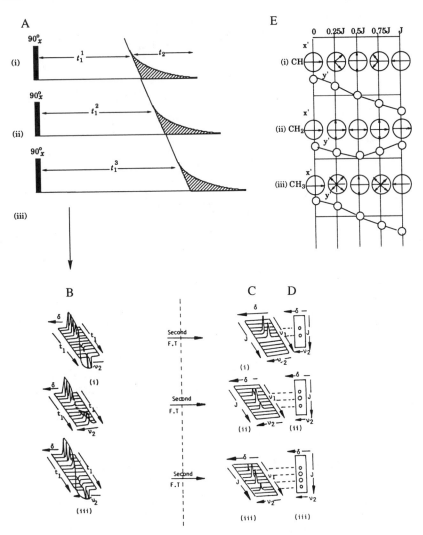

Figure 5.3 (A) Many FIDs are recorded at incremented evolution periods, t_1, as the first step involved when recording 2D spectra. (B) The first series of Fourier transformations affords a corresponding series of 1D spectra in which the individual peaks exhibit periodic changes in signal amplitudes at various incremented t_1 values. The second series of Fourier transformations yield the 2D plots as stacked plots (C) or as contour plots (D). In the 2D J-resolved spectra thus obtained, one axis defines the chemical shift (δ), while the other axis defines the coupling constant (J). (E) Different positions of magnetization vectors of (i) CH, (ii) CH$_2$, and (iii) CH$_3$ carbons with changing evolution time t_1. The periodic changes in the positions of the magnetization vectors results in corresponding periodic changes in signal amplitudes, as shown in (i), (ii) and (iii).

puter so that the spectra are arranged in rows, one behind the other. Fourier transformation is then again carried out along columns of modulated peaks to produce the 2D *J*-resolved spectrum in which chemical shift information lies along one axis and coupling information lies along the other axis. This is illustrated in Fig. 5.3C and D.

In the case of a CH_2 group, since there are now two protons coupled to the ^{13}C nucleus, it will be split into a triplet. If we again position the reference frequency at the center of the triplet, then the middle vector will appear stationary, the outer vectors will precess away in opposite directions with angular velocities of $+J$ and $-J$, and the resultant along the y'-axis will again oscillate but with a different periodicity (depending on δ and J) along the y'-axis (Fig. 5.3E, ii). Similarly, in the case of a CH_3 group, if the reference frequency is positioned at the center of the quartet, the two vectors corresponding to the outer peaks will possess angular velocities of $+\frac{3}{2}J$ and $-\frac{3}{2}J$, while the middle vectors will have angular velocities of $+J/2$ and $-J/2$ (Fig. 5.3E, iii). Clearly, the periodic perturbation of CH_3, CH_2, and CH carbons with varying t_1 values will be characteristically different, depending on the number of vectors and their respective angular velocities.

A simplifying assumption was made in the earlier discussion that the frame was rotating at the chemical shift frequency of the ^{13}C nucleus so that as the vectors diverged from one another, the position of the *net* magnetization remained unchanged, aligned with the y'-axis, and only its magnitude changed. If, however, we consider that the rotating frame is rotating at the chemical shift frequency of the TMS reference frequency, then both of the ^{13}C vectors of the CH group will rotate away in the same direction but with angular velocities determined by the chemical shift frequency of the CH group and that differ from each other by J Hz. This will be due to the difference in chemical shift frequency from the reference frequency of the rotating frame. The net magnetization will therefore actually not remain static along the y'-axis, but will precess in the $x'y'$-plane with a phase shift of v_0/J. This is illustrated in Fig. 5.4. The net magnetization is represented by the dotted line at angle α to the y'-axis. As the vectors rotate in the $x'y'$-plane, this angle α increases; when it reaches a value of 180°, the net magnetization points towards the $-y'$-axis. Thus the *resultant* along the y'-axis oscillates periodically, having a maximum positive value along the $+y$-axis immediately after the 90°_x pulse (when both vectors are aligned along the $+y'$-axis) and maximum negative amplitude when both vectors are aligned with the $-y'$-axis. This oscillation of the resultant is detected as corresponding oscillations of the NMR signal at different values of the evolution period, t_1.

A   B   C

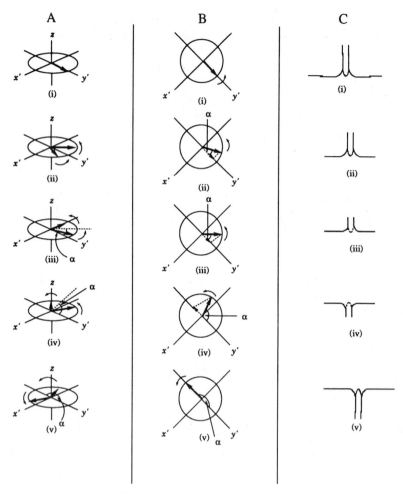

Figure 5.4 (A) The two ^{13}C vectors (representing the doublet of a CH group) move away from each other as well as away from the y'-axis (y'-axis is assumed to be rotating with angular velocity equal to TMS). The distance between the two vectors therefore grows with the delay till their angle of separation reaches 180° before it starts decreasing. The vectors therefore progressively go out-of phase and come-in phase. (B) The corresponding positions of the *net* magnetization vectors along the y'-axis. (C) Signal amplitudes with respect to the position of magnetization in the $x'y'$-plane.

Figure 5.5 shows the heteronuclear 2D J-resolved spectrum of camphor. The broad-band decoupled ^{13}C-NMR spectrum is plotted alongside it. This allows the multiplicity of each carbon to be read without difficulty, the F_1 dimension containing only the coupling information and the F_2 dimension only the chemical shift information. If, however, proton broad-band decoupling is applied in the evolution period t_1, then the 2D spectrum obtained again contains only the coupling information in the F_1 domain, but the F_2 domain now contains *both* the chemical shift and the coupling information (Fig. 5.6). Projection of the peaks onto the F_1 axis therefore gives the ^{1}H-decoupled ^{13}C spectrum; projection onto the F_2 axis produces the fully proton-coupled ^{13}C spectrum.

The most common way to record heteronuclear 2D J-resolved spectra is the *gated decoupler method,* so called because the decoupler is "gated," i.e., switched on during the preparation period (for nOe) during the first

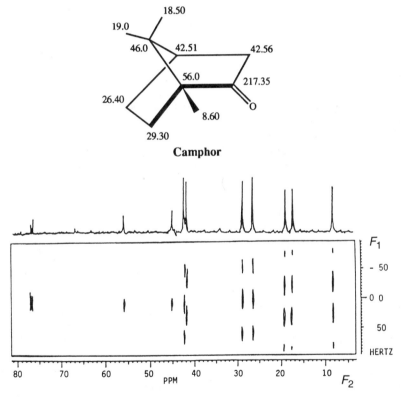

Camphor

Figure 5.5 A heteronuclear 2D J-resolved spectrum of camphor, along with a broad-band decoupled spectrum.

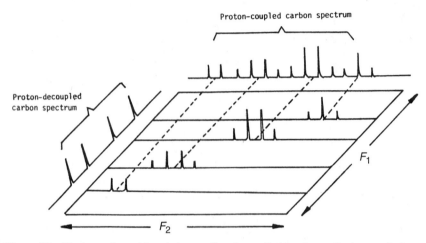

Figure 5.6 If proton broad-band decoupling is applied in the evolution period, t_1, then the resulting 2D spectrum contains only chemical shift information in the F_1 domain, while both chemical shift and coupling information is present in the F_2 domain. Projection onto the F_1-axis therefore gives the ^{1}H-decoupled ^{13}C spectrum, whereas projection along F_2 gives the fully coupled ^{13}C spectrum.

half of the evolution period, and switched off during the second half of the evolution period (Fig. 5.7). Since broad-band decoupling is carried out during the first half of the evolution period, the components of the ^{13}C multiplets do not diverge from one another, but rotate in the $x'y'$-plane with the same angular velocity, $\Omega/2\pi$. In the second half of the evolution period, the decoupler is switched off so that the ^{13}C multiplet components (doublet in CH, triplet in CH_2, quartet in CH_3) diverge and precess away from one another. *The extent of the divergence depends on the magnitude of their coupling constants and the duration of the time $t_1/2$ for which they are allowed to precess.* At the end of the evolution period, broad-band decoupling is turned on, so the vectors collapse to a single resultant signal for each carbon that is detected. These signals acquired at incremented t_1 values are modulated as a function of the coupling constant J so that data transposition and

A

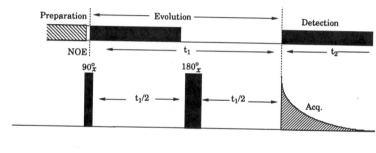

B

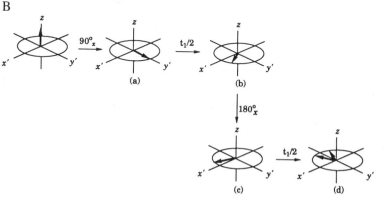

Figure 5.7 (A) Pulse sequence for gated decoupled J-resolved spectroscopy. It involves decoupling only during the first half of the evolution period t_1, which is why it is called "gated." (B) Positions of ^{13}C magnetization vectors at the end of the pulse sequence in (d) depend on the evolution time t_1 and the magnitude of the coupling constant, J. The signals are therefore said to be "J-modulated."

Fourier transformation gives 2D J-resolved spectra, with the chemical shifts appearing on the F_2 axis and the coupling constants along the F_1 axis.

Many variations of this experiment are known. Some of the pulse sequences used for recording heteronuclear 2D J-resolved spectra are shown in Fig. 5.8. In a *modified gated decoupler sequence* (Fig. 5.8b), the decoupler is off during the first half of the evolution period t_1 and is switched on during the second half. Any ^{13}C resonances that are folded over in the F_1 domain may be removed by employing the *fold-over corrected gated decoupler sequence* (FOCSY) (Fig. 5.8c) or the *refocused fold-over corrected decoupler sequence* (RE-FOCSY) (Fig. 5.8d).

In the *spin-flip method,* 180_x° pulses are applied simultaneously to both ^1H and ^{13}C nuclei at the midpoint of the evolution period so that the ^{13}C

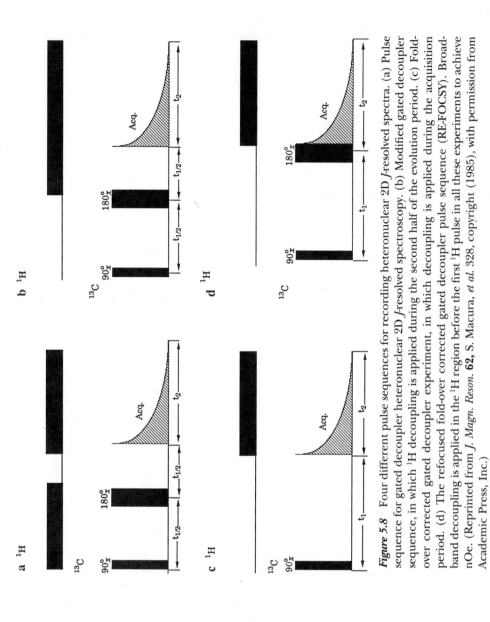

Figure 5.8 Four different pulse sequences for recording heteronuclear 2D *J*-resolved spectra. (a) Pulse sequence for gated decoupler heteronuclear 2D *J*-resolved spectroscopy. (b) Modified gated decoupler sequence, in which ^1H decoupling is applied during the second half of the evolution period. (c) Fold-over corrected gated decoupler experiment, in which decoupling is applied during the acquisition period. (d) The refocused fold-over corrected gated decoupler pulse sequence (**RE-FOCSY**). Broadband decoupling is applied in the ^1H region before the first ^1H pulse in all these experiments to achieve nOe. (Reprinted from *J. Magn. Reson.* **62,** S. Macura, *et al.* 328, copyright (1985), with permission from Academic Press, Inc.)

magnetization vectors do not converge after these pulses but continue to diverge from one another (Fig. 5.9). One disadvantage of this method is that if the 180° ^{1}H pulse is not applied accurately, then a component of the ^{13}C magnetization will be created that will not be *J*-modulated, giving rise to an artifact peak at $F_1 = 0$ on the F_1 axis.

In the selective spin-flip method (Fig. 5.10) developed by Bax and co-workers a selective 180°_x ^{1}H pulse is applied to one or more selected protons. The 2D *J*-resolved spectrum then shows heteronuclear coupling only with the irradiated proton(s). This is illustrated in the case of *β*-methyl cellobioside, a 1–4 linked disaccharide (Gidley, 1985). Irradiation of H-1 allowed the measurement of long-range coupling constants of H-1 with other distant protons, with the large one-bond coupling constant J_{CH} being suppressed by the 180°_x pulse in the middle of the evolution period. Measurement of the coupling constants across the glycosidic linkage (${}^{3}J_{C4,H-1'}$, ${}^{3}J_{C-1',H4}$) allowed a deduction of the torsional angles θ and Ψ, since a Karplus relationship exists (Fig. 5.11).

Another related experiment involves a BIRD (bilinear rotation decoupling) pulse sequence to flip distant protons selectively so the long-range J_{CH} interactions can be observed (Fig. 5.12) (Rutar, 1984). Alternatively, by reversing the phase of the last proton pulse, we can achieve the opposite effect, i.e., elimination of the long-range couplings so the one-bond coupling effects can be measured (Fig. 5.13).

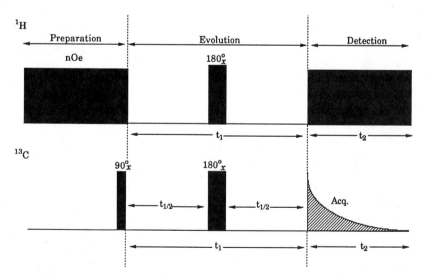

Figure 5.9 Pulse sequence for spin-flip heteronuclear *J*-resolved spectroscopy.

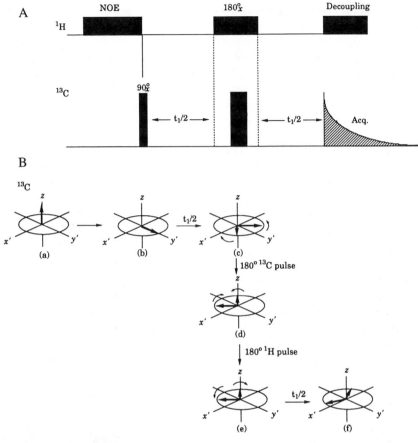

Figure 5.10   (A) Selective spin-flip pulse sequence for recording heteronuclear 2D *J*-resolved spectra. (B) Its effect on ^{13}C magnetization vectors. The selective 180_x° pulse in the middle of the evolution period eliminates the large one-bond coupling constants, $^1J_{CH}$.

In another related procedure, the protons that are bound directly to ^{13}C nuclei can be flipped selectively so that geminal couplings between nonequivalent protons can be measured (Fig. 5.14). This is known as *selective indirect J-spectroscopy* (Rutar *et al.*, 1984a).

Polarization transfer techniques like INEPT and DEPT have been used to enhance sensitivity in heteronuclear 2D *J*-resolved spectra. In combination with the semiselective sequence just described, INEPT has been used to suppress long-range J_{CH} couplings and to measure the one-bond couplings (Fig. 5.15) (Rutar, 1984). Driven equilibrium pulses for fast restora-

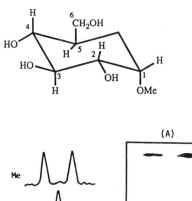

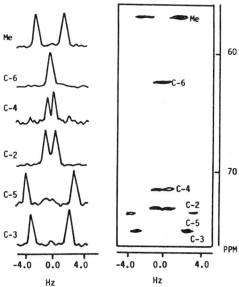

Figure 5.11 (A) Heteronuclear 2D *J*-resolved spectrum (selective spin-flip method) of α-methylglucoside after selective irradiation of H-1. The slices for various signals are shown on the left. (B) The peak separations represent the coupling constants of H-1 (proton irradiated) with various carbons. (Reprinted from *J. Chem. Soc. Chem. Comm.* M. J. Gidley and S. M. Bociek, 220, copyright (1985), with permission from The Royal Society of Chemistry, Thomas Graham House, Science Park, Milton Road, Cambridge CB4 4WF, U.K.).

tion of *z*-magnetization have also been employed for obtaining heteronuclear 2D *J*-resolved spectra (Becker *et al.,* 1969; Wang *et al.,* 1982; Wang and Wong, 1985).

Pure 2D absorption line shapes are readily obtained in heteronuclear 2D *J*-resolved spectra. The incorrect setting of 90° and 180° pulses can, however, cause "ghost" peaks that can be removed by a phase cycling procedure, appropriately named "Exorcycle" (Rutar, 1984b). A

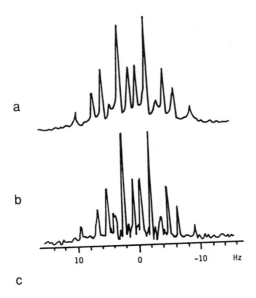

a

b

10   0   -10   Hz

c

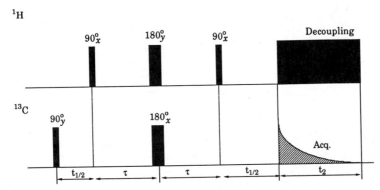

¹H

90°ₓ   180°ᵧ   90°ₓ   Decoupling

¹³C

90°ᵧ   180°ₓ   Acq.

t₁/₂   τ   τ   t₁/₂   t₂

Figure 5.12   (a) The upper trace shows elimination of one-bond couplings, which allows a decrease in bandwidth so that a significant increase in resolution is attained with a smaller number of data sets. (b) The cross-section in the ν_1 dimension of the central peak of the "triplet" (of the middle carbon of $ClCH_2CH_2CH_3$) is obtained by the spin manipulation method. The increase in resolution allows the two-bond coupling interactions to be measured with greater accuracy. (c) Pulse sequence for manipulation of spin in 2D NMR spectroscopy. (Reprinted from *J. Magn. Reson.* **56,** V. Rutar, 87, copyright (1984), with permission from Academic Press, Inc.)

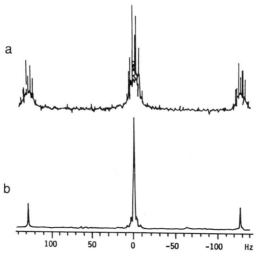

Figure 5.13 The normal 1D spectrum of the "triplet," corresponding to the central carbon of $ClCH_2CH_2CH_3$. The fine structure due to long-range couplings makes the one-bond coupling constant difficult to measure. The long-range couplings are eliminated by modifying the spin manipulation procedure (i.e., reversing the phase of the last pulse), which results in the disappearance of the fine structures. Only the one-bond couplings can now be measured, since the directly attached protons are flipped selectively. (Reprinted from *J. Magn. Reson.* **56,** V. Rutar, 87, copyright (1984), with permission from Academic Press, Inc.)

low-resolution heteronuclear 2D J-resolved spectrum, preferably obtained by the gated decoupler method, can quickly reveal the multiplicity of CH and CH_3 carbons, and it can therefore compete with INEPT and DEPT for the multiplicity determination of carbons.

5.1.2 Homonuclear 2D J-Resolved Spectroscopy

In the heteronuclear 2D J-resolved spectra described earlier, the chemical shifts of the ^{13}C nuclei were located along one axis and the $^1H/^{13}C$ couplings along the other. If the coupled nuclei of the *same* type are required to be observed, then homonuclear 2D J-resolved spectra are obtained. For instance, if the nuclear species being observed is ^1H, then the ^1H chemical shifts will be spread along the F_2 axis, and the coupling constants of the ^1H nuclei with other neighboring ^1H nuclei will be spread along the F_1 axis. This offers a powerful procedure for unraveling complex overlapping multiplets.

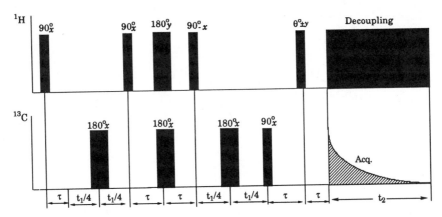

Figure 5.14 Pulse sequence for selective indirect *J*-spectroscopy. The three proton pulses at the center of the evolution period flip attached protons selectively, resulting in decoupling between distant and attached protons. (Reprinted from *J. Magn. Reson.* **60**, V. Rutar, *et al.*, 333, copyright (1984), with permission from Academic Press, Inc.)

Let us consider, for instance, an overlapping triplet and a quartet, so that seven peaks integrating for 2H appear in a certain region of the NMR spectrum (Fig. 5.16). If the coupling constants of the two multiplets are not significantly different, it may be difficult to recognize the overlapping signals readily.

It is very unusual to find that such overlapping multiplets have *exactly* the same chemical shifts. This difference in chemical shifts may be exploited if we could somehow rotate the triplet and the quartet about their respective chemical shifts by 90° so that, instead of lying flat in a horizontal plane, they come to lie above and below the plane, perpendicular to it. Each multiplet is then positioned vertically so it passes through the plane of the paper at its chemical shift. Clearly, if the two multiplets have different chemical shifts, the overlap present before the 90° rotation will disappear, with the chemical shifts lying along one axis and the coupling information appearing along a perpendicular axis. This is illustrated in Fig. 5.17.

The pulse sequence used in homonuclear 2D *J*-resolved spectroscopy is shown in Fig. 5.18. Let us consider a proton, A, coupled to another proton, X. The 90° pulse bends the magnetization of proton A to the y'-axis. During the first half of the evolution period, the two vectors $\mathbf{H}_F$ (faster vector) and $\mathbf{H}_S$ (slower vector) of proton A precess in the $x'y'$-plane with angular velocities of $\Omega/2 + (J/2)$ and $\Omega/2 - (J/2)$, respectively. The 180°_x ^{1}H pulse results in (a) the vectors' adopting mirror image positions across the x'-axis so that $\mathbf{H}_S$ comes to lie ahead of $\mathbf{H}_F$ when viewed in a clockwise fashion. However, since the 180° ^{1}H pulse was not selective, it also affects

A

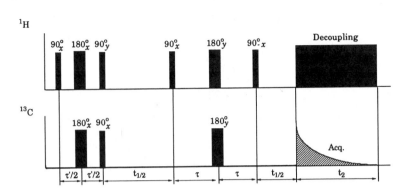

B

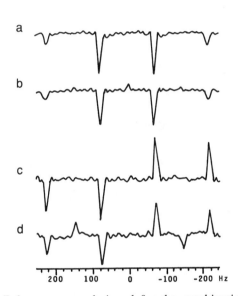

Figure 5.15 (A) Pulse sequence designed for the combination of polarization transfer and selective flip of attached protons. The long-range couplings are suppressed, and only one-bond heteronuclear couplings are observed in this experiment. (B) (a) Quartet of nitromethane obtained by using the basic pulse sequence of the heteronuclear 2D J-resolved experiment. (b) The same quartet, but with misset delays, which results in spurious peaks. (c) Quartet obtained by the combination of INEPT with spin manipulation. Peak intensification is due to the polarization transfer effect. (d) Effect of missetting of pulse widths in (c). (Reprinted from *J. Magn. Reson.* **58,** V. Rutar, 132, copyright (1984), with permission from Academic Press, Inc.)

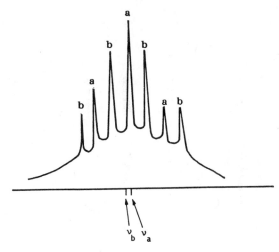

Figure 5.16 Overlapping triplet a and quartet b, respectively. ν_a and ν_b are the chemical shifts of the two protons.

neighboring proton X to which proton A was coupled. Thus, if $\mathbf{H}_F$ was coupled to the α state of proton X before the 180° pulse, it exchanges its identity with $\mathbf{H}_S$ and becomes coupled to the β state of proton X. This exchange of identity (or "relabeling") of the vectors $\mathbf{H}_F$ and $\mathbf{H}_S$ has $\mathbf{H}_F$ again becoming positioned ahead of $\mathbf{H}_S$ (Fig. 5.18B, C), so in the second half of the evolution period t_1, the two vectors continue to diverge from

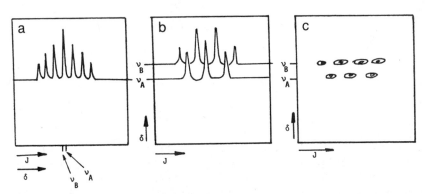

Figure 5.17 (a) Overlapping triplet and quartet of protons A and B, with the chemical shifts and coupling constants along the same axis. (b) Triplet and quartet of the same protons, shown with the chemical shifts lying along the vertical axis and coupling constants along the horizontal axis separately. (c) Overhead view of the same peaks as in (b) but in the form of contours.

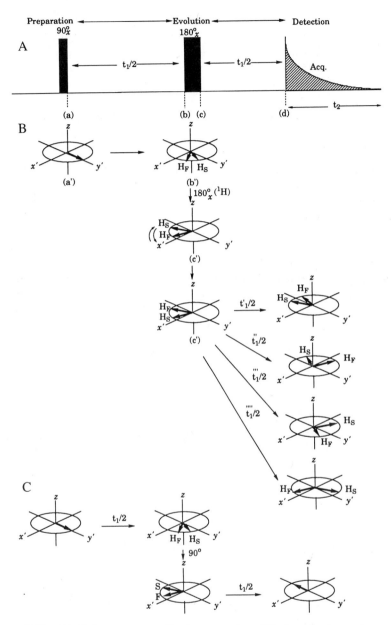

Figure 5.18 (A) Pulse sequence for homonuclear 2D J-resolved spectroscopy. (B) Effect of 90_x° ^{1}H and 180_x° ^{1}H pulses on an ^{1}H doublet. (C) In the absence of coupling, the vectors are refocused by the 180_x° ^{1}H pulse after t_1. This serves to remove any field inhomogeneities or chemical shift differences.

each other. The extent of this divergence depends both on the time t_1 for which they are allowed to diverge and on the coupling constant J. The first series of Fourier transformations give the 1D spectra in v_1 and F_2. Transformation of the data so they are arranged in rows, one behind the other, and Fourier transformation of the individual columns of oscillating signals along the F_2 (vertical) axis leads to the 2D plots in the F_1 and F_2 frequency dimensions, with the chemical shifts being defined by the F_2 axis and the coupling frequencies by the F_1 axis. The J-modulation of doublets and triplets with changing evolution time is shown in Fig. 5.19.

In homonuclear 2D J-resolved spectra, couplings are present during t_2; in heteronuclear 2D J-resolved spectra, they are removed by broad-band decoupling. This has the multiplets in homonuclear 2D J-resolved spectra appearing on the diagonal, and not parallel with F_1. If the spectra are plotted with the same Hz/cm scale in both dimensions, then the multiplets will be tilted by 45° (Fig. 5.20). So if the data are presented in the absolute-value mode and projected on the chemical shift (F_2) axis, the normal, fully coupled 1D spectrum will be obtained. To make the spectra more readable, a tilt correction is carried out with the computer (Fig. 5.21) so that F_1 contains only J information and F_2 contains only δ information. Projection

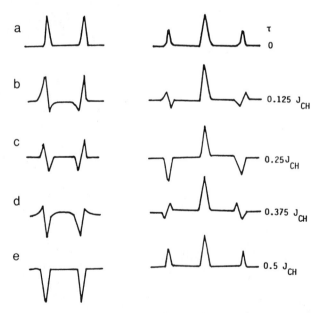

Figure 5.19 Modulation effects of a doublet and triplet by a spin-echo pulse sequence, $90^\circ_x–\tau–180^\circ_y–\tau–$echo.

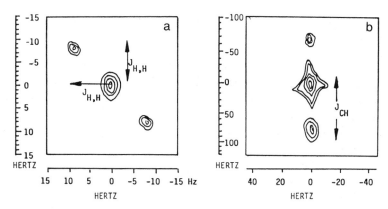

Figure 5.20 (a) Homonuclear J-spectrum. Since the coupling is both in ν_1 and ν_2 dimensions, a 45° tilt is observed. (b) Heteronuclear J-spectrum in which coupling appears only in one dimension.

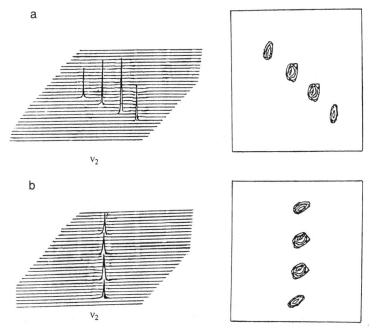

Figure 5.21 (a) The methylene protons of ethyl alcohol appear as a quartet with a 45° tilt in the homonuclear 2D J-resolved spectrum. (b) The same, but after tilt correction. The stacked plot is presented on the left; the corresponding contour plot appears on the right.

onto the F_2 (chemical shift) axis would then produce a fully decoupled spectrum.

Peaks in homonuclear 2D *J*-resolved spectra have a "phase-twisted" line shape with equal 2D absorptive and dispersive contributions. If a 45° projection is performed on them, the overlap of positive and negative contributions will mutually cancel and the peaks will disappear. The spectra are therefore presented in the absolute-value mode.

Homonuclear 2D *J*-resolved spectra often show artifact signals due to second-order effects appearing in strongly coupled systems (Fig. 5.22). Such artifact signals generally do not appear in weakly coupled spin systems ($\Delta\delta/J > 4$), which are largely first-order. They are usually readily recognized, for they appear at chemical shifts midway between strongly coupled protons, and they occur as wavy contours spread across the spectrum. Such

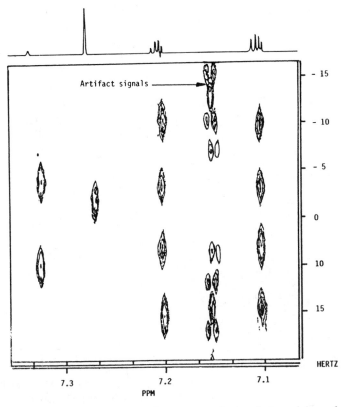

Figure 5.22 Artifact signals appear due to strongly coupled nuclei, as shown by the vertical lines of contours at about δ 7.16 in the spectrum.

artifact signals may disappear if the spectra are recorded at higher fields, due to the greater first-order character of the spectra recorded on high-field instruments. A common artifact in J-resolved spectra, as with other 2D spectra, is the presence of t_1-noise ridges, particularly when sharp, intense signals are present, like those due to methyl groups. These noise ridges appear parallel to F_1 before the tilt correction is performed; but after the tilt correction they appear at 45°, provided both dimensions are recorded with the same scale. Symmetrization can be carried out to remove the t_1-noise.

✦ *PROBLEM 5.1*

What are the essential ingredients of 2D J-resolved NMR spectroscopy?

─── ✦

✦ *PROBLEM 5.2*

What is the difference between homo- and heteronuclear 2D J-resolved spectroscopy?

─── ✦

5.2 HOMONUCLEAR AND HETERONUCLEAR SHIFT-CORRELATION SPECTROSCOPY

5.2.1 *Homonuclear Shift-Correlation Spectroscopy (COSY)*

Before the advent of 2D NMR spectroscopy, the classical procedure for determining proton–proton connectivities was by homonuclear proton spin decoupling experiments. Such experiments can still serve to determine some $^1H/^1H$ connectivities in simple molecules.

In the previous section we were concerned with 2D J-resolved spectroscopy; i.e., chemical shifts of the nuclei were presented along one axis and their couplings along the other axis. The data thus obtained provided information about *multiplicity*.

A more useful type of 2D NMR spectroscopy is shift-correlated spectroscopy (COSY), in which both axes describe the chemical shifts of the coupled nuclei, and the cross-peaks obtained tell us which nuclei are coupled to which other nuclei. The coupled nuclei may be of the same type—e.g., protons coupled to protons, as in *homonuclear* 2D shift-correlated experiments—or of different types—e.g., protons coupled to ^{13}C nuclei, as in *heteronuclear* 2D shift-correlated spectroscopy. Thus, in contrast to J-resolved spectroscopy, in which the nuclei were being modulated (i.e., undergoing a repetitive change in signal amplitude) by the *coupling frequencies* of the nuclei to which they were coupled, in shift-correlated spectroscopy the nuclei are modulated by the *chemical shift frequencies* of the nuclei to which they are coupled. So, if a proton H_A (having a chemical shift of 2 ppm, i.e., 200 Hz from TMS on a 100-MHz instrument) is coupled to a proton H_B (having a chemical shift of 4 ppm, i.e., 400 Hz from TMS on the same instrument) with a coupling constant of 8 Hz, then in *J*-resolved spectroscopy the protons H_A and H_B will both be modulated as a function of 8 Hz, while in shift-correlated spectroscopy H_A will be modulated as a function of δ_{HB}—i.e., 400 Hz—while H_B will be modulated as a function of δ_{HA}—i.e., 200 Hz. The variation of signal amplitude with changing evolution time t_1 is shown in Fig. 5.23. If the signal amplitude is modulated by more than one frequency, then a different pattern would result. Fig. 5.24 shows the amplitude being modulated by two different chemical shift frequencies.

When a Jeener pulse sequence (90°_x–$t_1/2$–90°_x–$t_1/2$–Acq.) is applied to coupled protons, the second 90°_x pulse causes the magnetization arising from one proton transition during t_1 to be distributed among all other transitions associated with it. For instance, in an AX spin system, the 1D spectrum will show two lines for proton A, corresponding to the transitions A_1 and A_2, and two lines for proton X, corresponding to the transitions X_1 and X_2. The energy-level diagram for such a spin system in which only single-quantum transitions are shown is given in Fig. 5.25. The second 90°_x pulse causes the magnetization arising from each transition to be distributed among all the four transitions. Thus the magnetization arising from the A_1 transition will be partly retained in the A_1 transition and partly distributed

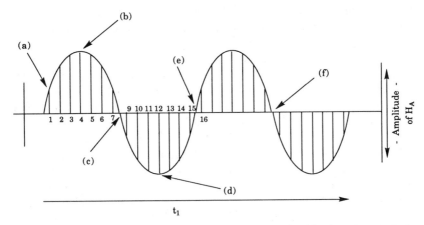

Figure 5.23 Variation in the amplitude of a proton H_A with changing evolution time t_1.

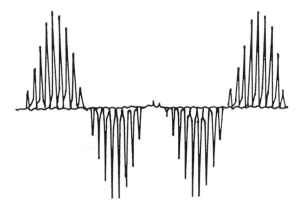

Figure 5.24 The amplitude of the lines is seen to be modulated by at least two different frequencies.

among the A_2, X_1, and X_2 transitions. A similar redistribution of magnetizations will occur for the A_2, X_1, and X_2 transitions. A line arising from the A_1 transition will therefore have its amplitude modulated by the frequency components of all the other transition lines of the AX spin system, giving rise to corresponding cross-peaks in the COSY spectrum.

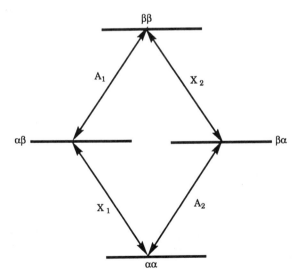

Figure 5.25 Energy levels for an AX spin system representing only single-quantum transitions.

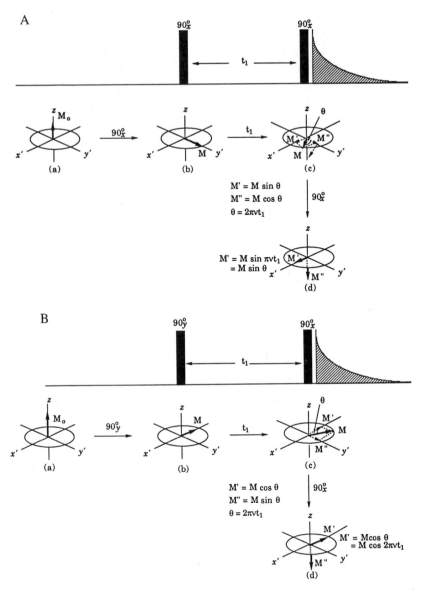

Figure 5.26 (A) The two 90°_x pulses are applied with an intervening variable t_1 period, which is systematically incremented. The first 90°_x pulse bends the magnetization from its equilibrium position along the z-axis in (a) to the y′-axis, as in (b). During the t_1 interval, the magnetization vector precesses in the x′y′-plane, and it may be considered to be composed of two component vectors, **M′** and **M″**, along the x′- and y′-axes, respectively. The magnitudes of the component vectors **M′** and

As is evident from Fig. 5.26, we can have either a sinusoidal modulation or a cosinusoidal modulation of the signals, depending on whether the two 90°_x pulses have the same phase or different phases. In Fig. 5.26A, both pulses have the same phase. This means that in the first t_1 period, the magnetization vector precesses in the $x'y'$-plane, so if it has precessed by

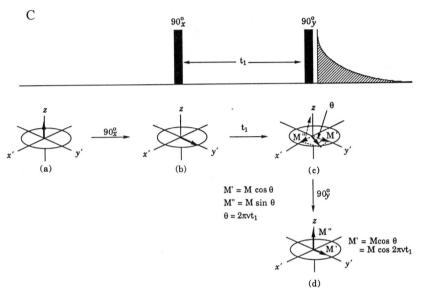

Figure 5.26 (*Continued*) $\mathbf{M}''$ are $\mathbf{M}' = \mathbf{M} \sin \theta$ and $\mathbf{M}'' = \mathbf{M} \cos \theta$, where θ is the angle of deviation of $\mathbf{M}$ from the y'-axis. The second 90°_y pulse bends the component $\mathbf{M}''$ lying along the y'-axis to the $-z$-axis so that it is no longer detectable, leaving the sinusoidal modulations of the $\mathbf{M}'$ vector along the x'-axis to be recorded. Therefore, if both the 90° pulses have the same phase, a *sinusoidal* modulation of the signals is observed. (B) Effect of two 90° pulses that are 90° out-of-phase with each other on the magnetization vector, $\mathbf{M}_0$. (a) The first 90°_y pulse bends $\mathbf{M}_0$ to the $-x'$-axis. (b) In the subsequent delay, t_1, the magnetization $\mathbf{M}$ precesses in the $x'y'$-plane. After time t_1 it will have precessed away from the $-x'$-axis by an angle θ, and its two components along the x'- and y'-axes will have magnitudes $\mathbf{M}' = \mathbf{M} \cos \theta$ and $\mathbf{M}'' = \mathbf{M} \sin \theta$, respectively. (c) The second 90°_x pulse removes the vector component along the y'-axis and bends it to the $-z$-axis. This leaves only one vector component along the $-x'$-axis ($\mathbf{M}' = \mathbf{M} \cos \theta$) to be measured. Consinusoidal modulation of the magnetization is therefore observed during different t_1 increments when the two 90° pulses applied are 90° out-of-phase with one another. (C) If the first pulse is a 90°_x pulse and the second pulse a 90°_y pulse, then a *cosinusoidal* modulation of the signals is observed. Now the second 90°_y pulse bends the component $\mathbf{M}''$ ($= \mathbf{M} \sin \theta$) to the z axis, leaving only the consinusoidally modulated component $\mathbf{M}'$ ($= \mathbf{M} \cos \theta$) in the transverse ($x'y'$) plane to be detected.

an angle θ away from the y'-axis by the end of the t_1 period, it will have two components, **M'** and **M''** along the x'- and y'-axes, respectively, with magnitudes **M'** = **M** sin θ and **M''** = **M** cos θ. The second $90°_x$ pulse bends the component **M''** lying along the y'-axis to the $-z$-axis so it will no longer be detected. This will leave the component vector along the x'-axis (with magnitude M sin θ) to be detected. In other words, *sinusoidal* modulations of the magnetization vector **M'** will be observed if *both* the first and the second 90° pulses have the same phase. If, however, they have opposite phases (Figs. 5.26B and 5.26C), then the second 90° pulse will remove the sinusoidally modulated component to the z-axis, leaving the cosinusoidally modulated component to be detected. Hence, with appropriate phase cycling, it is possible to obtain *phase-sensitive COSY spectra* in which the absorption- and dispersion-mode signals are nicely separated; i.e., the peaks appearing on or near the diagonal will differ in phase by 90° from the cross-peaks. Normally we adjust the phasing so as to show up the cross-

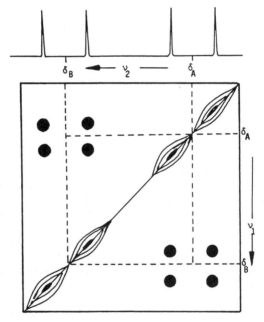

Figure 5.27 Schematic diagram of a COSY spectrum illustrating how intrinsically different cross-peaks are from diagonal peaks. The circles represent the cross-peaks in an AB coupled system. The diagonal peaks have dispersion shape, and the cross-peaks have an absorption shape though they alternate in sense. The diagonal peaks are centered at (δ_A, δ_A) and (δ_B, δ_B); the cross-peaks appear at (δ_A, δ_B) and (δ_B, δ_A).

peaks in the absorptive mode and the diagonal peaks in the dispersive mode, since it is the cross-peaks that are of real interest (Fig. 5.27).

An advantage of phase-sensitive COSY spectra, as compared to absolute-mode spectra, is that when recorded at enough digital resolution they allow us to determine coupling constants. Let us consider an AMX spin system. A schematic drawing of a phase-sensitive COSY spectrum of an AMX spin system is shown in Fig. 5.28. The coupling between H_A and H_M gives rise to the cross-peak (1) marked A/M. This is designated as the *active coupling* and contains information about J_{AM}. However, H_A is also coupled to H_X. And the fine structure in the A/M cross-peak (1) will also contain information about J_{AX} (*passive coupling*). Similarly, cross-peak (2), arising from the coupling of A with X, will contain the active coupling information J_{AX} as well as the fine structure for J_{AM}. The fine splitting can be seen if the phase-sensitive spectrum is recorded at a high digital resolution (Fig. 5.29). The active coupling shows an antiphase disposition of cross-peaks, while passive couplings appear in-phase. This is illustrated in the schematic drawing of a cross-peak obtained at ν_A, ν_X in an AMX spin system (Fig. 5.30). If we read the multiplet horizontally, the distance between the center points of peaks 1 and 2 (or 3 and 4) corresponds to the active coupling J_{AX} (note

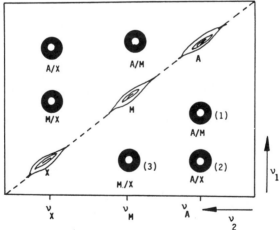

Figure 5.28 A schematic drawing of an AMX spin system representing coupling interactions recorded at low digital resolution so that no fine splittings are visible. Note the symmetrical appearance of the cross-peaks on either side of the diagonal.

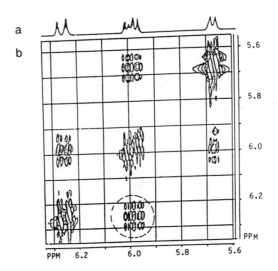

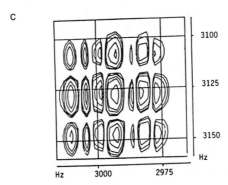

Figure 5.29 (a) The ¹H-NMR spectrum of ethyl acrylate showing signals for olefinic protons. (b) A phase-sensitive COSY spectrum recorded at higher digital resolution. (c) Expansion of the downfield cross-peak identified by a dashed circle in (b).

antiphase disposition). Similarly, the distance between peaks 1 and 3 (or 2 and 4) represents J_{AM}. If we read the multiplet vertically, the distance between peaks 1 and 5 again corresponds to J_{AX}, while the distance between peaks 1 and 9 provides J_{MX}. If we read horizontally, J_{AX} is active and J_{AM} passive, if we read vertically, J_{AX} is active and J_{MX} passive. If the antiphase character of the cross-peaks is not readily discernible due to peak overlap, then this may be read from the slices taken at different values of ν_1 and ν_2, which allows the phase patterns to be seen with greater clarity. This is illustrated in Fig. 5.31.

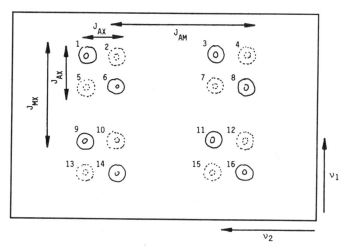

Figure 5.30 Drawing of a spectrum with cross-peak at ν_A, ν_X in an AMX system. The active couplings display an antiphase disposition of peaks. If we read horizontally, J_{AX} is active and J_{AM} passive; if we read vertically, J_{AX} is active and J_{MX} passive.

5.2.1.1 COSY-45° SPECTRA

If the width of the second pulse in the COSY pulse sequence is reduced from 90° to 45°, the mixing of populations between transitions that do not share the same energy level (*unconnected transitions*) is reduced. Only directly connected transitions then show cross-peaks, resulting in a simplification of the multiplet structure and, to some extent, a decrease in the area occupied by the multiplet. The intensity of the diagonal peaks is also reduced. In such spectra, the direction in which the cross-peak responses are tilted gives the relative signs of the coupling constant. This can be useful in distinguishing positive vicinal coupling constants from negative geminal couplings (Fig. 5.32).

In general, the ratio of intensities of cross-peaks arising from connected transitions to those arising from unconnected transitions is given by $\cot^2(\alpha/2)$, where α is the flip angle of the last mixing pulse. With a mixing pulse of 45°, the connected transitions give five to eight times stronger cross-peaks than unconnected transitions. A better signal-to-noise ratio will result if the mixing pulse has an angle of 60° (a COSY-60° experiment), with the cross-peaks from connected transitions appearing threefold stronger than those from unconnected transitions.

Since many of the signals in COSY spectra are in antiphase, they may not show up as cross-peaks due to the intrinsic nature of the polarization transfer experiment. The intensities of cross-peaks in COSY spectra may be represented by an antiphase triangle (Fig. 5.33B), in contrast to multiplet

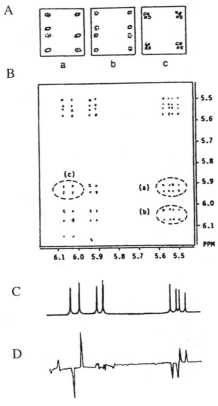

Figure 5.31 (A) The circled cross-peaks in (B) are expanded in (a), (b), and (c) to show up the fine structures. (B) Phase-sensitive COSY spectrum of acrylonitrile. (C) 1D ^{1}H-NMR spectrum of the compound. (D) Cross section (slice) along the column marked B in the COSY spectrum.

intensities in 1D spectra, in which the multiplicities are represented by a Pascal triangle (Fig. 5.33A). This is illustrated by a schematic drawing of the coupling of H_4 and H_6 with H_5 in *m*-dibromobenzene (Fig. 5.34). The COSY cross-peaks are missing at the center of the triplet since its intensity ratio in the polarization transfer experiment is represented by +1, 0 and −1.

✦ *PROBLEM 5.3*

How can we correlate protons that are coupled to each other (proton coupling network)?

─── ✦

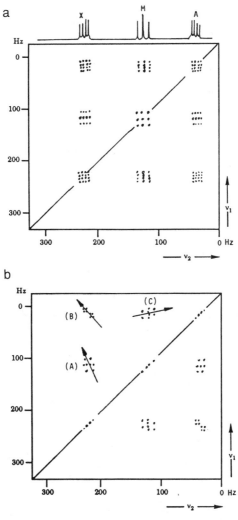

Figure 5.32 (a) COSY-90° spectrum of 2,3-dibromopropionic acid. (b) COSY-45° spectrum of the same compound. In cross-peak (B), only directly connected transitions appear. Each cross-peak consists of two overlapping square patterns. The sign of the slope of the line joining their centers gives the relative signs of the coupling constants, J. (Reprinted from *J. Magn. Reson.* **44,** A. Bax, *et al.* 542, copyright (1981), with permission from Academic Press, Inc.)

✦ *PROBLEM 5.4*

How do the peaks appearing on the diagonal differ from the "off-diagonal" cross-peaks in COSY spectra? How do they arise?

✦

(A)   **(B)**

Figure 5.33 (A) Pascal triangle. (B) Antiphase triangle.

As in 1D spectra, the folding of signals may occur in 2D spectra if the spectral width chosen is too small, giving rise to artifacts. This can be readily corrected by recording the spectra with a wider spectral width. Such artifacts as fold-over peaks or t_1-noise lines running across the spectrum can be corrected by the process of *symmetrization* (Fig. 5.35). Since genuine cross-peaks appear as symmetrically disposed signals on either side of the diagonal, the process of symmetrization will remove all random noise and other nonsymmetrically disposed signals on the two sides of the diagonal. Of course, the cosmetic improvement due to symmetrization must be undertaken with caution, for any noise signals that appear symmetrically on either side of the diagonal will not be removed and they will show up as a "cross-peak." The application of shaping functions for the improvement of 2D spectra has already been discussed in Section 3.2.2 and will therefore not be repeated here.

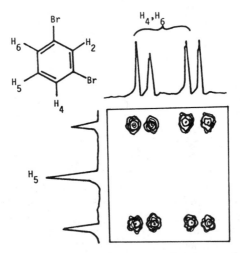

Figure 5.34 Schematic representation of the coupling interactions of the H_4, H_5, and H_6 protons of *m*-dibromobenzene. The H_4 and H_6 protons are split into double doublets due to their couplings with H_5 and H_2; H_5 is split into a triplet by the two *ortho* protons. Note that the COSY cross-peaks are missing at the central peak of the triplet due to the intensity ratio of the triplet in the polarization transfer experiment being $+1$, 0, -1.

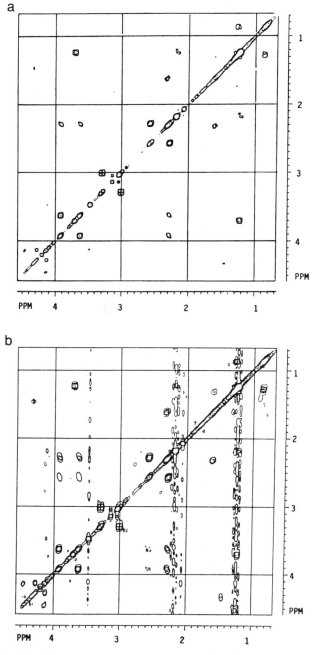

Figure 5.35 (a) Symmetrized version of a COSY spectrum in which noise lines are eliminated. (b) Unsymmetrized version of the same spectrum.

✦ *PROBLEM 5.5*

Following are the 1D ¹H-NMR and COSY-45° spectra of ethyl acrylate.
Interpret the spectrum.

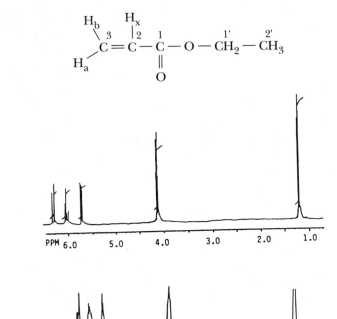

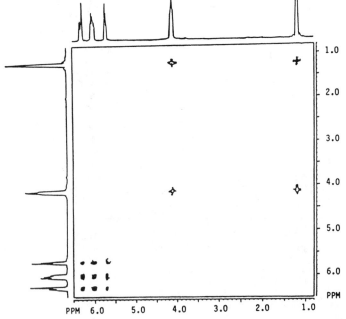

✦ *PROBLEM 5.6*

Is it possible to obtain proton–proton coupling information from a 1D ^{1}H-NMR experiment?

―――――――――――――――――――――――――――――― ✦

✦ *PROBLEM 5.7*

Give the advantages of the COSY-45° experiment over the COSY-90° experiment.

―――――――――――――――――――――――――――――― ✦

5.2.1.2 DECOUPLED COSY SPECTRA

When there is serious overlapping of cross-peaks because the coupled protons lie close to one another, then we can narrow the area of the cross-peaks by decoupling in the ν_1 dimension. This is illustrated in Fig. 5.36. All the protons appear as fully decoupled signals at their respective chemical shifts along the ν_1 axis, while the projection on the ν_2 axis corresponds to the normal coupled spectrum. This provides a useful procedure for obtaining the fully decoupled NMR spectrum and for obtaining the chemical shifts as well as the number of different protons in a molecule—information that may not be readily obtained from the 1D NMR spectrum because of the overlapping of multiplets. Since the cross-peaks are now fairly thin, the information about all the coupling partners of a given proton (cross-section along ν_2) are also obtained much more readily. Many other modifications of the COSY experiment have been developed; some of the more important ones will be presented here.

✦ *PROBLEM 5.8*

How is decoupling in the ν_1 dimension achieved in decoupled COSY spectra, and what are the advantages of this technique?

―――――――――――――――――――――――――――――― ✦

5.2.1.3 DOUBLE-QUANTUM FILTERED COSY

One problem associated with COSY spectra is the dispersive character of the diagonal peaks, which can obliterate the cross-peaks lying near the diagonal. Moreover, if the multiplets are resolved incompletely in the cross-peaks, then because of their alternating phases an overlap can weaken their intensity or even cause them to disappear. In *double-quantum* filtered COSY spectra, both the diagonal and the cross-peaks possess antiphase character, so they can be phased simultaneously to produce pure 2D absorption line

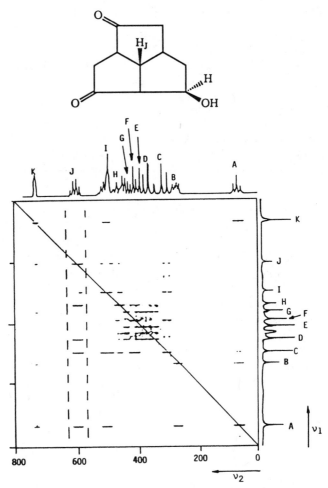

Figure 5.36 Effect of decoupling in ν_1 dimension on the COSY spectrum of 9-hydroxytricyclodecan-2,5-dione. The 1D ^{1}H-NMR spectrum on ν_2 is fully coupled, while along ν_1 it is fully decoupled. The region between the dashed lines represents a band of signals associated with proton H_J and its connectivities with other protons by projection onto the ν_1 axis. (Reprinted from *J. Magn. Reson.* **44**, A. Bax, *et al.*, 542, copyright (1981), with permission from Academic Press, Inc.)

shapes in both. In practice this is done by applying a third 90°_x pulse immediately after the second (mixing) pulse. This third pulse converts the double-quantum coherence generated by the second pulse into detectable single-quantum coherence. *Only* signals due to double-quantum coherence are inverted by the third 90° pulse. And if the receiver phase is correspondingly inverted, then only signals due to double-quantum coherence will be detected. (In the case of triple-quantum coherence spectra, the signal inversion will be achieved by a 60° phase shift.) Since all other coherences are filtered out, only the desired double-quantum coherences will be modulated as a function of t_1 and yield cross-peaks in the 2D spectrum. This will significantly simplify the COSY spectrum. Thus, in a double-quantum filtered spectrum, couplings due to AB and AX spin systems will show up, but solvent signals—which cannot generate double-quantum coherence— or signals due to triple-quantum coherence (as in a three-spin system) will not appear. Direct connectivities are represented in double-quantum spectra by pairs of signals situated symmetrically on the two sides of the diagonal; remotely connected or magnetically equivalent protons will appear as lone multiplets.

The sensitivity of multiple-quantum spectra is less than that of unfiltered COSY spectra. In the case of double-quantum filtered (DQF) COSY spectra, the sensitivity is reduced by a factor of 2, though this is usually not a serious drawback, since in homonuclear COSY spectra we are normally concerned with the more-sensitive $^1H/^1H$ coupling interactions. Figure 5.37 presents the phase-sensitive COSY and DQF-COSY spectra. The singlet lines due to the solvent signal are largely suppressed, and the peaks lying near the diagonal are no longer obliterated in the phase-sensitive DQF spectrum.

5.2.1.4 DELAYED-COSY

In normal COSY spectra, the stronger coupling interactions appear as prominent cross-peaks and the weaker long-range couplings may not appear. Introducing a fixed-delay Δ before acquisition results in the weakening (and often even the disappearance) of the stronger coupling interactions and an intensification of the smaller long-range couplings. The pulse sequence used is shown in Fig. 5.38. The value of the fixed delay is optimized to between 0.3 and 0.5. To improve the sensitivity, the last mixing pulse is set at 50°–60°.

✦ *PROBLEM 5.9*

When is the delayed-COSY experiment advantageous over the standard COSY experiment?

———————————————————————————————✦

5.2.1.5 E. COSY

The measurement of coupling constants is possible in multiple-quantum COSY spectra with phase-sensitive display. The recording of spec-

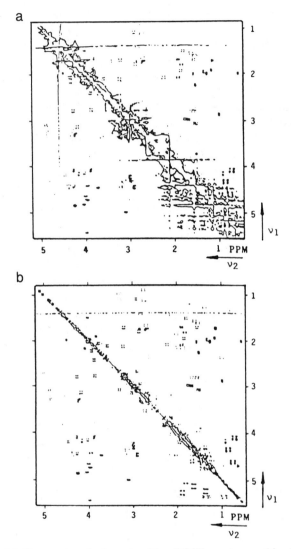

Figure 5.37 (a) Conventional phase-sensitive COSY spectrum of basic pancreatic trypsin inhibitor. (b) Double-quantum filtered (DQF) phase-sensitive COSY spectrum of the same trypsin inhibitor, in which singlet resonances and solvent signal are largely suppressed. Notice how clean the spectrum is, especially in the region near the diagonal line. (Reprinted from *Biochem. Biophys. Res. Comm.* **117,** M. Rance, *et al.,* 479, copyright (1983) with permission from Academic Press, Inc.)

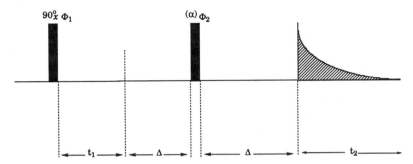

Figure 5.38 Pulse sequence for delayed COSY—a modification of the COSY experiment. The fixed delays at the end of the evolution period t_1 and before the acquisition period t_2 allow the detection of long-range couplings between protons.

tra in the absolute-value mode is necessary to remove the phase-twist caused by the mixing of absorption and dispersion lines. Pseudoecho weighting functions can be employed, but with some loss of sensitivity. Many procedures have been developed (States *et al.*, 1982) for measuring coupling constants in 2D spectra; only one will be mentioned here. This is a modification of the COSY experiment that incorporates a mixing pulse β of variable angle in the pulse sequence $90^\circ_x-t_1-(\beta)y-t_2$. In practice it is more convenient to use the variable parameter β as a phase shift in the equivalent sequence $(90^\circ)_\beta-t_1-(90^\circ)_\beta(90^\circ)_{-x}-t_2$. In this *exchange correlation spectroscopy* (E. COSY), many experiments with different values of β are combined, so multiplet components between unconnected transitions are eliminated and only multiplet components arising from connected transitions are detected. Thus, in contrast to conventional COSY spectra, in which spin systems comprising N mutually coupled spins give 2^{2N-2} multiplet components, in E. COSY spectra only 2^N correlations will appear from connected transitions, resulting in a simplification of the multiplet components.

✦ *PROBLEM 5.10*

How are coupling constants measured from the E. COSY spectrum?

　　　　　　　　　　　　　　　　　　　　　　　　　　　　✦

5.2.1.6 SECSY

Another 2D homonuclear shift-correlation experiment that provides the coupling information in a different format is known as SECSY (spin-echo correlation spectroscopy). It is of particular use when the coupled nuclei lie in a narrow chemical shift range and nuclei with large chemical shift differences are not coupled to one another. The experiment differs from the conventional COSY experiment in that acquisition is not immediately after the second mixing pulse but is delayed by time $t_1/2$ after the mixing pulse (i.e., the mixing pulse is placed in the middle of the t_1 period). Because the observation now begins at the top of the coherence transfer echo, the experiment is called "spin-echo correlation spectroscopy".

The untransferred magnetization now appears along a horizontal line (instead of along a diagonal, as in COSY). The cross-peaks appear above and below the horizontal line at distances equal to half the chemical shift differences between the two coupled nuclei. The identification of cross-peaks is facilitated by the fact that they occur on slanted parallel lines that make the same angle with the horizontal line and by the fact that each pair of cross-peaks is equidistant above and below the horizontal line. To determine the connectivities, we drop lines vertically from the cross-peaks onto the horizontal line; the points at which they meet the horizontal line give the chemical shifts of the coupled protons. This is illustrated in the SECSY spectrum of majidine (Fig. 5.39). The coupling of the anomeric proton resonating at δ 5.56 is observed with the C-2″ proton (δ 4.62). The C-4″ α and β methylenic protons in the sugar moiety show interactions between geminal protons resonating at δ 3.78 and 4.22, respectively. Similarly, the other α and β methylenic protons resonating at δ 3.17 and 3.57 also display geminal coupling with each other. The C-2‴ proton resonating at δ 3.08 affords a cross-peak with the C-1‴ proton (δ 4.38). The downfield C-5 proton (δ 7.61) also shows through-bond interaction with the C-8 proton resonating at δ 6.90. The various interactions based on the SECSY experiment are presented around the structure (Fig. 5.39).

✦ PROBLEM 5.11

What are the major differences between the COSY and SECSY experiments?

——————————————————————————————————— ✦

5.2.2 Heteronuclear Shift-Correlation Spectroscopy

In *homonuclear* shift-correlation experiments like COSY we were concerned with the correlation of chemical shifts between nuclei of the same nuclear species, e.g., ^1H with ^1H. In *heteronuclear* shift-correlation experiments, however, the chemical shifts of nuclei belonging to different nuclear species are determined (e.g., ^1H with ^{13}C). These may be one-bond chemical shift correlations, e.g., between directly bound ^1H and ^{13}C nuclei, or they may be long-range chemical shift correlations, in which the interactions

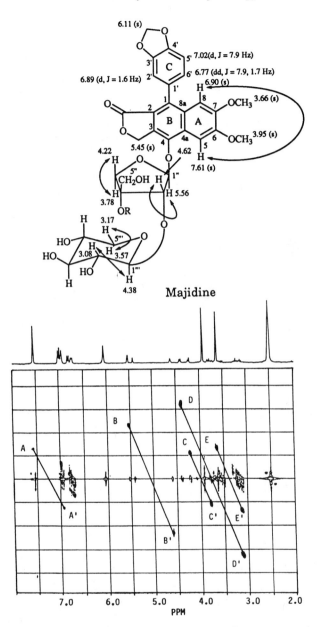

Figure 5.39 The structure and SECSY spectrum of majidine. Cross-peaks represent coupling interactions between various protons. These interactions are also presented around the structure.

between protons with carbon atoms that are 2, 3, or 4 bonds away are detected.

The basic pulse sequence employed in the heteronuclear 2D shift-correlation (or HETCOR) experiment is shown in Fig. 5.40. The first 90°_x ^{1}H pulse bends the ^{1}H magnetization to the y'-axis. During the subsequent evolution period this magnetization precesses in the $x'y'$-plane. It may be considered to be made up of two vectors corresponding to the lower (α) and higher (β) spin states of carbon to which ^{1}H is coupled. These two

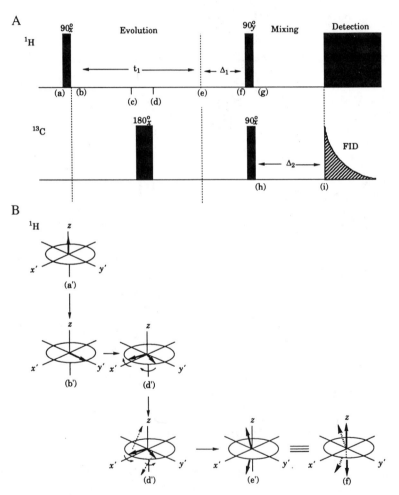

Figure 5.40 (A) Pulse sequence for the 2D heteronuclear shift-correlation experiment. (B) Effect of the pulse sequence in (A) on ^{1}H magnetization vectors of CH.

vectors diverge in the first half of the evolution period due to their differing angular velocities. The 180° ^{13}C pulse applied at the center of the evolution period reverses their direction of rotation so that during the second half of the evolution period they converge and are refocused at the end of the t_1 period. The angle α by which the vector has diverged from the y'-axis corresponds to its chemical shift, with the downfield protons having a larger angle α than the upfield protons. A constant mixing delay period Δ_1 occurs after the variable evolution period t_1, and the length of the mixing delay is set to the average value of J_{CH}. After the first $\frac{1}{2}J$ period, the vectors diverge and become aligned in opposite directions. The second 90°_x mixing ^{1}H pulse is then applied, which rotates the two vectors towards the $+z$- and $-z$-axes. One vector now points toward the $+z$-axis (equilibrium position), while the other points toward the $-z$-axis (nonequilibrium position).

What we have succeeded in doing is to invert *one* of the proton lines of the CH doublet (corresponding to the vector pointing toward the $-z$-axis), amounting to a selective population inversion, as in the INEPT experiment described earlier. This correspondingly intensifies the ^{13}C signals due to the polarization transfer. A 90° ^{13}C pulse is then applied that samples the evolved spin states. Clearly the extent of the polarization transfer will depend on the disposition of the proton multiplet components when the second 90° mixing pulse is applied (which in turn depends on the ^{1}H chemical shifts) as well as on the duration of the evolution period t_1. The amplitude of the ^{13}C signals detected during t_2 is therefore modulated as a function of t_1 by the frequencies of ^{1}H spins, giving rise to ^{1}H/^{13}C shift correlation spectra.

If broad-band decoupling is not applied during acquisition, then a coupled spectrum is obtained in both dimensions. This complicates the spectrum, and the splitting of the signals into multiplets also results in a general lowering of intensity. Since the multiplet components are in antiphase, an overlap of multiplets can also cause their mutual cancellation. It is therefore usual to eliminate the heteronuclear couplings in both dimensions. In the case of nonequivalent geminal protons of CH$_2$ groups, two different cross-peaks will appear at the same carbon chemical shift but at two different proton chemical shifts, thereby facilitating their recognition.

The HETCOR spectrum of a naturally occurring isoprenylcoumarin is shown in Fig. 5.41. The spectrum displays one-bond heteronuclear correlations of all protonated carbons. These correlations can easily be determined by drawing vertical and horizontal lines starting from each peak. For example, peak A represents the correlation between a proton resonating at δ 1.9 and the carbon at δ 18.0. Similarly, cross-peaks E and F show that the protons at δ 4.9 and 5.1 are coupled to the same carbon, which resonates at δ 114.4; i.e., these are the nonequivalent protons of an exomethylenic

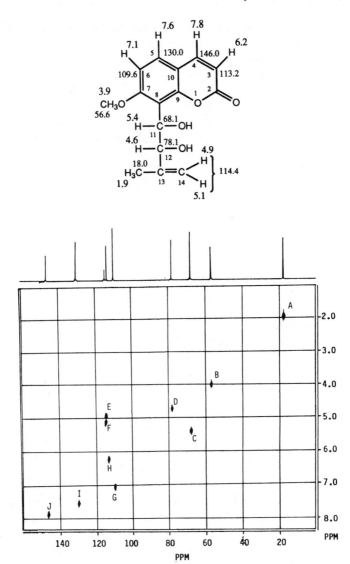

Figure 5.41   Heteronuclear shift correlation (HETCOR) spectrum of an isoprenyl-coumarin.

group. Cross-peak *J* represents the one-bond heteronuclear correlation between the proton at δ 7.8 and the carbon at δ 146.0. The chemical shift assignments to various protonated carbons based on the HETCOR experiment are presented around the structure in Fig. 5.41.

If the delays Δ_1 and Δ_2 are adjusted so they correspond to J_{CH} of 130–150 Hz, then only the direct one-bond couplings between sp^3- and sp^2-hybridized $^{13}C-^1H$ bonds will be observed. However, if we wish to observe long-range ^1H couplings with ^{13}C nuclei, then we adjust the value of the delay to correspond to J_{CH} of 5–10 Hz. The cross-peaks will then correspond to coupling interactions between ^1H and ^{13}C nuclei separated by two, three, or even four bonds. In such a *long-range hetero COSY* experiment, the one-bond magnetization transfer is suppressed and long-range heteronuclear couplings observed.

Many improved modifications of the basic heteronuclear shift-correlation experiment have been developed (Garbow *et al.*, 1982; Bax, 1983; Bauer *et al.*, 1984; Kessler *et al.*, 1984a,b,1985,1986; Schenker and Philipsborn, 1985; Wimperis and Freeman, 1984, 1985; Welti, 1985; Zektzer *et al.*, 1986, 1987a,b; Quast *et al.*, 1987; Wernly and Lauterwein, 1985). A discussion of these is beyond the scope of this text. For long-range shift correlations, COLOC (correlation spectroscopy via long-range coupling) has proved particularly useful (Kessler *et al.*, 1984a,b,1985,1986). However, if facilities for carrying out inverse polarization transfer experiments are available, then the HMQC (heteronuclear multiple-quantum coherence) experiment should be the experiment of choice for determining one-bond C-H shift correlations, and the HMBC (heteronuclear multiple-bond connectivity) experiment should be employed for determining long-range C–H connectivities, particularly when the sample quantity is limited so that maximum sensitivity is required. These inverse experiments are discussed in a later section.

✦ *PROBLEM 5.12*

How can we determine the carbon–carbon connectivities in a molecule through a combination of homo- and heteronculear shift-correlation experiments?

─── ✦

5.3 TWO-DIMENSIONAL NUCLEAR OVERHAUSER SPECTROSCOPY

5.3.1 NOESY

A large number of 1D nOe experiments may have to be performed if the spatial relationships among many protons in a molecule are to be determined. In such cases, instead of employing multipulse 1D nOe experiments, we can opt for the nOe spectroscopy (NOESY) experiment. If many protons have close chemical shifts, then NOESY may be particularly advantageous, since it is difficult to carry out irradiation in 1D experiments selectively without affecting other protons nearby. In the NOESY experiment, all interproton nOe effects appear simultaneously, and the spectral overlap is minimized due to the spread of the spectrum in two dimensions.

The pulse sequence for the NOESY experiment is shown in Fig. 5.42. Let us assume that there are two close-lying protons A and X that are not scalar-coupled together. The first 90°_x pulse bends the longitudinal z-magnetization of nucleus A to the y-axis; in the subsequent t_1 time period, this magnetization precesses in the $x'y'$-plane so that after time t_1 it will have traveled by an angle $\Omega_A t_1$, where Ω_A is the precession frequency of nucleus A. The second 90°_x pulse will rotate this magnetization from the $x'y'$-plane to the $x'z$-plane. At this time the x'-component of magnetization $\mathbf{M}'$ will be proportional to sin $(\Omega_A t_1)$, and the $-z$-component $\mathbf{M}''$ will be proportional to $-\cos (\Omega_A t_1)$. The $-z$-component is of primary interest since it is this component that will be converted (after modulation during the mixing period) to detectable transverse magnetization by the third $90_{\pm x}$ pulse. During the mixing period, magnetization transfer occurs between spins A and X so that the z-magnetizations of nuclei A and X are modulated by each other. Thus, spin A magnetization will be modulated by cos $(\Omega_X t_1)$

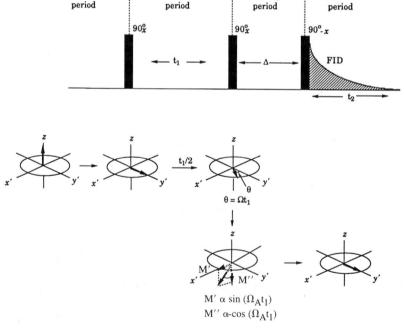

$$M' \propto \sin (\Omega_A t_1)$$
$$M'' \propto -\cos (\Omega_A t_1)$$

Figure 5.42   Pulse sequence and vector representation of a NOESY experiment.

and spin X magnetization will be modulated by cos $(\Omega_A t_1)$. These modulations will appear as cross-peaks on either side of the diagonal, i.e., at the coordinates $(\Omega_1, \Omega_2) = (\Omega_A, \Omega_X)$ and (Ω_X, Ω_A), thereby establishing the existence of nOe between them. The unmodulated signals appear on the diagonal, as in COSY spectra. A typical NOESY spectrum is shown in Fig. 5.43.

The diagonal and cross- (off-diagonal) peaks in the NOESY spectrum have either the same phase or exactly opposite phases, depending on the magnitude of the molecular correlation time, τ_c, so NOESY spectra are normally recorded in the pure-absorption mode. Some "COSY peaks" (i.e., peaks due to coherent magnetization transfer from scalar coupling) can appear as artifacts in NOESY spectra; therefore either we cycle the phases of the first two pulses, keeping the receiver phase constant, or we cycle the last pulse together with the receiver phase to remove COSY-type signals.

5.3.2 Phase-Sensitive NOESY

Phase-sensitive NOESY spectra are generally significantly superior in resolution to absolute-mode spectra. As in phase-sensitive COSY spectra,

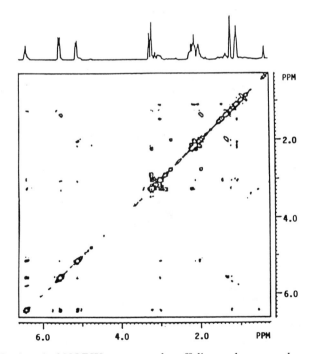

Figure 5.43 A typical NOESY spectrum; the off-diagonal cross-peaks represent the nOe interactions between various nuclei.

discussed in Section 5.2.1, there are "four phase quadrants" arising from the various combinations of real (R) and imaginary (I) parts corresponding to quadrants RR (real F_1, real F_2), RI (real F_1, imaginary F_2), IR (imaginary F_1, real F_2), and II (imaginary F_1, imaginary F_2). Phasing of the 2D NOESY spectrum is carried out to obtain the pure double-absorption phase in the RR quadrant; the other three quadrants are then discarded. Figure 5.44 shows a phase-sensitive absorption-mode spectrum compared to an absolute-value-mode spectrum.

5.3.3 CAMELSPIN, or ROESY (Rotating-Frame Overhauser Enhancement Spectroscopy)

When molecules have the molecular correlation time τ_c close to the reciprocal of the angular Larmor frequency ($\tau_c = 1/\upsilon_L$), then nOe signals may be completely absent or too weak to be detected. In such cases, it is advisable to record the CAMELSPIN (or ROESY) experiment instead of the NOESY experiment, since it always gives a positive value for the nOe effect, even when $\tau_c = 1/\upsilon_L$. Another advantage of the ROESY experiment is that cross-peaks due to relayed nOe (i.e., magnetization of one spin relayed via an intermediate spin to another spin) have the same phase as the diagonal peaks, so they can be readily distinguished from direct nOe signals which have the opposite phase relative to the diagonal signals.

The pulse sequence used for the ROESY experiment is shown in Fig. 5.45. A 90°_x pulse bends the magnetization by 90°. At the end of the evolution period t_1, a strong Rf field is applied for the duration of the mixing time τ_m, so the magnetization vectors precessing in the $x'y'$-plane become "spin-locked" with the applied Rf field and precess synchronously with it. As soon as the vectors of two different nuclei have acquired an identical precessional frequency, a transfer of *transverse* magnetization can occur between them, i.e., an exchange of magnetization in the $x'y'$-phase takes place. This is in contrast to the nOe experiment, in which an exchange of longitudinal magnetization occurs (i.e., in nOe-difference or NOESY experiments, a change of z-magnetization of one nucleus changes the z-magnetization of another, close-lying nucleus, which is observed). Hence, ROESY is a 2D *transverse* nOe experiment. The ROESY spectrum of thermopsine, a lupine base isolated from *Thermopsis turcica*, is shown in Fig. 5.46.

5.3.4 Heteronuclear nOe Spectroscopy (HOESY)

Heteronuclear nOe experiments are similar to homonuclear experiments, discussed earlier. Normally, protons are irradiated and the enhancement of the heteronucleus is recorded, since if the reverse were done, the

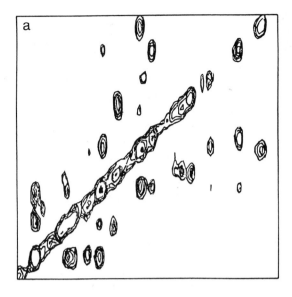

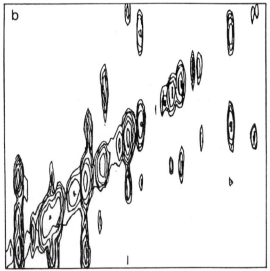

Figure 5.44 (a) Phase-sensitive absorption-mode NOESY spectrum of bovine phospholipase A_2 in D_2O. (b) The same spectrum in the absolute-value mode. The phase-sensitive spectra are generally superior in resolution to absolute-value-mode spectra. (Reprinted from David Neuhaus and Michael Williamson, The Nuclear Overhauser Effect in Structural and Confirmational Analysis, copyright (1989), 278, with permission from VCH Publishers Inc.)

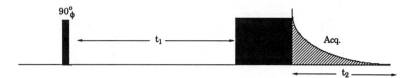

Figure 5.45 Pulse sequence used in the ROESY experiment. The data obtained from odd and even scans are stored separately.

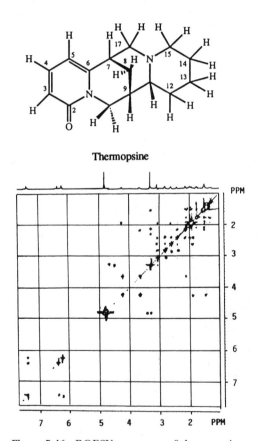

Figure 5.46 ROESY spectrum of thermopsine.

enhancements would usually be very small (the magnetogyric ratios of most nuclei are much smaller than those of ^1H). An exception to this is ^{19}F, so ^1H {^{19}F} experiments have been reported.

5.3.5 Relayed nOe Experiments

Both homonuclear and heteronuclear versions of relayed nOe experiments are known. The homonuclear *relayed* NOESY experiment involves both an *incoherent* transfer of magnetization between two spins H_k and H_l that are not coupled but close in space, and a *coherent* transfer of magnetization between two spins H_l and H_m that are *J*-coupled together. The magnetization pathway may be depicted as

$$H_k \overset{nOe}{\to} H_l \overset{J}{\to} H_m.$$

In a heteronuclear nOe experiment, the first step may be identical to the homonuclear nOe experiment (i.e., involving an incoherent transfer of coherence from H_k to H_l), while the second step could involve a coherence transfer from ^1H to ^{13}C nucleus by an INEPT sequence. These methods suffer from poor sensitivity and have therefore not been used extensively.

5.4 TWO-DIMENSIONAL CHEMICAL EXCHANGE SPECTROSCOPY

It is possible to transfer magnetization between nuclei that are undergoing slow exchange. When the exchange process is slow on the NMR time scale, then each form contributing to the exchange process will produce a separate signal, and irradiation of one nucleus will cause the magnetization to be transferred to other nuclei by an exchange process—a phenomenon known as *transfer of saturation* (or *saturation transfer*). This is exemplified by the classical case of *N,N*-dimethylacetamide (Fig. 5.47). The methyl groups attached to the nitrogen are nonequivalent and slowly undergo exchange by rotation about the amide bond. They appear at differing chemical shifts, and a 2D exchange spectrum would show up in the exchange process as cross-peaks between them. Again, as in the NOESY experiment, it is the modulations during the mixing period that give rise to the cross-peaks.

If we designate the two methyl groups as A and X, then that proportion of the X resonance that has been exchanged with the A resonance will suffer the modulation. The cross-peaks that appear at (ν_X, ν_A) therefore represent that portion of the total magnetization that has undergone a change in precessional frequency during the mixing period; but the unmod-

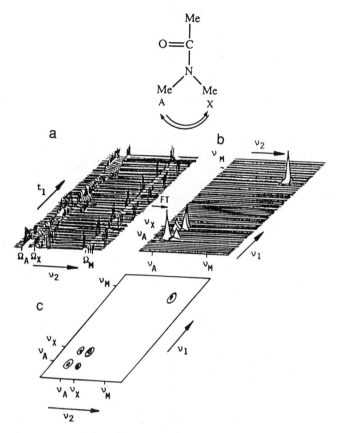

Figure 5.47 Two-dimensional exchange spectrum of *N,N*-dimethylacetamide and its generation. (a) The first set of spectra results from the first series of Fourier transformations with respect to t_2. The modulation of signals as a function of t_1 is observed. (b) The second set of spectra is obtained by the second series of Fourier transformations. The unmodulated signals appear on the diagonal at (ν_A, ν_A), (ν_X, ν_X), and (ν_M, ν_M), whereas the modulations due to exchange show up as cross-peaks on either side of the diagonal at (ν_A, ν_X) and (ν_X, ν_A). (c) A contour plot representation of (b). (Reprinted from *Science* **232,** A. Bax, *et al.,* 960, copyright (1986), with permission from Science-AAAS, c/o Direct Partners Int., P.O. Box 599, 1200 AN Hilversum, The Netherlands)

ulated signals (i.e., those experiencing no change in precessional frequency) appear on the diagonal.

5.5 HOMONUCLEAR HARTMANN-HAHN SPECTROSCOPY (HOHAHA), OR TOTAL CORRELATION SPECTROSCOPY (TOCSY)

When protons are subjected to *strong* irradiation, then the chemical shift differences between them are effectively scaled to zero and *isotropic mixing* occurs. In other word, when the two scalar-coupled nuclei are subjected to identical field strengths through the application of a coherent Rf field, the Zeeman contributions are removed and the Hartmann-Hahn matching condition is established, so an oscillatory exchange of "spin-locked" magnetization (with a periodicity of $1/J$) occurs. Such nuclei experience strong coupling effects, even if there are several intervening nuclei between them, as long as they are a part of the same spin network. The magnetization diffuses to the neighboring nuclei within the spin network at a rate dependent on the magnitude of the scalar coupling (or dipolar coupling in solids). At short mixing times ($<0.1J$), normally only the vicinal and geminal protons are affected. And cross-peaks are found largely between them, so a COSY-like spectrum is obtained. With longer mixing intervals the magnetization initially transferred to the immediately neighboring protons is *relayed* further down the chain within that spin network. Net magnetization transfer occurs, and a phase-sensitive 2D spectrum is obtained.

In practice it is convenient to record HOHAHA spectra at, say, 20, 40, 60, and 100 ms. Slices are then taken at specific chemical shifts at the various time intervals. This yields highly informative plots of the spread of magnetization from key points in the molecule to adjoining regions in the same spin system, thereby providing a powerful tool for structure elucidation. It is thus possible to divide an NMR spectrum into many subspectra, with each subspectrum corresponding to the NMR spectrum of a separate spin system. This simplifies the interpretation greatly, since the overlap between protons belonging to different spin systems is removed in such subspectra. Thus we can rather neatly divide a molecule into various fragments corresponding to discrete subspectra. The pulse sequence employed is shown in Fig. 5.48.

Figure 5.49 shows the four HOHAHA spectra obtained at 20, 40, 60, and 100 ms of hyocyamine, an alkaloid isolated from *Datura fastusa*. The various short-range and long-range homonuclear connectivities established on the basis of HOHAHA spectra of hyoscyamine provided a powerful tool for structure elucidation. The spectrum obtained with a mixing time of

Figure 5.48 The pulse sequence employed in 2D homonuclear Hartmann-Hahn spectroscopy. An extended MLEV-16 sequence is used. Trim pulses are applied before and after this pulse sequence in order to refocus the magnetization not parallel to the x-axis.

20 ms closely resembled the COSY-45° spectrum, showing mainly direct connectivities (geminal and vicinal). With longer mixing intervals, the magnetization is seen to spread to more distant protons within the individual spin systems. This spread of magnetization with increase in the mixing delay can be seen in the projections of the HOHAHA spectra, in which slices have been taken at δ 1.30, 3.50, and 5.16 (Fig. 5.50).

5.5.1 Heteronuclear Hartmann-Hahn Spectroscopy

The cross-polarization in the rotating frame described earlier is not confined to nuclei of the same nuclear species. Indeed, the concept was originally introduced to transfer polarization between different nuclear species in solids. In such cases, two strong Rf fields are applied simultaneously to the two nuclear species, I and S, at their precessional frequencies, and the nuclei I and S then exchange their spin energies at a rate dependent on the magnitude of the Rf fields. This is done in practice by applying a $90°_y$ pulse to create transverse magnetization of the I spins, and then "spin-locking" it in the rotating frame by applying a long Rf pulse that is phase-shifted by 90°. Simultaneously, a long Rf pulse is applied on the S spins so that the Hartmann-Hahn matching condition is reached and an oscillatory exchange of spin energy takes place. The two spins I and S are now said to be "in contact." Once an equilibrium state has been established, they are said to possess the same "negative spin temperature." During this contact period, polarization transfer occurs from the I spin (the abundant spin, with a high gyromagnetic ratio) to the S spin (the dilute spin, with a low gyromagnetic ratio) in the IS two-spin system. If the contact period $\tau = 1/J_{IS}$, then complete transfer of polarization takes place, resulting in a sensitivity enhancement that is determined by the ratio of the gyromagnetic ratios of the two spins. This produces an enhancement of about 4 if polarization transfer is occurring from ^1H to ^{13}C nuclei and an enhancement of about 10 if polarization is being transferred from ^1H to ^{15}N nuclei.

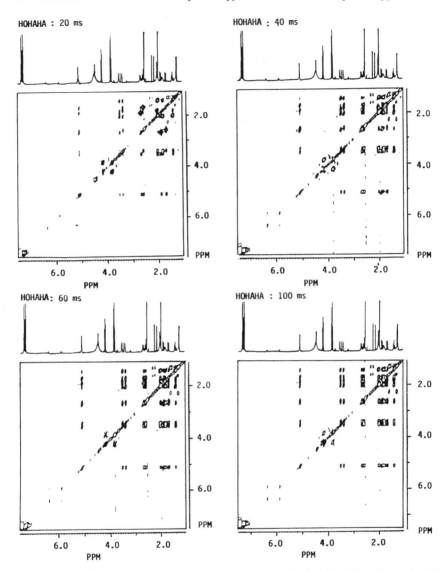

Figure 5.49 HOHAHA spectra of hyoscyamine recorded with mixing intervals of 20, 40, 60, and 100 ms.

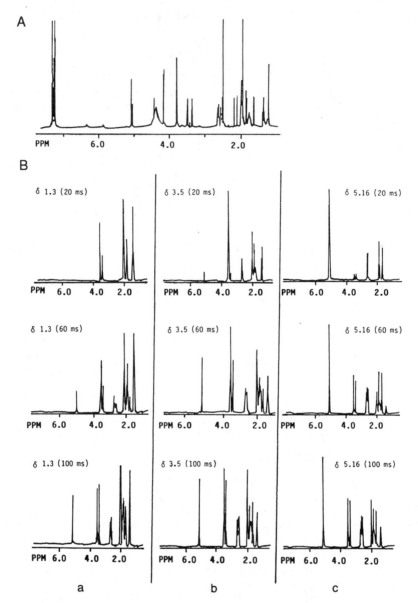

Figure 5.50 (A) 1D ¹H-NMR spectrum of hyoscyamine, an alkaloid isolated from *Datura fastusa*. (B) HOHAHA spectra of the same compound recorded at 20, 60, and 100 ms. The projection spectra are taken at (a) δ 1.30, (b) δ 3.50, and (c) δ 5.16.

5.6 "INVERSE" NMR SPECTROSCOPY

5.6.1 ^{1}H-Detected Heteronuclear Multiple-Quantum Coherence (HMQC) Spectra

A considerable enhancement of sensitivity is obtainable when nuclei with a low gyromagnetic ratio, like ^{13}C, are detected through their effects on the more sensitive nuclei, like ^{1}H. Protons not coupled to the ^{13}C nuclei are suppressed by an initial bilinear (BIRD) pulse, so only protons directly bonded to the ^{13}C nuclei produce cross-peaks in the 2D spectrum. The heteronuclear multiple-quantum coherence (HMQC) experiment (L. Muller, 1979) has provided a highly sensitive procedure for determination of one-bond proton-carbon shift correlations. The pulse sequence for the HMQC spectrum is shown in Fig. 5.51. The BIRD pulse is followed by a delay τ whose magnitude is adjusted so the inverted magnetizations of protons not bound to the ^{13}C nuclei pass through zero amplitude (as they change from the original to the final positive amplitude) when the first 90°_x pulse is applied. The relaxation time T_1 of protons directly bonded to ^{13}C nucleus is short, because of the presence of adjacent ^{13}C nuclei. So such nuclei relax efficiently, and the time between successive scans can be kept quite short (about $1.3T_1$ of the fastest-relaxing protons).

The HMQC experiment is about 16 or 100 times more sensitive for ^{13}C and ^{15}N detections, respectively, than the conventional heteronuclear shift-correlated experiment. The sensitivity advantage in this experiment (as well as in the HMBC experiment described later) is due to the fact that it relies on the equilibrium magnetizations derived from *protons*. Since this magnetization is proportional to the larger population difference in *proton energy levels* (rather than the smaller population difference of ^{13}C energy levels), a stronger NMR signal is obtained. Moreover, since the strength of the NMR signal

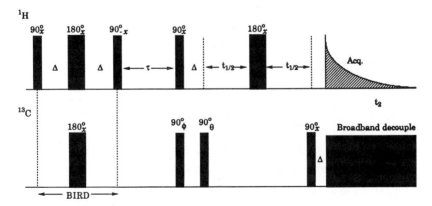

Figure 5.51 The pulse sequence employed for the HMQC experiment.

increases with the frequency of observation, a larger signal will be obtained at the higher proton observation frequencies than at the lower ^{13}C observation frequencies. An HMQC spectrum of thermopsine is shown in Fig. 5.52.

The HMQC spectrum of thermopsine contains one-bond heteronuclear shift correlations. The C-14, C-8, C-12, C-13, C-10, C-15, and C-17 carbons (δ 25.4, 26.4, 28.3, 30.9, 46.3, 57.4, and 64.4, respectively) show one-bond interactions with the following methylenic protons: H-14α,β (δ 1.30 and 1.8), H-8α,β (δ 1.4 and 1.6), H-12α,β (δ 1.85 and 2.05), H-13α,β (δ 1.6 and 1.8), H-10α,β (δ 4.25 and 3.65), H-15α,β (δ 1.95 and 2.65), and H-17α,β (δ 2.85

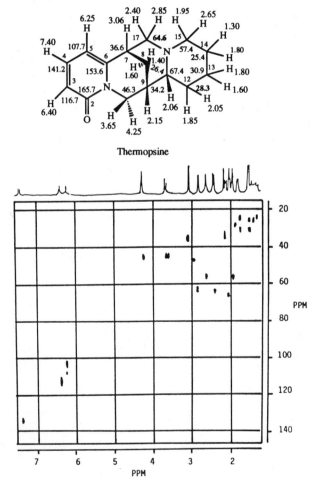

Thermopsine

Figure 5.52 HMQC spectrum of thermopsine.

and 2.40), respectively. The C-5, C-3, and C-4 carbons (δ 107.7, 116.7, and 141.2) show heteronuclear connectivities with their respective protons resonating at δ 6.25, 6.40, and 7.40. The remaining methine protons at δ 2.15, 3.06, and 2.06 exhibit correlations with the C-9, C-7, and C-11 carbons (δ 34.2, 36.6, and 67.4), respectively. The one-bond heteronuclear shift interactions based on the HMQC spectrum are shown on the structure.

5.6.2 Heteronuclear Multiple-Bond Connectivity (HMBC) Spectra

In contrast to the HMQC experiment, which provides connectivity information about directly bonded ^{1}H–^{13}C interactions (i.e., one-bond coupling), the HMBC (heteronuclear multiple-bond connectivity) experiment provides information about long-range ^{1}H coupling interactions with ^{13}C nuclei. Like the HMQC experiment, the HMBC experiment is also an "inverse" experiment (i.e., the ^{13}C population differences are measured indirectly through their effects on ^{1}H nuclei), and it is therefore significantly more sensitive than the conventional long-range heteronuclear shift-correlated or 2D COLOC experiments. The pulse sequence used is shown in Fig. 5.53. The first 90° ^{13}C pulse serves to remove the one-bond $^1J_{CH}$ correlations so that cross-peaks due to direct connectivities do not appear, allowing long-range ^{1}H–^{13}C connectivities to be recorded. The second 90° ^{13}C pulse creates zero- and double-quantum coherences, which are interchanged by the 180° ^{1}H pulse. After the last 90° ^{13}C pulse, the ^{1}H signals resulting from ^{1}H–^{13}C multiple-quantum coherence are modulated by ^{13}C chemical shifts and homonuclear proton couplings.

It is easier to obtain a higher digital resolution in 2D experiments in the F_2 domain than in the F_1 domain, since doubling the acquisition time t_2 results in little overall increase in the experiment time. This is so because

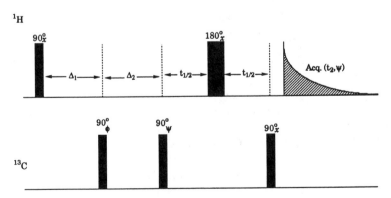

Figure 5.53 The pulse sequence for the HMBC experiment.

we can compensate for the increase by decreasing the relaxation delay between successive experiments. However, if we double the maximum time reached during t_1, we have to double the number of t_1 increments, thereby not only increasing the total time taken for the experiment but also suffering a loss of signal due to relaxation during t_1. This problem is avoided in the inverse experiment, because C–H correlation is detected via the proton spectrum, and the crowded proton region (requiring a higher resolution) now lies in the F_2 dimension, which has a high digital resolution, whereas the better-dispersed ^{13}C spectrum (which requires a lower digital resolution) lies in the F_1 dimension.

The HMBC experiment, like the 2D COLOC experiment, is particularly useful for locating the quaternary carbons by identifying the various protons interacting with them through two-bond ($^2J_{CH}$), three-bond ($^3J_{CH}$), and occasionally four-bond ($^4J_{CH}$) interactions. An HMBC spectrum of griffithine, a new alkaloid isolated in our laboratory from *Sophora griffithii* (Atta-ur-Rahman *et al.*, 1992), is presented in Fig. 5.54. The spectrum was particularly helpful in joining individual structural fragments (established on the basis of COSY, HOHAHA, etc.) into a complete structure. It showed that the quaternary carbon at δ 148.05 was coupled to the protons at δ 6.04 (H-5), 7.25 (H-4), 3.34 (H-13α), 3.12 (H-7α), and 2.05 (H-8). This quaternary carbon was therefore assigned to C-6, adjacent to C-5 and C-7. Such arguments led to the unraveling of the structure of griffithine. Other examples of the HMBC experiment are presented as problems at the end of this chapter.

5.7 INADEQUATE

The NMR techniques discussed so far provide information about proton–proton interactions (e.g., COSY, NOESY, SECSY, 2D *J*-resolved), or they allow the correlation of protons with carbons or other hetero atoms (e.g., hetero COSY, COLOC, hetero *J*-resolved). The resulting information is very useful for structure elucidation, but it does not reveal the carbon framework of the organic molecule directly. One interesting 2D NMR experiment, INADEQUATE (Incredible Natural Abundance Double Quantum Transfer Experiment), allows the entire carbon skeleton to be deduced directly via the measurement of ^{13}C–^{13}C couplings.

Carbon is a mixture of two isotopes in its natural abundance. The major isotope, ^{12}C, occurs in a natural abundance of 98.9%, but it is insensitive to the NMR experiment. The minor isotope, ^{13}C, occurs in a natural abundance of 1.1%; it is with this isotope that we are concerned in the INADEQUATE and many other NMR experiments. Thus there will be only about one molecule in a hundred with a particular carbon bearing the ^{13}C isotope. To find two adjacent carbons bearing ^{13}C isotope would be even less likely

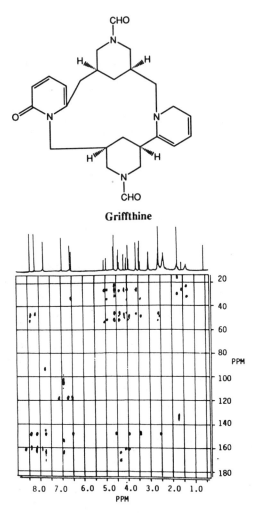

Griffthine

Figure 5.54 HMBC spectrum of griffithine.

(i.e., 1.1% × 1.1% ≅ 1 in 10,000 molecules). The INADEQUATE experiment, which is based on the detection of ^{13}C–^{13}C interactions in such molecules, is therefore very insensitive, requiring about 500 mg of a compound of molecular weight 400–500, and scanning for 24–36 h before an experiment with an acceptable signal-to-noise ratio can be recorded.

The INADEQUATE experiment is a homonuclear NMR technique in which ^{13}C–^{13}C coupling satellites are detected after suppressing the stronger ^{13}C–^{1}H couplings. Historically, this technique evolved out of ideas con-

Figure 5.55   Single-spin energy-level diagram.

cerned with determining the $^{13}C–^{13}C$ coupling constants at natural abundance. But the concept has now emerged to be more general and useful. The theory behind the technique can be explained by exploiting two particular properties of the coherence processes: (a) Multiple-quantum coherence, or MQC, can occur only in a coupled-spin system. (b) Multiple-quantum coherences have significantly different phase properties than single-quantum coherences.

MQC can be understood by analogy with the more familiar single-quantum coherence. For instance, a spin $-½$ nucleus (such as ^1H or ^{13}C) has two energy levels, α and β. The energy difference ΔE between these two levels can be related to the absorption of frequency, $\Delta E = h\nu$ (Fig. 5.55). When the population is promoted from the lower to the upper level after absorption of energy, the process is called a *transition*. In the NMR process we do not detect such transitions directly. Instead, we detect the evolution of magnetization in the *xy*-plane. This *single-quantum coherence* can be created by the application of a pulse on the equilibrium *z*-magnetization.

In contrast, in a two-spin system the two nuclei coupled with each other by the coupling constant, *J*, will have four energy levels available for transitions (Fig. 5.56). Such a system not only has single-quantum coher-

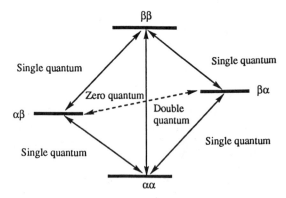

Figure 5.56   Energy-level diagram representing two nuclei in a two-spin system coupled with each other.

ences ($\alpha\alpha \rightarrow \beta\alpha$; $\alpha\alpha \rightarrow \alpha\beta$; $\alpha\beta \rightarrow \beta\beta$; $\beta\alpha \rightarrow \beta\beta$) but also has a double-quantum transition ($\alpha\alpha \rightarrow \beta\beta$) and a zero-quantum transition ($\alpha\beta \rightarrow \beta\alpha$). These transitions cannot be detected directly but can be detected by indirect procedures: forcing the magnetization to evolve at the double-quantum (or zero-quantum) frequency during a certain time period before it is converted by a mixing pulse into a single-quantum coherence for detection.

A two-dimensional INADEQUATE experiment therefore requires a pulse sequence capable of producing the following four changes in the spin and energy states:

(a) It should create double-quantum coherence through an appropriate pulse sequence.
(b) It should allow the double-quantum coherence to evolve during the incremented delays.
(c) The evolution is then converted into observable magnetization by a "mixing" or "detect" pulse.
(d) The FID's are acquired and processed through double Fourier transformation in the usual fashion.

In the 2D INADEQUATE spectrum, the two adjacent ^{13}C nuclei (say, A and X) appear at the following coordinates:

$$\delta_A; \qquad \delta_A + \delta_X \text{ and } \delta_X; \qquad \delta_A + \delta_X.$$

Thus, if the vertical axis representing the double-quantum frequencies ν_{DQ} is defined by ν_1, and the horizontal axis representing the chemical shifts of the two carbons A and X is defined by ν_2, then both pairs of ^{13}C satellites of nuclei A and X will lie on the same row (at $\nu_A + \nu_X$), equidistant from the diagonal. This greatly facilitates identification. The advantage is that no overlap can occur, even when J_{CC} values are identical, since each pair of coupled ^{13}C nuclei appear on different horizontal rows.

The basic INADEQUATE pulse sequence is:

$$90°_{\phi1}-\tau-180°_{\phi2}-\tau-90°_{\phi1}-t_1-\theta_{\phi2}-Acq.$$

The delay is generally kept at $\frac{1}{4}J_{CC}$. The coupling constant J_{CC} for directly attached carbons is usually between 30 and 70 Hz. The first two pulses and delays ($90°_{\phi1}-\tau-180°_{\phi2}-\tau$) create a spin echo, which is subjected to a second $90°_x$ pulse (i.e., the second pulse in the pulse sequence), which then creates a double-quantum coherence for all directly attached ^{13}C nuclei. Following this is an incremented evolution period t_1, during which the double quantum-coherence evolves. The double-quantum coherence is then converted to detectable magnetization by a third pulse $\theta_{\phi2}$, and the resulting FID is collected. The most efficient conversion of double-quantum coherence can be achieved through a flip angle of 90°, but this makes the determination of the sign of the double-quantum coherence frequency difficult. Selection of the optimum double-quantum coherence transfer signal can also be made by cycling phase ϕ_2 of the pulse θ, but it requires a phase shift of 45°, which is difficult to implement. We can, however, detect the desired type of response selectively by designing the appropriate phase cycle.

The 2D INADEQUATE spectrum of α-picoline (Fig. 5.57) provides carbon–carbon connectivity information. The pairs of coupled carbons

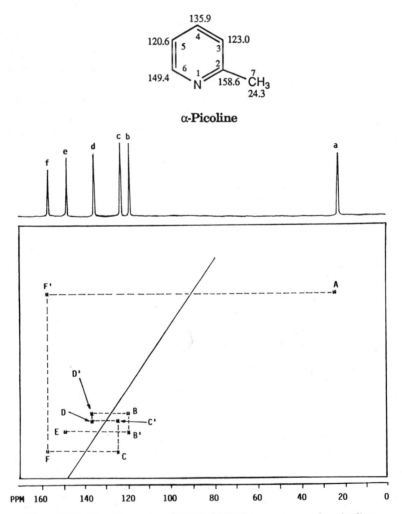

Figure 5.57 Two-dimensional INADEQUATE spectrum of α-picoline.

appear on the same horizontal line, allowing them to be identified readily. A diagonal line is drawn between the satellite peaks, which occur at equal distances on either side of the diagonal.

In order to interpret the spectrum, let us start with the most downfield carbon resonance and trace the subsequent carbon–carbon connectivities. For instance, a horizontal line is drawn from satellite peak F, which corresponds to the most downfield carbon resonating at δ 158.6 (signal "f"), to another carbon resonance C at δ 123.0 (signal "c"), establishing the connectivities between the C-2 and C-3 carbons. A vertical line is then drawn upward from C to satellite peak C', which has another satellite peak, D, at a mirror image position across the diagonal line on the horizontal axis. This establishes the connectivity of the C-3 carbon with C-4 (δ 135.9). Similarly, satellite peak D', lying on the same vertical axis, shows connectivity with another satellite peak, B, corresponding to the carbon resonating at δ 120.6. By dropping a vertical line from B, we reach the point B', which has a mirror image partner, E, on the other side of the diagonal line, indicating connectivities of the C-5 and C-6 carbons (δ 120.6 and 149.4, respectively).

A line is then drawn vertically upward from F to satellite peak F'. The mirror image partner of F', appearing on the same horizontal axis, is A, which establishes the connectivity of the C-2 quaternary carbon with the attached methyl carbon (δ 24.3). The carbon–carbon connectivity assignments based on the 2D INADEQUATE experiment are presented around the structure.

✦ *PROBLEM 5.13*

¹H/¹H Homonuclear Shift Correlations by COSY-45° Experiment

The COSY-45° spectrum of podophyllotoxin and its ¹H-NMR data are shown. Assign and interpret the ¹H/¹H cross-peaks in the COSY spectrum.

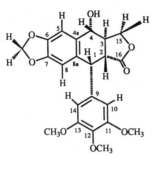

δ_H	No. of protons	J -Values
2.70 m	1H	-
2.95 m	1H	-
3.70 s	6H	-
3.77 s	3H	-
4.10 m	1H	-
4.52 m	1H	-
4.55 d	1H	3.6
4.70 d	1H	8.6
5.90 dd	2H	1.32, 5.09
6.35 s	2H	-
6.46 s	1H	-
7.07 s	1H	-

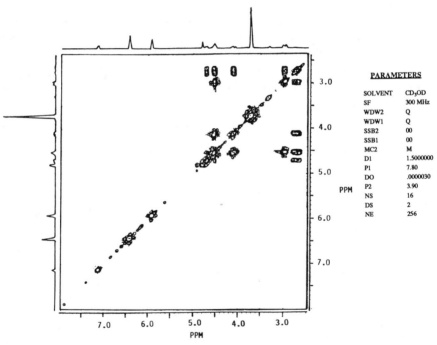

PARAMETERS

SOLVENT	CD₃OD
SF	300 MHz
WDW2	Q
WDW1	Q
SSB2	00
SSB1	00
MC2	M
D1	1.5000000
P1	7.80
DO	.0000030
P2	3.90
NS	16
DS	2
NE	256

COSY-45° spectrum of podophyllotoxin.

✦

✦ PROBLEM 5.14

¹H/¹H Homonuclear Shift Correlation by COSY-45° Experiment

The COSY-45° spectrum of vasicinone and ¹H-NMR data are shown. Assign various protons on the basis of cross-peaks in the COSY-45° spectrum.

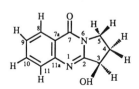

δ_H	No. of protons	J -values
2.20 m	1H	-
2.70 m	1H	-
4.05 m	1H	-
4.21 ddd	1H	-
5.10 dd	1H	6.9, 7.4
7.52 ddd	1H	1.4, 7.0, 8.0
7.72 ddd	1H	0.6, 1.4, 8.3
7.85 m	1H	-
8.25 ddd	1H	0.6, 1.5, 8.0

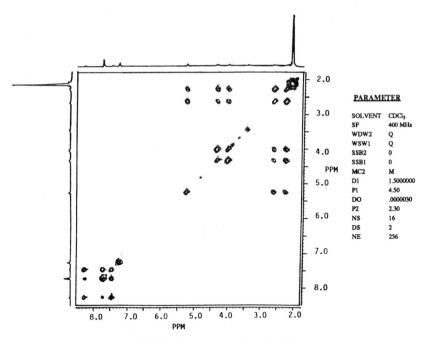

PARAMETER

SOLVENT	CDCl₃
SF	400 MHz
WDW2	Q
WSW1	Q
SSB2	0
SSB1	0
MC2	M
D1	1.5000000
P1	4.50
DO	.0000030
P2	2.30
NS	16
DS	2
NE	256

COSY-45° spectrum of vasicinone.

✦

✦ PROBLEM 5.15

¹H/¹H Homonuclear Shift Correlations by DQF-COSY Experiment

The DQF (double-quantum filtered)-COSY spectrum of an isoprenyl coumarin along with ¹H-NMR data are shown. Determine the ¹H/¹H homonuclear interactions in the DQF-COSY spectrum.

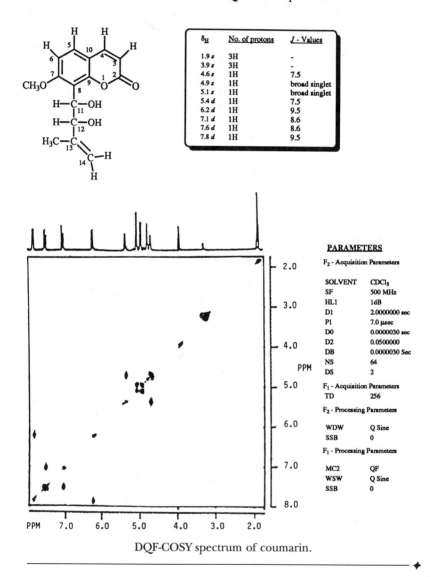

δ_H	No. of protons	J - Values
1.9 s	3H	-
3.9 s	3H	-
4.6 s	1H	7.5
4.9 s	1H	broad singlet
5.1 s	1H	broad singlet
5.4 d	1H	7.5
6.2 d	1H	9.5
7.1 d	1H	8.6
7.6 d	1H	8.6
7.8 d	1H	9.5

PARAMETERS

F₂ - Acquisition Parameters

SOLVENT	CDCl₃
SF	500 MHz
HL1	1dB
D1	2.0000000 sec
P1	7.0 μsec
DO	0.0000030 sec
D2	0.0500000
DB	0.0000030 Sec
NS	64
DS	2

F₁ - Acquisition Parameters
TD	256

F₂ - Processing Parameters

WDW	Q Sine
SSB	0

F₁ - Processing Parameters

MC2	QF
WSW	Q Sine
SSB	0

DQF-COSY spectrum of coumarin.

✦ PROBLEM 5.16

¹H/¹H Homonuclear Shift Correlations by Long-Range COSY-45° Experiment

The LR COSY-45° spectrum and ¹H-NMR chemical shifts of an isoprenyl coumarin are given below. Determine the long-range ¹H/¹H homonuclear correlations based on the long-range COSY-45° spectrum. Demonstrate with reference to problem -5.15 how they can be helpful in interconnecting different spin systems?

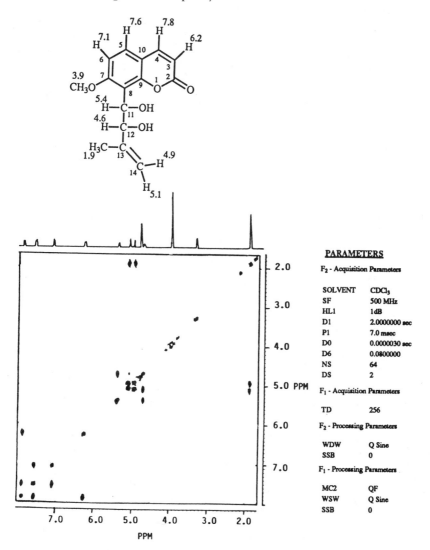

Long-range COSY-45° spectrum of coumarin.

✦

✦ PROBLEM 5.17

¹H/¹H Homonuclear Shift Correlations by SECSY Experiment

The SECSY spectrum of buxapentalactone (a triterpenoid isolated from *Buxus papillosa*) along with the chemical shift assignments are given. The SECSY spectrum, a variant of COSY, allows the assignment of vicinal and geminal couplings between various protons. The display mode of SECSY spectra is different from that of normal COSY spectra. The spectrum comprises a horizontal line at δ 0.0 (unmodulated signals) and cross-peaks lying equidistant above or below it. When connecting such cross-peaks, the following considerations must be accounted for: (a) the cross-peaks to be connected should lie at equal distances above and below the horizontal line, and (b) all the connected cross-peaks should lie on *parallel* slanted lines. By dropping vertical lines from such connected cross-peaks to the horizontal line, we arrive at the chemical shifts of the coupled protons. By this method the coupled protons can easily be identified. Interpret the spectrum, and determine the homonuclear correlations between various protons based on the SECSY spectrum.

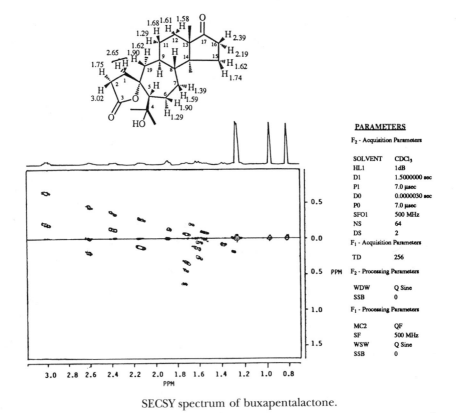

SECSY spectrum of buxapentalactone.

✦ *PROBLEM 5.18*

$^1H/^1H$ Homonuclear Shift Correlations by SECSY Experiment

The SECSY spectrum of an isoprenyl coumarin along with the ^{1}H-NMR chemical shifts are shown. Determine the homonuclear shift correlations between various protons based on the SECSY spectrum.

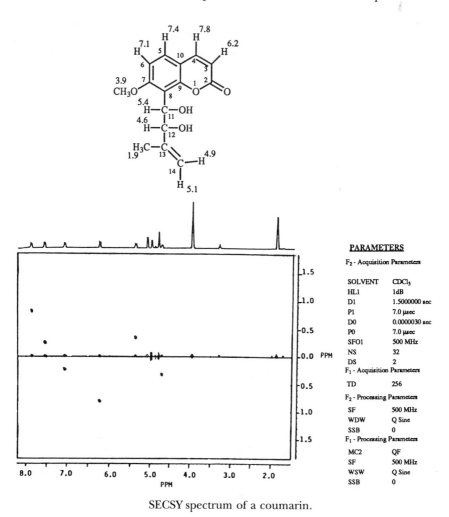

SECSY spectrum of a coumarin.

✦ PROBLEM 5.19

1H/1H Long-Range Total Homonuclear Shift Correlations by HOHAHA Experiment

The HOHAHA spectrum (100 ms) of podophyllotoxin is presented. The HOHAHA, or TOCSY (total correlation spectroscopy), spectrum (100 ms) shows coupling interactions of all protons within a spin network, irrespective of whether they are directly coupled to one another or not. As in COSY spectra, peaks on the diagonal are ignored as they arise due to magnetization that is not modulated by coupling interactions. Podophyllotoxin has only one large spin system, extending from the C-1 proton to the C-4 and C-15 protons. Identify all homonuclear correlations of protons within this spin system based on the cross-peaks in the spectrum.

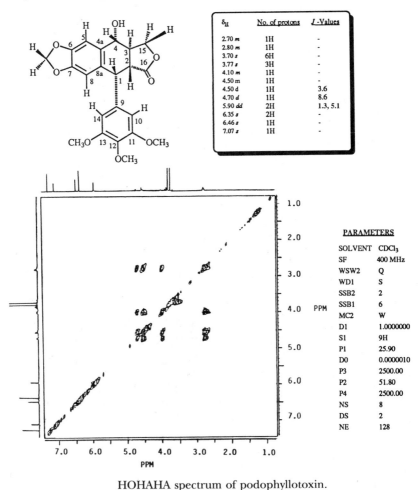

δ_H	No. of protons	J-Values
2.70 m	1H	-
2.80 m	1H	-
3.70 s	6H	-
3.77 s	3H	-
4.10 m	1H	-
4.50 m	1H	-
4.50 d	1H	3.6
4.70 d	1H	8.6
5.90 dd	2H	1.3, 5.1
6.35 s	2H	-
6.46 s	1H	-
7.07 s	1H	-

PARAMETERS

SOLVENT	CDCl₃
SF	400 MHz
WSW2	Q
WD1	S
SSB2	2
SSB1	6
MC2	W
D1	1.0000000
S1	9H
P1	25.90
D0	0.0000010
P3	2500.00
P2	51.80
P4	2500.00
NS	8
DS	2
NE	128

HOHAHA spectrum of podophyllotoxin.

✦ *PROBLEM 5.20*

¹H/¹H Long-Range Total Homonuclear Shift Correlations by HOHAHA Experiment

The HOHAHA spectrum (100 ms) of vasicinone is shown. Interpret the spectrum, and determine the different spin systems from the cross-peaks in the spectrum.

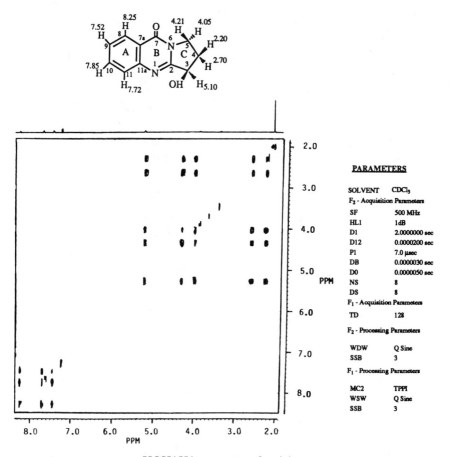

HOHAHA spectrum of vasicinone.

✦

✦ PROBLEM 5.21

¹H/¹³C One-Bond Shift Correlations by Hetero-COSY (HETCOR) Experiment

The one-bond HETCOR spectrum and ^{13}C-NMR data of podophyllo-toxin are shown. The one-bond heteronuclear shift correlations can readily be made from the HETCOR spectrum by locating the positions of the cross-peaks and the corresponding δ_H and δ_C chemical shift values. The ^{1}H-NMR chemical shifts are labeled on the structure. Assign the ^{13}C-NMR resonances to the various protonated carbons based on the heteronuclear correlations in the HETCOR spectrum.

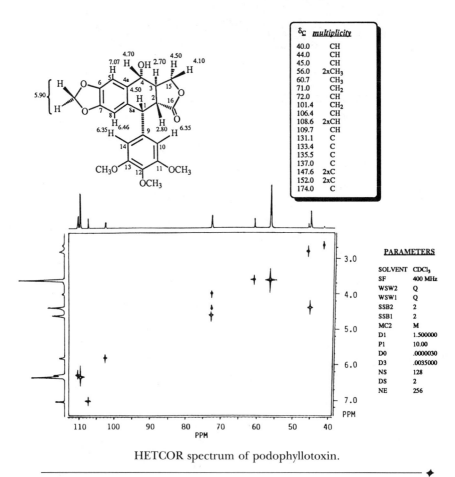

HETCOR spectrum of podophyllotoxin.

✦ *PROBLEM 5.22*

¹H/¹³C One-Bond Shift Correlation by Hetero-COSY Experiment

The hetero-COSY spectrum of vasicinone along with the ¹H-NMR assignments and ¹³C-NMR data are shown. Assign the ¹³C-NMR chemical shifts to the various protonated carbons based on the one-bond heteronuclear shift correlations.

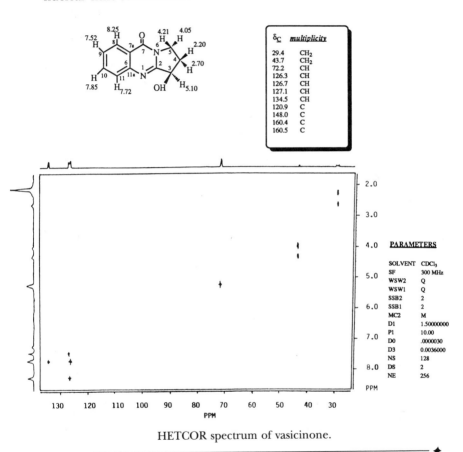

HETCOR spectrum of vasicinone.

✦

✦ PROBLEM 5.23

¹H/¹³C One-Bond Shift Correlations by Hetero-COSY Experiment

The hetero-COSY spectrum of 7-hydroxyfrullanolide, $C_{15}H_{20}O_3$, a sesquiterpene lactone, along with the ¹H-NMR and ¹³C-NMR data are
shown. Assign the ¹³C-NMR shifts to various protonated carbons based
on the cross-peaks in the hetero-COSY spectrum.

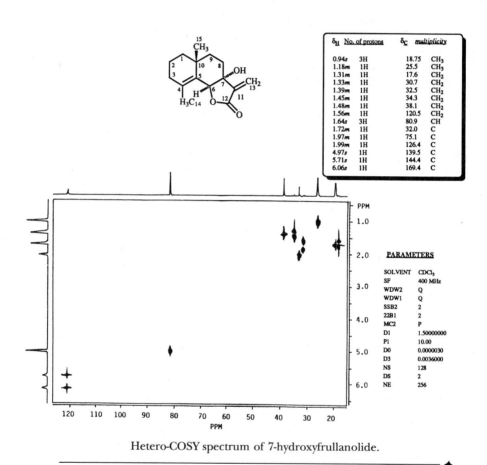

δ_H	No. of protons	δ_C	multiplicity
0.94s	3H	18.75	CH₃
1.18m	1H	25.5	CH₃
1.31m	1H	17.6	CH₂
1.33m	1H	30.7	CH₂
1.39m	1H	32.5	CH₂
1.45m	1H	34.3	CH₂
1.48m	1H	38.1	CH₂
1.56m	1H	120.5	CH₂
1.64s	3H	80.9	CH
1.72m	1H	32.0	C
1.97m	1H	75.1	C
1.99m	1H	126.4	C
4.97s	1H	139.5	C
5.71s	1H	144.4	C
6.06s	1H	169.4	C

PARAMETERS

SOLVENT	CDCl₃
SF	400 MHz
WDW2	Q
WDW1	Q
SSB2	2
22B1	2
MC2	P
D1	1.50000000
P1	10.00
D0	0.0000030
D3	0.0036000
NS	128
DS	2
NE	256

Hetero-COSY spectrum of 7-hydroxyfrullanolide.

✦ *PROBLEM 5.24*

$^1H/^{13}C$ One-Bond Shift Correlations from the Hetero-COSY Experiment

The HETCOR spectrum, ^{13}C-NMR data, and ^1H-NMR chemical shift assignments of buxoxybenzamine ($C_{35}H_{50}N_2O_5$) are presented below. Assign the ^{13}C-NMR chemical shifts to the various protonated carbons using the HETCOR plot.

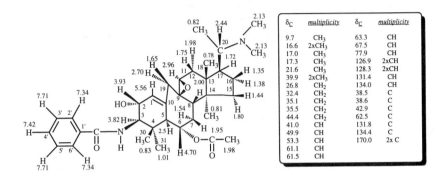

δ_C	multiplicity	δ_C	multiplicity
9.7	CH_3	63.3	CH
16.6	$2xCH_3$	67.5	CH
17.0	CH_3	77.9	CH
17.3	CH_3	126.9	2xCH
21.6	CH_3	128.3	2xCH
39.9	$2xCH_3$	131.4	CH
26.8	CH_2	134.0	CH
32.4	CH_2	38.5	C
35.1	CH_2	38.6	C
35.5	CH_2	42.9	C
44.4	CH_2	62.5	C
41.0	CH	131.8	C
49.9	CH	134.4	C
53.3	CH	170.0	2x C
61.1	CH		
61.5	CH		

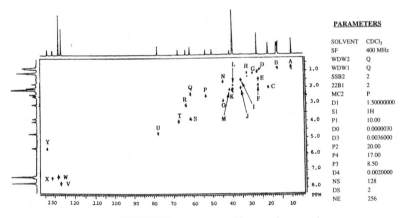

PARAMETERS	
SOLVENT	CDCl$_3$
SF	400 MHz
WDW2	Q
WDW1	Q
SSB2	2
22B1	2
MC2	P
D1	1.50000000
S1	1H
P1	10.00
D0	0.0000030
D3	0.0036000
P2	20.00
P4	17.00
P3	8.50
D4	0.0020000
NS	128
DS	2
NE	256

HETCOR spectrum of buxoxybenzamine

✦

+ PROBLEM 5.25

¹H/¹³C-NMR Chemical Shift Correlations by HMQC Experiments

The heteronuclear multiple-quantum coherence (HMQC) spectrum, ¹H-NMR chemical shift assignments, and ¹³C-NMR data of podophyllotoxin are shown. Determine the chemical shifts of various carbons and connected protons. The HMQC spectra provide information about the one-bond correlations of protons and attached carbons. These spectra are fairly straightforward to interpret. The correlations are made by noting the position of each cross-peak and identifying the corresponding δ_H and δ_C values. Based on this technique, interpret the following spectrum.

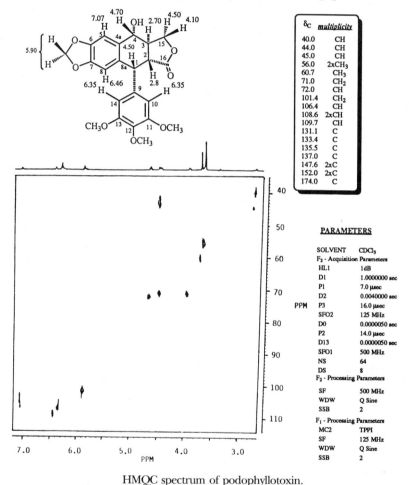

HMQC spectrum of podophyllotoxin.

✦ PROBLEM 5.26

¹H/¹³C-NMR Chemical Shift Correlations by HMQC Experiments

The HMQC spectrum, ¹H-NMR chemical shift assignments, and ¹³C-NMR data of vasicinone are shown. Consider the homonuclear correlations obtained from the COSY spectrum in Problem 5.14, and then determine the carbon framework of the spin systems.

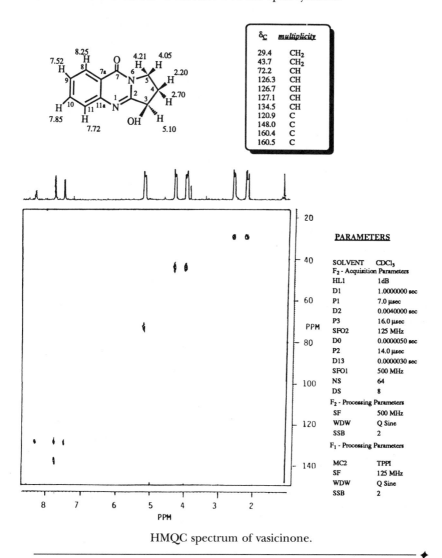

HMQC spectrum of vasicinone.

✦ *PROBLEM 5.27*

¹H/¹³C Long-Range Interactions from the HMBC Spectrum

The HMBC spectrum of podophyllotoxin is shown. The cross-peaks in the HMBC spectrum represent long-range heteronuclear $^1H/^{13}C$ interactions within the same substructure or between different substructures. Interpretation should start with a readily assignable carbon (or proton), and then you identify the proton/s (or carbon/s) with which it has coupling interactions. Then proceed from these protons, and look for the carbon two, three, or, occasionally, four bonds away. One-bond heteronuclear interactions may also appear in HMBC spectrum.

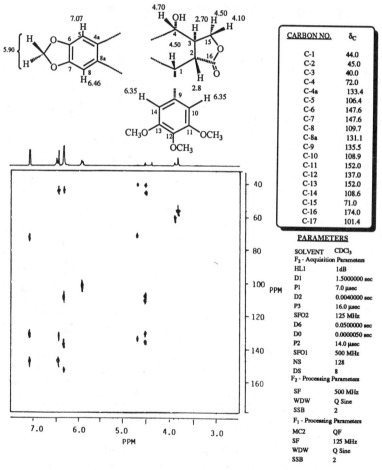

CARBON NO.	δ_C
C-1	44.0
C-2	45.0
C-3	40.0
C-4	72.0
C-4a	133.4
C-5	106.4
C-6	147.6
C-7	147.6
C-8	109.7
C-8a	131.1
C-9	135.5
C-10	108.9
C-11	152.0
C-12	137.0
C-13	152.0
C-14	108.6
C-15	71.0
C-16	174.0
C-17	101.4

PARAMETERS

SOLVENT	CDCl₃
F₂ - Acquisition Parameters	
HL1	1 dB
D1	1.5000000 sec
P1	7.0 μsec
D2	0.0040000 sec
P3	16.0 μsec
SFO2	125 MHz
D6	0.0500000 sec
D0	0.0000050 sec
P2	14.0 μsec
SFO1	500 MHz
NS	128
DS	8
F₂ - Processing Parameters	
SF	500 MHz
WDW	Q Sine
SSB	2
F₁ - Processing Parameters	
MC2	QF
SF	125 MHz
WDW	Q Sine
SSB	2

HMBC spectrum of podophyllotoxin.

The ^{1}H-NMR and ^{13}C-NMR chemical shifts have been assigned and substructures have been deduced on the basis of COSY-45° (Problem 5.13) and other spectroscopic observations. Interpret the HMBC spectrum and identify the heteronuclear long-range coupling interactions between the ^{1}H and ^{13}C nuclei.

✦ *PROBLEM 5.28*

$^1H/^{13}C$ Long-Range Interactions from the HMBC Spectrum

The HMBC spectrum of vasicinone along with the ^{1}H-NMR assignments are shown. Determine the ^{1}H/^{13}C long-range heteronuclear shift correlations based on the HMBC experiment, and explain how HMBC correlations are useful in chemical shift assignments of nonprotonated quaternary carbons.

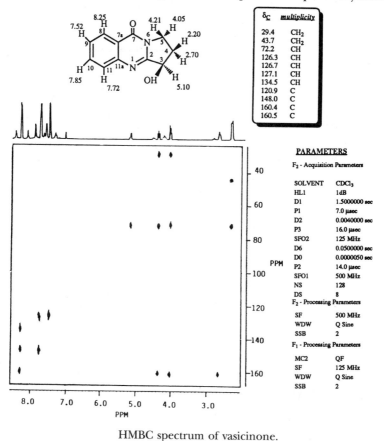

δ_C	*multiplicity*
29.4	CH$_2$
43.7	CH$_2$
72.2	CH
126.3	CH
126.7	CH
127.1	CH
134.5	CH
120.9	C
148.0	C
160.4	C
160.5	C

PARAMETERS

F$_2$ - Acquisition Parameters

SOLVENT	CDCl$_3$
HL1	1 dB
D1	1.5000000 sec
P1	7.0 µsec
D2	0.0040000 sec
P3	16.0 µsec
SFO2	125 MHz
D6	0.0500000 sec
DO	0.0000050 sec
P2	14.0 µsec
SFO1	500 MHz
NS	128
DS	8

F$_2$ - Processing Parameters

SF	500 MHz
WDW	Q Sine
SSB	2

F$_1$ - Processing Parameters

MC2	QF
SF	125 MHz
WDW	Q Sine
SSB	2

HMBC spectrum of vasicinone.

✦ PROBLEM 5.29

$^{1}H/^{13}C$ Long-Range Shift Correlation by COLOC Experiment

Like the HMBC, the COLOC experiment provides long-range hetero-nuclear chemical shift correlations. The COLOC spectrum, ^{1}H-NMR, and ^{13}C-NMR data of 7-hydroxyfrullanolide are presented here. Use the data to assign the quaternary carbons.

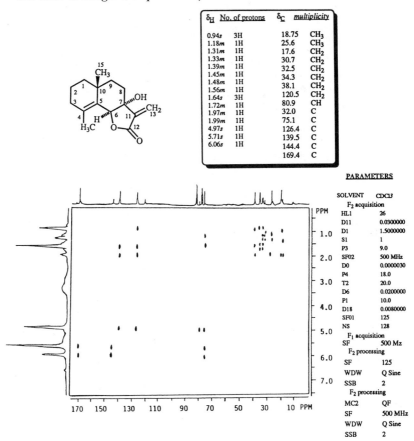

δ_H	No. of protons	δ_C	multiplicity
0.94s	3H	18.75	CH_3
1.18m	1H	25.6	CH_3
1.31m	1H	17.6	CH_2
1.33m	1H	30.7	CH_2
1.39m	1H	32.5	CH_2
1.45m	1H	34.3	CH_2
1.48m	1H	38.1	CH_2
1.56m	1H	120.5	CH_2
1.64s	3H	80.9	CH
1.72m	1H	32.0	C
1.97m	1H	75.1	C
1.99m	1H	126.4	C
4.97s	1H	139.5	C
5.71s	1H	144.4	C
6.06s	1H	169.4	C

PARAMETERS

SOLVENT	CDCl3
F₂ acquisition	
HL1	26
D11	0.0300000
D1	1.5000000
S1	1
P3	9.0
SF02	500 MHz
D0	0.0000030
P4	18.0
T2	20.0
D6	0.0200000
P1	10.0
D18	0.0080000
SF01	125
NS	128
F₁ acquisition	
SF	500 Mz
F₂ processing	
SF	125
WDW	Q Sine
SSB	2
F₂ processing	
MC2	QF
SF	500 MHz
WDW	Q Sine
SSB	2

COLOC spectrum of 7-hydroxyfrullanolide.

✦ PROBLEM 5.30

Stereochemical Correlations by nOe Experiment

The nOe difference and 1D ^{1}H-NMR spectra (in C_5D_5N) of podophyllo-toxin are shown. Justify the stereochemical assignments of the C-2, C-3, and C-4 asymmetric centers based on the nOe data, assuming that the C-4 proton is β-oriented.

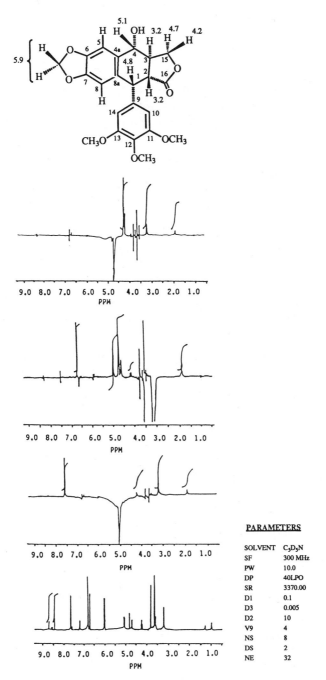

nOe spectra of podophyllotoxin.

✦ *PROBLEM 5.31*

nOe Difference Spectrum and Its Use in Structure Elucidation

Two possible structures, A and B, were initially considered for a naturally occurring coumarin isolated from *Murraya paniculata*. Its nOe difference and ^{1}H-NMR spectra are presented. The relevant ^{1}H chemical shift values are given around structure A. On the basis of the nOe data, identify the correct structure of the coumarin.

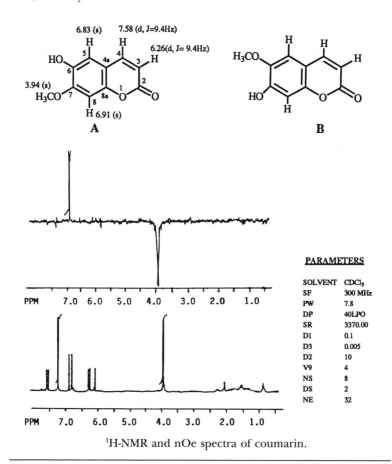

^{1}H-NMR and nOe spectra of coumarin.

✦ PROBLEM 5.32

Structural Information from NOESY Experiment

The NOESY spectrum appears in a similar format as that of COSY spectra; i.e., unmodulated signals lie along the diagonal, whereas the chemical shift frequencies of a proton modulated by the chemical shift of another proton that lies spatially close to it (dipolar coupling) gives rise to an off-diagonal NOESY cross-peak. This cross-peak appears at the junction of the chemical shift frequencies of the two protons. Three possible structures, A, B and C, were considered for buxatenone, $C_{22}H_{30}O_2$, a triterpene isolated from *Buxus papillosa*. The relevant ^1H-NMR chemical shifts have been assigned to the various protons. Assign the correct structure based on the NOESY interactions.

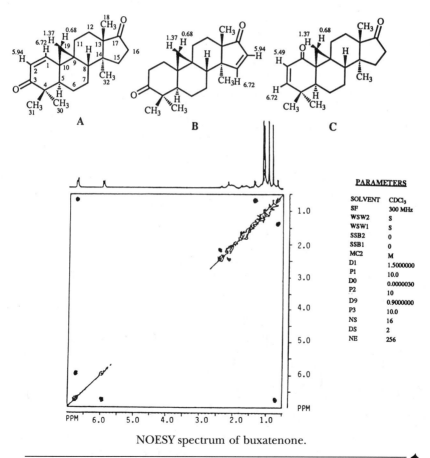

NOESY spectrum of buxatenone.

✦ PROBLEM 5.33

Structural Information by NOESY Experiment

The NOESY spectrum and ¹H-NMR chemical shift assignments of 7-hydroxyfrullanolide are shown. Interpret the NOESY spectrum. What conclusions can you draw about the stereochemistry at C-6 and C-10?

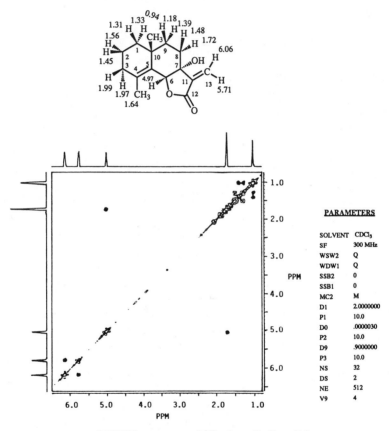

NOESY spectrum of 7-hydroxyfrullanolide.

✦ PROBLEM 5.34

Stereochemical Correlations by ROESY Experiment

The ROESY spectrum affords homonuclear *transverse* nOe interactions as cross-peaks between the various dipolarly coupled hydrogens. This may be compared with NOESY spectra, which arise due to a decrease in the *longitudinal magnetization* (z-magnetization) of one nucleus (the irradiated nucleus) that causes an increase in the z-magnetization of another nucleus that lies spatially close to the irradiated nucleus. ROESY spectra, on the other hand, involve an exchange of *transverse magnetization*, i.e., magnetization in the $x'y'$-plane. The ROESY spectrum and ¹H-NMR data of podophyllotoxin are shown. Deduce the stereochemistry and ¹H–¹H couplings in this compound.

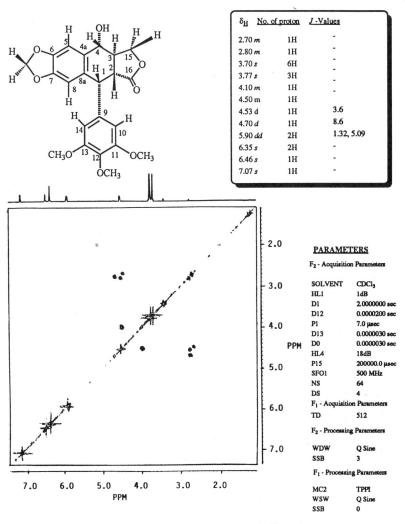

δ_H	No. of proton	J-Values
2.70 *m*	1H	-
2.80 *m*	1H	-
3.70 *s*	6H	-
3.77 *s*	3H	-
4.10 *m*	1H	-
4.50 m	1H	-
4.53 d	1H	3.6
4.70 d	1H	8.6
5.90 *dd*	2H	1.32, 5.09
6.35 *s*	2H	-
6.46 *s*	1H	-
7.07 *s*	1H	-

PARAMETERS

F₂ - Acquisition Parameters

SOLVENT	CDCl₃
HL1	1dB
D1	2.0000000 sec
D12	0.0000200 sec
P1	7.0 μsec
D13	0.0000030 sec
D0	0.0000030 sec
HL4	18dB
P15	200000.0 μsec
SFO1	500 MHz
NS	64
DS	4

F₁ - Acquisition Parameters

TD	512

F₂ - Processing Parameters

WDW	Q Sine
SSB	3

F₁ - Processing Parameters

MC2	TPPI
WSW	Q Sine
SSB	0

ROESY spectrum of podophyllotoxin.

✦ PROBLEM 5.35

Homonuclear 2D J-Resolved Assignments

Homonuclear 2D J-resolved spectra are generally used to determine the coupling constants and multiplicity of complex proton signals. In these spectra the chemical shifts of all the protons are displayed along one axis and the coupling constants along the other axis, thereby eliminating the overlap between close-lying multiplets. The homonuclear 2D J-resolved spectrum of ethyl acrylate is shown. The ^{1}H-NMR chemical shifts are assigned on the structure. Determine the approximate coupling constants and multiplicities for each signal.

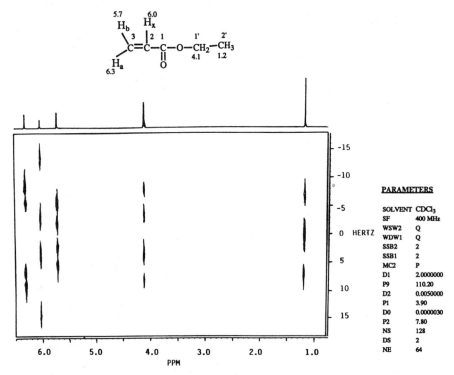

PARAMETERS	
SOLVENT	CDCl$_3$
SF	400 MHz
WSW2	Q
WDW1	Q
SSB2	2
SSB1	2
MC2	P
D1	2.0000000
P9	110.20
D2	0.0050000
P1	3.90
D0	0.0000030
P2	7.80
NS	128
DS	2
NE	64

Homonuclear 2D J-resolved spectrum of ethyl acrylate.

✦

✦ PROBLEM 5.36

Carbon–Carbon Connectivity Assignment by 2D INADEQUATE Experiment

The 2D INADEQUATE spectrum contains satellite-peaks representing direct coupling interactions between adjacent ^{13}C nuclei. The 2D INADEQUATE spectrum and ^{13}C-NMR data of methyl tetrahydrofuran are shown. Assign the carbon–carbon connectivities using the 2D INADEQUATE plot.

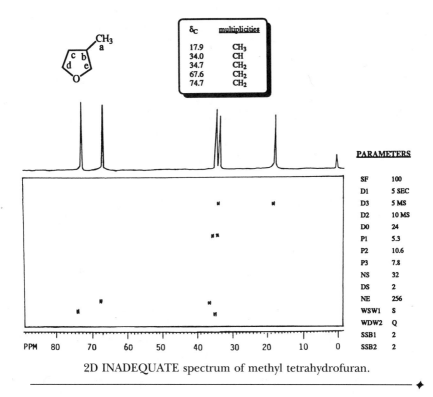

2D INADEQUATE spectrum of methyl tetrahydrofuran.

✦ PROBLEM 5.37

Carbon–Carbon Connectivity Assignment by 2D INADEQUATE Experiment

The 2D INADEQUATE spectrum and ^{13}C-NMR data of 7-hydroxyfrulla-nolide are given. Determine the carbon–carbon connectivities, and subsequently build the entire carbon skeleton based on the 2D INADE-QUATE experiment.

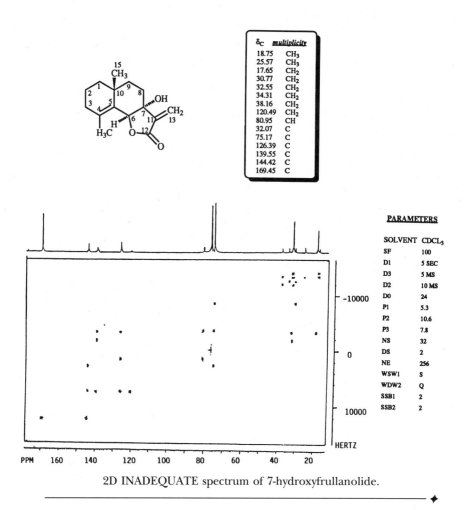

2D INADEQUATE spectrum of 7-hydroxyfrullanolide.

SOLUTIONS TO PROBLEMS

✧ *5.1*

2D J-Resolved NMR experiments are based on the elimination of chemical shift modulations during t_1, allowing the observation of J-coupling only in the F_1 domain. Dephasing in the $x'y'$-plane after the initial 90°_x pulse occurs due to both chemical shift effects and J-couplings in the system. The 180° refocusing pulse at the end of t_1 reverses only the dephasing of chemical shifts, whereas the effect of J-coupling continues to modulate the observed nucleus during t_2. This is then recorded as the second dimension in the 2D plot.

✧ *5.2*

In both experiments, chemical shifts and coupling constants are separated along the two axes. In homonuclear 2D J-resolved spectra, the chemical shifts are along one axis and the couplings with the same type of nucleus lie along the other axis (e.g., ^{1}H chemical shifts along one axis and ^{1}H/^{1}H couplings along the other). In heteronuclear 2D J-resolved spectra, however, one axis would contain the chemical shifts of one type of nucleus, e.g., ^{13}C, while the other axis would contain the coupling of the ^{13}C nuclei with another type of nucleus, e.g., ^{1}H, i.e., δ_C along one axis and J_{CH} along the other. Both experiments are based on the same principle of J-modulation during the evolution period t_1, but the pulse sequences are substantially different. This can be understood by considering the behavior of the magnetization vectors in a spin-echo experiment. Let us first consider vector dephasing due to spin-spin coupling and chemical shifts in a homonuclear AX system. Application of the single 180° pulse at this stage will not only affect the nucleus (A) which is being observed, but will also affect the coupling partner nucleus (X). The vector components will therefore continue to diverge from each other at a rate governed by J, so the amplitudes of the signals detected during t_2 will be modulated according to their respective J values ("J-modulation"). In the case of heteronuclear AX spin coupling, application of simultaneous 180° pulses in the frequency range of both nuclei generates similar J_{AX} modulation effects. The chemical shifts will be refocused at the end of t_1 by the 180° pulse, while heteronuclear J-modulation effects are retained.

✧ *5.3*

Homonuclear shift-correlation spectroscopy (COSY) is a standard method for establishing proton coupling networks. Diagonal and off-diagonal peaks appear with respect to the two frequency dimensions,

F_1 and F_2. The off-diagonal peaks (cross-peaks) represent the direct coupling interactions between protons. Working through cross-peaks, one can easily correlate protons that are coupled to each other. Several versions of the COSY experiment have been designed to get optimum performance in a variety of situations (such as DQF COSY, COSY-45°, and COSY-60°).

✧ *5.4*

The 2D shift-correlation technique is based on Jeener's original experiment and has now become the standard method for establishing proton–proton coupling networks and proton–carbon connectivity. When we excite protons with a 90° pulse and allow them to precess in the $x'y'$-plane, they acquire phases that differ according to the differences in their respective chemical shifts and J-coupling interactions. The second 90°_x pulse causes the mixing of spin states within a spin system. The $x'y'$-magnetization of one proton can thus be transferred to another coupled proton by a pulse applied along an axis in the $x'y'$-plane. The effect of this second 90°_x pulse on the magnetization vector will vary, depending on its position in the $x'y'$-plane. Fourier transformation of these data will then yield a spectrum with two frequency axes, F_1 and F_2. The "off-diagonal" peaks in the spectrum represent the spin-spin couplings between the protons. The peaks on the diagonal, however, represent unperturbed magnetization.

✧ *5.5*

In the COSY-45° spectrum, diagonal peaks are not important. The off-diagonal cross-peaks are the ones that arise due to vicinal and geminal coupling interactions between coupled protons of a spin system. In the COSY spectrum of ethyl acrylate, four cross-peaks are clearly visible. Cross-peak A represents the coupling interactions between the signals at δ 1.2 and 4.1, assigned to the C-2′ methyl and C-1′ methylene protons of the ethyl substituent. Cross-peak B is due to the COSY interaction of the proton at δ 5.7 (C-3H$_b$) with that at δ 6.0 (C-2H$_x$). The same proton (C-3H$_b$) is further coupled to δ 6.3 (C-3H$_a$), as is evident from cross-peak C. Cross-peak D represents the interaction between C-2H$_x$ and C-3H$_a$.

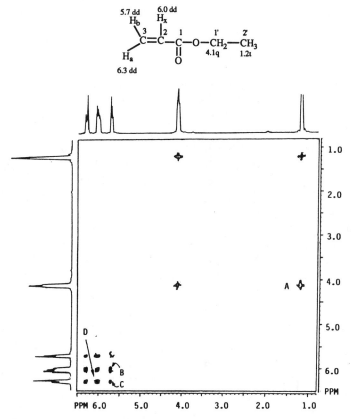

✧ 5.6

One-dimensional double-resonance or homonuclear spin-spin decoupling experiments can be used to furnish information about the spin network. However, we have to irradiate each proton signal sequentially and to record a larger number of 1D ^{1}H-NMR spectra if we wish to determine all the coupling interactions. Selective irradiation (saturation) of an individual proton signal is often difficult if there are protons with close chemical shifts. Such information, however, is readily obtainable through a single COSY experiment.

✧ 5.7

The COSY-45° experiment has the following major advantages over the basic COSY-90° experiment.

(i) The COSY-45° spectrum has reduced intensity of the component signals near the diagonal. This simplifies this region, thereby making it possible to identify correlations that would otherwise be hidden in the cluster of peaks close to the diagonal.

(ii) The cross-peaks visible in the COSY-45° spectrum represent essentially the couplings between directly connected transitions. This allows us to determine the relative signs of the coupling constants.

✧ *5.8*

Decoupling in the ν_1 dimension is carried out by fixing the interval between the two pulses to a value of t_d s and then applying a 180° refocusing pulse after a variable $t_{1/2}$ delay following the 90° pulse. Since the amplitude of the transferred magnetization depends on J_{t_d}, no J-modulation of the signal will occur during t_1. The signal will therefore be modulated by the chemical shift frequency. It will be at a maximum at $t_1 = 0$, zero at $t_1 = t_{d/2}$, and maximum again at $t_1 = t_d$. This results in the narrowing of cross-peaks along ν_1, and can facilitate the identification and assignments of the coupled protons, especially in cases where the protons lie close to one another. Decoupled-COSY is the method of choice in molecules in which the cross-peaks are not distinguishable due to overlap in normal coupled COSY spectra, since in decoupled COSY spectra all the protons produce narrow cross-peaks at their chemical shifts on the ν_1 axis.

✧ *5.9*

The pulse sequence of the delayed-COSY experiment has a fixed delay at the end of the evolution period t_1 and before the acquisition period t_2. This results in the weakening and often the disappearance of stronger direct-coupling interactions (cross-peaks) and intensification of the weaker long-range interactions. This allows long-range couplings to be seen with greater clarity.

✧ *5.10*

Exchange correlation spectroscopy (E. COSY), a modified form of COSY, is useful for measuring coupling constants. The pulse sequence of the E. COSY experiment has a mixing pulse β of variable angle. A number of experiments with different values of β are recorded that eliminate the multiplet components of unconnected transitions and leave only the multiplet components for connected transitions. This simplified 2D plot can then be used to measure coupling constants.

✧ *5.11*

SECSY (spin-echo correlated spectroscopy) is a modified form of the COSY experiment. The difference in the pulse sequence of the SECSY experiment is that the acquisition is delayed by time $t_1/2$ after the mixing pulse, while the mixing pulse in the SECSY sequence is placed in the middle of the t_1 period. The information content of the resulting SECSY spectrum is essentially the same as that in COSY, but the mode of presentation is different. If the individual nuclei in the SECSY spectrum are coupled to others with similar chemical shifts, then the F_1 spectral width may be less and the resolution consequently higher in SECSY spectra than in COSY spectra. However, for routine purposes the SECSY experiment does not offer any significant advantages over the COSY experiment.

✧ 5.12

COSY and HETCOR experiments are extremely useful in the structure elucidation of complex organic molecules. The geminal and vicinal protons and their one-bond C–H connectivities are first identified from the HETCOR spectrum, and then the geminal couplings are eliminated from the COSY spectrum, leaving vicinal connectivities. By careful interpretation of the COSY and the one-bond HETCOR spectra, it is then possible to obtain information about the carbon–carbon connectivities of the *protonated* carbons ("pseudo-INADEQUATE" information). In this way the carbon–carbon connectivity information of protonated carbons is obtainable through a combination of COSY and HETCOR experiments.

Long-range carbon–proton coupling interactions provide important information regarding coupling of protons with carbons across two, three, or four bonds ($^2J_{CH}$, $^3J_{CH}$, $^4J_{CH}$). Interaction of protons with quaternary carbons across two, three, or four bonds is particularly important in joining the various fragments together.

✧ 5.13

The COSY-45° spectrum of podophyllotoxin shows several cross-peaks (A–G). The obvious starting point in interpretation is the most downfield member of the spin network, i.e., the C-4β proton geminal to the hydroxy function, resonating at δ 4.70. This shows the coupling interaction with the C-3 proton (δ 2.70) represented by cross-peak E in the spectrum. Cross-peaks A, B, and C are due to the correlation of the C-3 proton (δ 2.70) with the C-2, C-15β, and C-15α protons resonating at δ 2.95, δ 4.10, and δ 4.52, respectively. Cross-peak F establishes that the protons at δ 4.10 and 4.52 are coupled with each other. Cross-peak D corresponds to the coupling of C-2βH (δ 2.95) with the C-1 proton (δ 4.55). The chemical shift assignments based on COSY cross-peaks are presented around the structure.

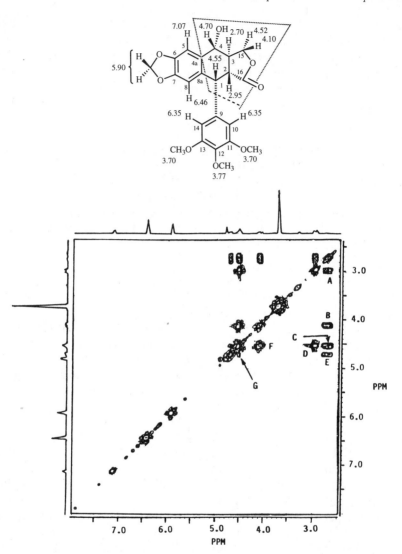

✧ *5.14*

The COSY-45° spectrum of vasicinone displays two distinct spin systems, as indicated by square brackets. Cross-peaks A–H represent coupling between five relatively upfield protons; cross-peaks I–K are due to couplings between the aromatic protons. We start from the most down-field proton of the upfield spin system, resonating at δ 5.10, and trace all the other coupling interactions. For instance, the peak at δ 5.10

for the hydroxyl-bearing C-3 methine proton shows interactions with protons resonating at δ 2.20 and 2.70 (cross-peaks D and G). These two protons are also coupled with each other through cross-peak A, and they can therefore be assigned to the geminally coupled C-4 methylene protons. The C-4 methylene protons also show couplings with a set of protons resonating at δ 4.05 and 4.21 (through cross-peaks B, C, E, F). The latter two protons are also found to be geminally coupled with each other through cross-peak H, and this can therefore be easily assigned to the C-5 methylene protons.

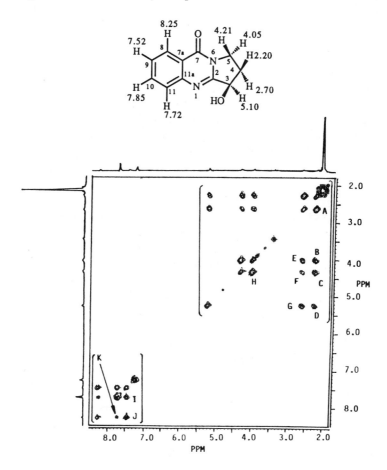

Signals falling in the second spin system represent aromatic methine protons. Starting from the most downfield signal, at δ 8.25, representing the C-8 methine proton (which appears downfield due to the deshielding effect of the electron-withdrawing C-7 *ortho* carbonyl group), we can trace back the coupling interactions between various members of the spin network. For example, cross-peaks K and J (δ 8.25/ 7.85 and δ 8.25/7.52) represent the coupling of the C-8 methine proton with the C-10 (*meta* coupling) and C-9 (*ortho* coupling) methine protons. Cross-peak I (δ 7.52/7.85) indicates the coupling between the C-9 and C-10 methine protons. The coupling between the C-10 (δ 7.85) and C-11 (δ 7.72) protons cannot be identified easily due to closeness of their chemical shifts. These COSY-45° interactions are not only useful for determining the spin network but also help to confirm chemical shift assignments to the various protons.

✧ *5.15*

The DQF-COSY spectrum of isoprenyl coumarin displays clean and well-formed cross-peaks, in contrast to the COSY-45° spectrum. Two distinct spin systems are clearly visible in the spectrum. The upfield spin system comprises cross-peaks A and B; the downfield spin system contains cross-peaks C and D. The upfield spin system represents the oxygen-bearing methine protons and the vinylic protons. Cross-peak A represents the geminal interaction between the vinylic methylene protons resonating at δ 5.1 and δ 4.9. Cross-peak B is due to the coupling between C-12H (δ 4.6) and C-11H (δ 5.4). The downfield spin system falls in the aromatic region. Cross-peak D represents coupling between the aromatic C-5 (δ 7.6) and C-6 (δ 7.1) protons, respectively; the coupling between the C-3 and C-4 vinylic protons is inferred by cross-peak C in the DQF-COSY spectrum.

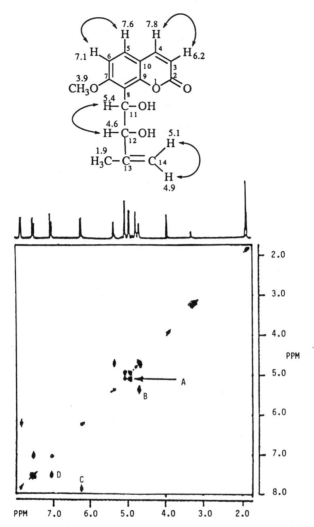

✧ 5.16

The LR COSY-45° spectrum of the coumarin exhibits several long-range correlations that are easily distinguishable in comparison to the DQF-COSY spectrum shown in Problem 5.15. Both spin systems have some additional peaks when compared with the DQF-COSY spectrum (Problem 5.15). Cross-peaks A, B, C, and H represent the long-range interactions; cross-peaks D, E, F, and G are due to direct interactions. This is a common phenomenon in the long-range COSY spectra; i.e., not only do the long-range coupling interactions appear, but the stronger vicinal

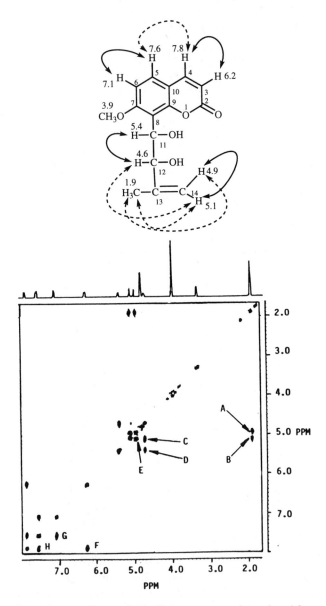

(or geminal) couplings also often appear, though with suppressed intensities in comparison to the standard COSY-45° spectrum. Cross-peaks A and B are due to the allylic interaction of the C-14 vinylic methylene protons (δ 5.1 and 4.9) with the methyl protons resonating at δ 1.9. Cross-peak C also represents the allylic interaction of the C-14 vinylic proton (δ 5.1) with the C-12 methine proton (δ 4.6). Cross-peak D represents vicinal coupling between the C-11 and C-12 oxygen-bearing methine protons, while cross-peak E is due to geminal coupling between the C-14 vinylic protons. Cross-peak H represents the long-range interaction between the C-4 vinylic proton (δ 7.8) and the C-5 aromatic proton (δ 7.6) in the molecule. The long-range interactions are presented as broken arrows; the direct couplings are displayed as solid arrows around the structure. As can be seen, the long-range interactions represented by broken lines can be used to connect different spin systems.

✧ 5.17

The SECSY spectrum of buxapentalactone displays a number of off-diagonal peaks. Pairs of interconnected cross-peaks can be identified

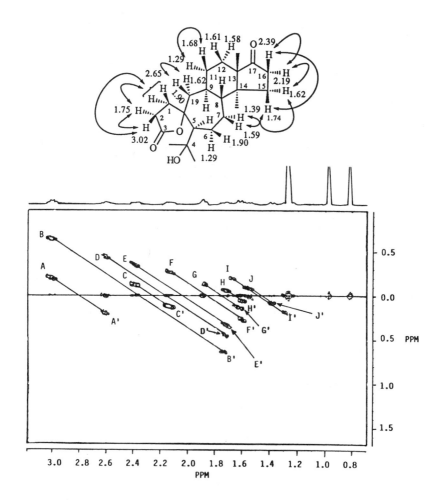

by connecting cross-peaks that lie at equal distances above and below the horizontal line at δ 0.0 on parallel slanted lines. For example, cross-peak A connects to cross-peak A' by the diagonal line, establishing the connectivity of the proton at δ 3.02 (C-2βH) with the protons at δ 2.65 (C-1H). Another cross-peak, B, of the same C-2βH shows connectivity via a line *parallel* to the first diagonal line to cross-peak B' for C-2αH at δ 1.75. Similarly, cross-peaks D and D' represent vicinal connectivity between the C-1 protons (δ 2.65) and the C-2α proton. Cross-peaks C and C' are due to the geminal correlations of the C-16β proton (δ 2.39) with the C-16α (δ 2.19) proton. The C-16β proton (δ 2.39) also shows vicinal interaction with the C-15β proton, as is apparent from the cross-peaks E and E', respectively, in the spectrum. On the other hand, the C-15α proton (δ 1.62) shows cross-peaks F' and H', representing coupling with the C-16α and C-15β protons resonating at δ 2.19 and 1.74, respectively. These SECSY interactions establish the respective positions of the various protons and are presented around the structure.

❖ *5.18*

The SECSY spectrum of the coumarin presents cross-peaks for various coupled nuclei. These cross-peaks appear on diagonal lines that are *parallel* to one another. By reading the chemical shifts at such connected cross-peaks we arrive at the chemical shifts of the coupled nuclei. For instance, cross-peaks A and A' exhibit connectivity between the vinylic C-4 and C-3 protons resonating at δ 7.8 and 6.2, respectively. The C-4 methine appears downfield due to its β-disposition to the lactone carbonyl. Similarly, cross-peaks B and B' show vicinal coupling between the C-5 and C-6 methine protons (δ 7.6 and 7.1, respectively) of the aromatic moiety. The signals C and C' represent the correlation between the oxygen-bearing C-11 (δ 5.4) and C-12 (δ 4.6) methine protons in the side chain. These interactions are presented around the structure.

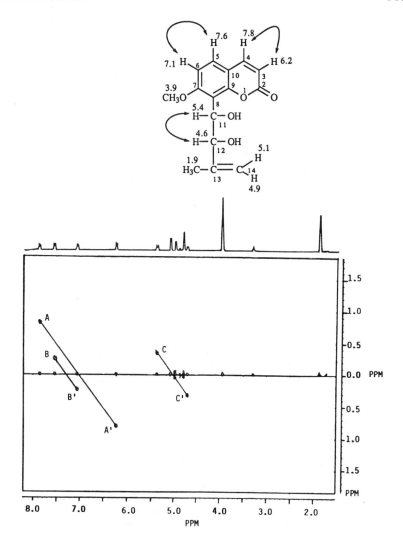

✧ *5.19*

The HOHAHA spectrum of podophyllotoxin shows interactions between all protons of the spin systems in rings B and C of the molecule.
Cross-peaks A, B, and C represent long-range interactions of the C-2
proton (δ 2.80) with the C-15 α and β protons (δ 4.50 and 4.10), as
well as with the C-4β proton (δ 4.70). Cross-peak D represents geminal
coupling between the C-15 α and β protons (δ 4.50 and 4.10). Cross-
peaks E and F are due to the interactions of the C-4 proton (δ 4.70)
with the C-15β (δ 4.10) and C-15α (δ 4.50) protons, respectively. Other

uncoupled protons only appear as diagonal peaks. The HOHAHA
interactions between various protons of a spin-system are shown as
broken lines on the structure.

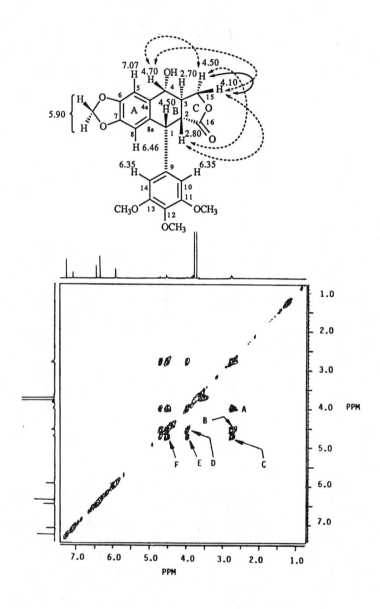

✧ *5.20*

The presence of two very distinct spin systems can easily be inferred from the HOHAHA spectrum of vasicinone. The 10 cross-peaks A–J represent an upfield spin system in which five protons of ring C are participating. Cross-peak A represents geminal interaction between C-4α and β protons (δ 2.20 and 2.70), while cross-peak J represents geminal coupling between C-5α and β protons (δ 4.05 and 4.21). Cross-peaks B, C, D, and E are due to vicinal coupling interactions of the C-4α and β protons with the C-5 methylene protons. Cross-peaks G and F indicate couplings between the oxygen-bearing C-3 methine proton (δ 5.10) and C-4α and β methylene protons. Cross-peaks H and

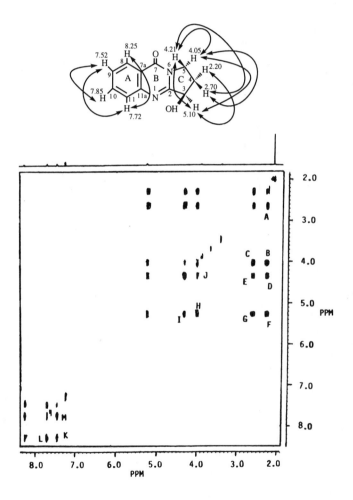

I, however, represent the only long-range interactions between C-5α and β protons, and C-3 proton in the spin system (ring C). The downfield second spin system is represented by cross-peaks K–M due to the aromatic protons. Even the most downfield C-8 proton (δ 8.25) is seen to show a strong cross-peak L with the C-11 proton (δ 7.72), a feature characteristic of HOHAHA spectra. The HOHAHA interactions are presented on the structure.

✧ *5.21*

Twelve cross-peaks (A–L) are visible in the HETCOR spectrum of podophyllotoxin, representing 13 protonated carbons in the molecule.

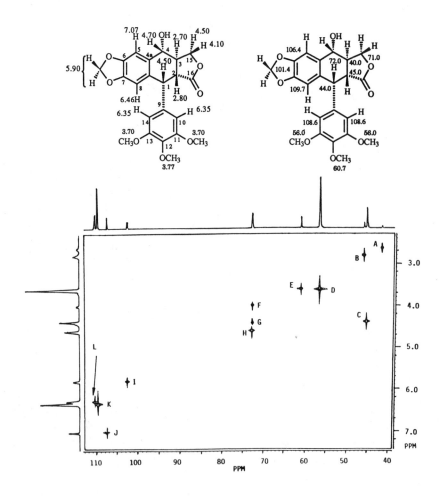

The most upfield cross-peak, A, is due to the interaction of the C-3 proton (δ 2.70) with the most upfield carbon at δ 40.0. Cross-peaks B and C display heteronuclear interactions of the C-2 proton (δ 2.80) with the carbon at δ 45.0 and the C-1 proton (δ 4.50) with the carbon at δ 44.0, respectively. Cross-peaks F and G are due to the one-bond interactions of the C-15 methylene protons (δ 4.10 and 4.50) with the same carbon resonating at δ 71.0, respectively. Cross-peak D represents heteronuclear interactions between protons of two magnetically equivalent methoxy groups (δ 3.70) with their respective carbons (δ 56.0), while cross-peak E displays heteronuclear coupling between the protons of the C-12 methoxy protons (δ 3.77) and its carbon, which resonates at δ 60.7. Coupling between the protons and carbon of the methylenedioxy group is represented by cross-peak I (linking δ_{1H} 5.90 with δ_{13C} 101.4). Cross-peak K is due to the coupling interactions between magnetically equivalent C-10 and C-14 methine protons and carbons; cross-peaks L and J represent heteronuclear couplings between H-5/C-5 and H-8/C-8, respectively. The ^{13}C-NMR chemical shift assignments based on heteronuclear one-bond correlations visible in the HETCOR experiment are displayed on the structure.

✧ 5.22

The HETCOR spectrum of vasicinone exhibits cross-peaks for all seven protonated carbons. Cross-peaks A–E represent heteronuclear couplings of ring C atoms, while cross-peaks F–H are due to heteronuclear couplings between the aromatic protons and carbons. The protons at δ 4.05 and 4.21 show one-bond heteronuclear interactions (cross-peaks A and B) with the carbon resonating at δ 43.7. This confirms that the carbon at δ 43.7 is a methylene carbon, since two different protons show one-bond coupling interactions with it. The presence of another methylene is indicated by heteronuclear interactions (cross-peaks C and D) of the protons at δ 2.20 and 2.70 with the carbon at δ 29.4. The downfield proton signal at δ 5.10 displays one-bond interactions with the carbon at δ 72.2 (cross-peak E), which is assigned to the oxygen-bearing C-3. The cross-peaks F–I are representative of ^{13}C/^{1}H couplings of the four protonated carbons of the aromatic ring. Cross-peak H represents coupling between the most downfield C-8 aromatic methine proton (δ 8.25) and C-8 (δ 126.7). The ^{13}C-NMR assignments based on the heteronuclear correlations are summarized around the structure.

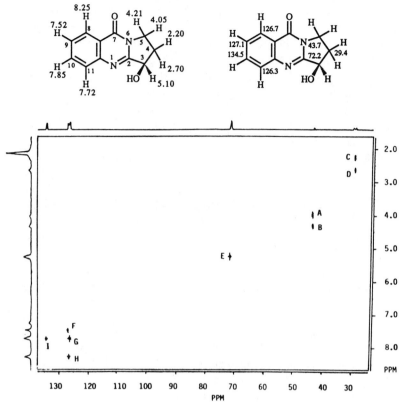

✧ *5.23*

The one-bond hetero-COSY spectrum of 7-hydroxyfrullanolide exhibits interactions for all nine protonated carbons. The most downfield cross-peaks, K and L, represent one-bond heteronuclear correlations of the two vinylic exomethylenic protons resonating at δ 5.71 and 6.06 with the C-13 carbon (δ 120.5). The C-6α proton, which resonates downfield at δ 4.97 due to the directly bonded oxygen atom, displays correlation with the carbon resonating at δ 80.9 (cross-peak D). Cross-peaks G and M represent $^1J_{CH}$ interactions of the C-1 methylene protons (δ 1.33 and 1.31, respectively) with C-1 (δ 38.1). Similarly, cross-peaks E and F display heteronuclear interactions of the C-8 methylenic protons (δ 1.48 and 1.72) with C-8 (δ 30.7), while cross-peak C couplings of C-3 methylene protons at δ 1.97 and 1.99 with C-3 (δ 32.5). Couplings between the C-1 methylene protons and C-1 (δ 38.1) can be inferred from cross-peak A, though in this case both the C-1 α and β protons resonate very close to each other (i.e., δ 1.31 and 1.33). Cross-peak C is due to C-9 methylene, while cross-peak I represents the C-15 methyl. The heteronuclear interactions between the most upfield C-2 methylenic protons and C-2 are visible as cross-peak J and B. Cross-peak H represents one-bond heteronuclear coupling between vinylic C-14 methyl protons and C-14 carbon. Various other one-bond heteronuclear interactions are delineated around the structure.

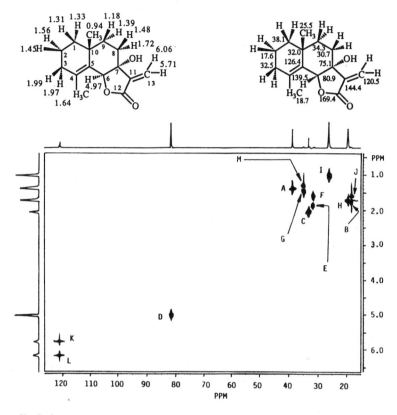

❖ *5.24*

The most downfield cross-peaks, V–Y, are due to heteronuclear couplings of the aromatic or vinylic protons and carbons. For instance, cross-peak Y represents heteronuclear interaction between the C-1 vinylic proton (δ 5.56) and a carbon resonating at δ 134.0 (C-1). The downfield cross-peaks, V and W, are due to the heteronuclear correlations of the *ortho* and *meta* protons (δ 7.34 and 7.71) in the aromatic moiety with the carbons resonating at δ 128.3 and 126.9, respectively. The remaining cross-peak X is due to the one-bond correlation of the C-4' aromatic proton (δ 7.42) with the C-4' carbon appearing at δ 131.4. The cross-peak U displays direct ¹H/¹³C connectivity between the carbon at δ 77.9 (C-6) and C-6 methine proton (δ 4.70). The cross-peak T is due to the one-bond heteronuclear correlation of carbon

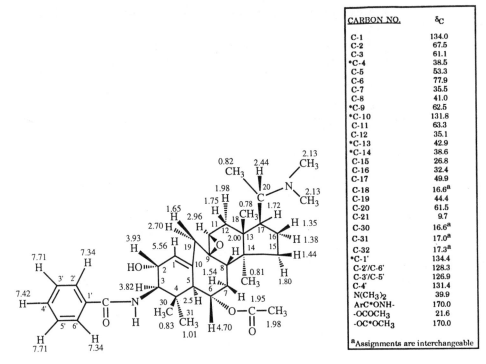

CARBON NO.	δ$_C$
C-1	134.0
C-2	67.5
C-3	61.1
*C-4	38.5
C-5	53.3
C-6	77.9
C-7	35.5
C-8	41.0
*C-9	62.5
*C-10	131.8
C-11	63.3
C-12	35.1
*C-13	42.9
*C-14	38.6
C-15	26.8
C-16	32.4
C-17	49.9
C-18	16.6[a]
C-19	44.4
C-20	61.5
C-21	9.7
C-30	16.6[a]
C-31	17.0[a]
C-32	17.3[a]
*C-1'	134.4
C-2'/C-6'	128.3
C-3'/C-5'	126.9
C-4'	131.4
N(CH$_3$)$_2$	39.9
ArC*ONH-	170.0
-OCOCH$_3$	21.6
-OC*OCH$_3$	170.0

[a]Assignments are interchangeable

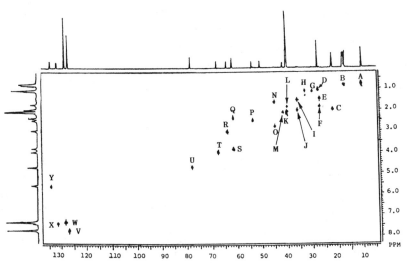

resonating at δ 67.5 (C-2) with the C-2 methine proton (δ 3.93). The downfield chemical shift values of C-2 and C-6 reflected the presence of geminal hydroxy and acetoxy functionalities at C-2 and C-6, respectively. The cross-peak R represents one-bond heteronuclear interaction of carbon resonating at δ 63.4 (C-11) with the C-11 proton (δ 2.96). Another downfield cross-peak S was due to the heteronuclear interactions of C-3 proton (δ 3.82) with a carbon resonating at δ 61.1. The cross-peak Q was due to the correlation of C-20 proton (δ 2.44) with the C-20 carbon resonating at δ 61.5. The cross-peak P indicates the interaction of C-5 proton (δ 2.50) with a carbon appearing at δ 53.4 (C-5). Cross-peaks N and O represent correlation of C-19 protons (δ 2.70 and 1.65) with C-19 methylene carbon (δ 44.4). Cross-peak K represents interaction between the $N_b(CH_3)_2$ protons (δ 2.13) with the attached carbons appearing at δ 39.9. Cross-peak M is due to the interaction of carbon appearing at δ 41.0 (C-8) with C-8 proton (δ 2.00). Cross-peaks I and J represent correlation of C-7 methylene protons (δ 1.95 and 1.54) with carbon resonating at δ 35.5. Cross-peaks H represent correlation of C-16 methylene protons (δ 1.38 and 1.35) with a carbon appearing at δ 32.4. Cross-peaks E and F are due to the correlations between C-15 methylene protons (δ 1.80 and 1.44) and C-15 (δ 26.8). The cross-peak C indicates $^1H/^{13}C$ correlation between acetyl methyl protons (δ 1.98) and acetyl methyl carbon (δ 21.6). Similarly chemical shifts of other protonated carbon was deduced from HETCOR plot. The chemical shift of nonprotonated (quaternary) carbons are marked with asterisks.

✧ *5.25*

The HMQC spectrum of podophyllotoxin shows heteronuclear cross-peaks for all 13 protonated carbons. Each cross-peak represents a one-bond correlation between the ^{13}C nucleus and the attached proton. It also allows us to identify the pairs of geminally coupled protons, since both protons display cross-peaks with the same carbon. For instance, peaks A and B represent the one-bond correlations between protons at δ 4.10 and 4.50 with the carbon at δ 71.0 and thus represent a methylene group (C-15). Cross-peak D is due to the heteronuclear correlation between the C-4 proton at δ 4.70 and the carbon at δ 72.0, assignable to the oxygen-bearing benzylic C-4. Heteronuclear shift correlations between the aromatic protons and carbons are easily distinguishable as cross-peaks J–L, while I represents C/H interactions between the methylenedioxy protons (δ 5.90) and the carbon at δ 101.5. The ^{13}C-NMR and ^1H-NMR chemical shift assignments based on the HMQC cross-peaks are summarized on the structure.

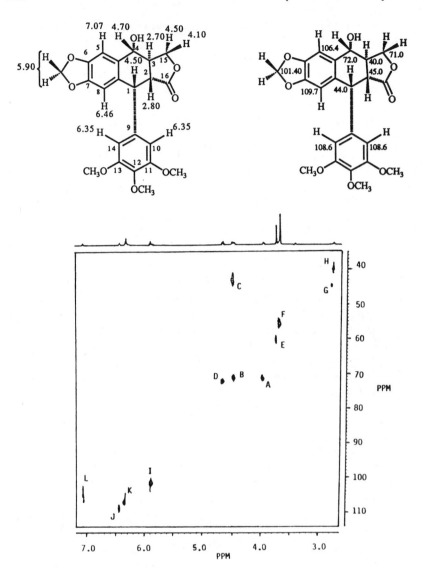

✧ *5.26*

The HMQC spectrum of vasicinone shows nine cross-peaks representing seven protonated carbons, since two of them (i.e., A and B, and C and D) represent two methylene groups. The C-4α and β methylene protons (δ 2.70 and 2.20) show one-bond heteronuclear correlations with the carbon resonating at δ 29.4 (cross-peaks A and B), while the C-5α and β methylene protons (δ 4.21 and 4.05) exhibit cross-peaks

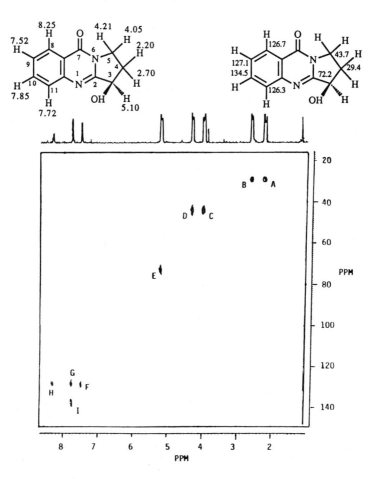

with the carbon signal at δ 43.7 (cross-peaks C and D). The oxygen-bearing C-3 proton at δ 5.10 displays interaction with the carbon at δ 72.2 (cross-peak E). The four protonated carbons of the benzene ring appear at δ 134.5, 126.7, 126.3, and 127.1 and display one-bond interactions with the protons at δ 7.85, 8.25, 7.72, and 7.52 (cross-peaks G, H, I, and F, respectively). After establishing the coupling relationships between the carbons and the attached protons, careful examination of the COSY-45° spectrum (Problem 5.14) provides indirect information about the carbon framework of the entire spin network based on [1]H–[1]H couplings.

For instance, the COSY interactions between the proton geminal to the oxygen at δ 5.10 (attached to the methine carbon at δ 72.2) with the protons at δ 2.20 and 2.70 (methylenic protons at the C-4

carbon appearing at δ 29.4) indicate that the oxygen-bearing C-3 methine (δ 72.2) is directly bonded to C-4 (δ 29.4). COSY interactions
between the C-4 methylene and the C-5 methylene protons establish
that the C-4 carbon (δ 29.4) is attached directly to a methylenic carbon
at δ 43.7 (C-5). A similar strategy can be followed to build the carbon
framework of the aromatic spin system. For instance, the C-8 proton
(δ 8.25) showed COSY interaction with C-9 proton (δ 7.52), which was
vicinally coupled with the C-10 proton (δ 7.85). The C-10 proton also
showed COSY coupling with the C-11 proton (Problem 5.14). Based
on coupling interactions between the various aromatic protons visible
in the COSY-45° spectrum and ^{1}H-^{13}C interactions obtained from the
HMQC spectrum, we can very simply connect the carbons resonating
at δ 126.7, 127.1, 134.5, and 126.3 in a framework. The ^{13}C-NMR assignments to various carbons based on HMQC cross-peaks are presented
around the structure.

✧ *5.27*

The HMBC spectrum of podophyllotoxin exhibits a number of cross-
peaks representing long-range heteronuclear interactions between the
various carbons and protons. For instance, the downfield oxygen-
bearing aromatic carbon(s) (δ 147.6) can be recognized as the
oxygen-bearing aromatic carbon(s), i.e., C-6 and C-7, and they exhibit
HMBC interactions with the protons resonating at δ 7.07 and 6.46
(cross-peaks A and D respectively). This establishes that the aromatic
methine carbons C-5 and C-8 are bound to the oxygen-bearing
quaternary carbons C-6 and C-7, respectively. Cross-peak B is due
to the heteronuclear coupling interaction between the C-5 proton
at δ 7.07 and the quaternary C-8a (δ 131.1). Similarly, long-range
heteronuclear relationships between the C-8 proton (δ 6.46) and C-
8a (cross-peak E) and between the C-1 proton (δ 4.50) and C-8a
(cross-peak N') are also visible in the HMBC spectrum. The C-4
proton geminal to the oxygen atom, which is also the terminus
of the largest substructure (i.e., spin system), exhibits long-range
heteronuclear coupling interaction with C-3 (δ 40.0; cross-peak K),
establishing connectivity in the spin system. Cross-peaks N, O, O',
P, and Q represent multiple-bond correlations of the C-1 proton (δ
4.50) with the C-9, C-8, C-10, C-2, and C-3 carbons, respectively.
Cross-peak L represents a one-bond interaction between the oxygen-
bearing C-4 proton (δ 4.70) and C-4 (δ 72.0), while cross-peak K
is due to the long-range interaction of the C-4 proton with the
quaternary C-4α. The C-2 carbon showed a one-bond interaction
with the C-2 proton (cross-peak T) and long-range heteronuclear

interactions with the C-15α and C-4 protons, respectively, as displayed by cross-peaks Q and M. The heteronuclear multiple-bond correlations in the HMBC plot are presented around the structure.

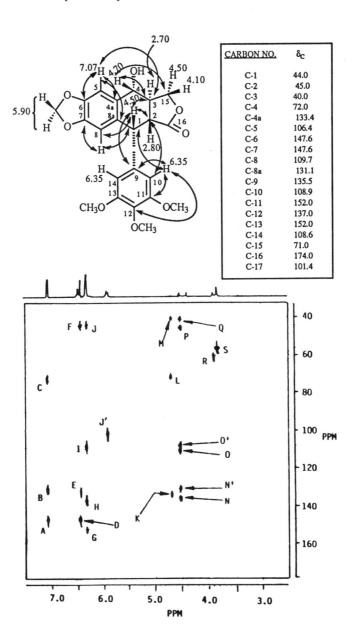

CARBON NO.	δ$_C$
C-1	44.0
C-2	45.0
C-3	40.0
C-4	72.0
C-4a	133.4
C-5	106.4
C-6	147.6
C-7	147.6
C-8	109.7
C-8a	131.1
C-9	135.5
C-10	108.9
C-11	152.0
C-12	137.0
C-13	152.0
C-14	108.6
C-15	71.0
C-16	174.0
C-17	101.4

✧ *5.28*

The HMBC spectrum of vasicinone displays long-range heteronuclear shift correlations between the various $^1H/^{13}C$ nuclei. These correlations are very helpful to determine the ^{13}C-NMR chemical shifts of quaternary carbons and allow the interlinking of the different substructures obtained.

Cross-peaks A and B represent interactions of the C-4 proton (H_α) resonating at δ 2.20 with the C-3 and C-5 carbons (δ 72.2 and 43.7, respectively). Cross-peak C corresponds to the coupling between the C-4 proton (δ 2.70) (H_β) with the C-2 iminic carbon (δ 160.4). Cross-peaks D, E, and F represent long-range correlations between the C-5α

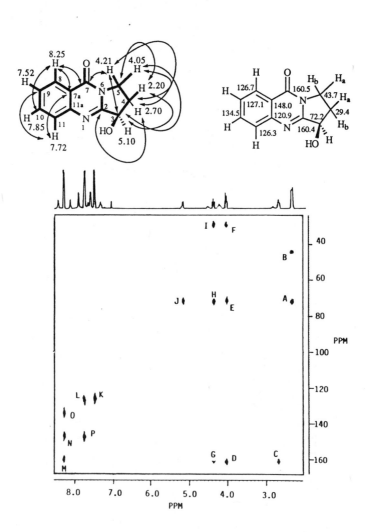

proton (δ 4.05) and the iminic C-2 (δ 160.39), the oxygen-bearing C-3 (δ 72.2), and the C-4 methylene (δ 29.4) carbons, respectively. Cross-peaks G, H, and I are due to the coupling of C-5Hβ (δ 4.21) with the same set of carbons (i.e., C-2, C-3, and C-4), thereby further confirming the ^{13}C-NMR chemical shift assignment of the C-2 quaternary iminic carbon. A one-bond correlation of the C-3 proton (δ 5.10) with the carbon resonating at δ 72.2 is also represented by cross-peak J in the spectrum. Correlations between the aromatic methine protons and carbons appear as cross-peaks K–M in the HMBC plot. For instance, cross-peaks M, N, and O represent long-range shift correlations between the C-8 proton (δ 8.25) and the C-7 carbonyl carbon (δ 160.0), quaternary C-7a (δ 148.0), and C-10 methine carbon (δ 134.5). These correlations not only help in the chemical shift assignments of nonprotonated C-7a and C-7, but also indicate that the aromatic spin system is linked to the aliphatic spin system through the amidic carbonyl at one end and the iminic nitrogen at the other. Various long-range heteronuclear shift correlations based on the HMBC experiment are shown on the structure.

✧ 5.29

The COLOC spectrum of 7-hydroxyfrullanolide shows long-range (^2J, ^3J, and ^4J) heteronuclear interactions of various carbons with protons that are two, three, or four bonds away and thus provides a tool to identify the signals of nonprotonated (quaternary) carbons. The ^{1}H-NMR and ^{13}C-NMR data along with the multiplicities of the signals (determined by DEPT and broad-band experiments) are essential for the interpretation of the COLOC plot. Cross-peaks A–C represent ^3J interactions of the C-1 methylene protons (δ 1.33 and 1.31) with the carbons at δ 34.3 (CH$_2$), 25.6 (CH$_3$), and 30.7 (CH$_2$) assigned to C-9, C-15, and C-8, respectively. The C-2 protons (δ 1.56 and 1.45) exhibit cross-peaks D and E with the carbon resonances at δ 32.5 (CH$_2$) and 32.0 ($-$C$-$). These carbon resonances are therefore assigned to the C-3 methylene carbon and the C-10 quaternary carbon, respectively. Cross-peak F is due to the heteronuclear interaction of the C-6 proton (δ 4.97) with a downfield quaternary carbon at δ 126.3, assigned to the olefinic C-5. The exomethylenic protons at δ 5.71 and 6.06 exhibit three sets of cross-peaks, G, H, and I, displaying their respective heteronuclear interactions with the quaternary carbons at δ 75.1, 144.4, and 169.4. These downfield quaternary signals are assigned to the oxygen-bearing C-7, olefinic C-11, and ester carbonyl C-12, respectively, based on their chemical shifts. The C-6 proton (δ 4.97) also shows a ^3J interaction (cross-peak J) with a quaternary carbon at δ 139.5, which may be ascribed to the C-4 olefinic carbon. The chemical shift assignments to various quaternary carbons based on long-range heteronuclear COLOC interactions are presented around the structure.

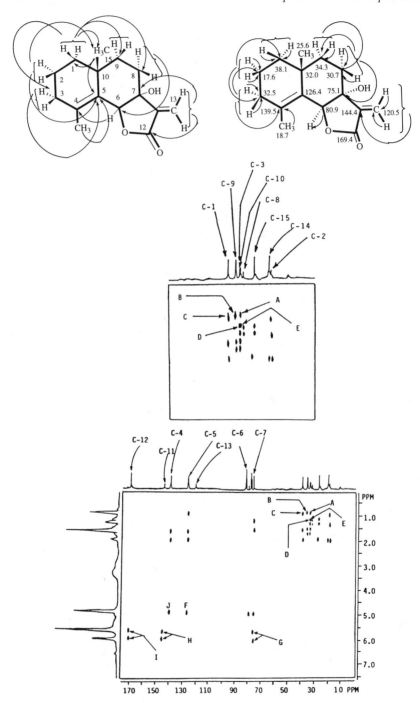

✧ *5.30*

Assuming a β-orientation of the C-1 methine proton (δ 4.8), the stereo-chemistry at other centers can be assigned. For instance, β-orientation of the C-2 methine proton (δ 3.2) is confirmed by nOe interaction between C-2 and C-1 protons. Selective irradiation of C-1 proton (δ 4.8) result an nOe on C-2 proton (δ 3.2) and C-15β proton (δ 4.2) (spectrum a). When overlapping signal of C-3 and C-2 protons (δ 3.2) was irradiated (spectrum b), nOe effects on C-4αH (δ 5.1), C-1αH (δ 4.8) and C-15αH (δ 4.7) were observed. The nOe between the C-4 proton (δ 5.1) and to C-2H (δ 3.2) indicates that they both are β-oriented (spectrum c).

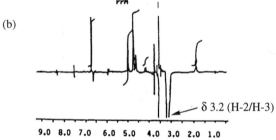

(a)

δ 4.8 (H-1)

9.0 8.0 7.0   6.0 5.0   4.0 3.0 2.0 1.0
                    PPM

(b)

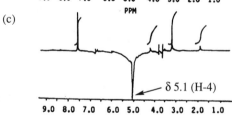

δ 3.2 (H-2/H-3)

9.0 8.0 7.0   6.0 5.0   4.0 3.0 2.0 1.0
                    PPM

(c)

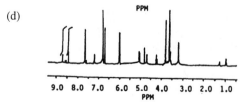

δ 5.1 (H-4)

9.0 8.0 7.0   6.0 5.0   4.0 3.0 2.0 1.0
                    PPM

(d)

9.0 8.0 7.0   6.0 5.0 4.0   3.0 2.0 1.0
                    PPM

✧ *5.31*

The nOe difference measurements not only help in stereochemical assignments but also provide connectivity information. A large nOe at δ 6.91 (C-8H), resulting from the irradiation of the methyl singlet of the methoxy group (δ 3.94), confirms their proximity in space. This nOe result is consistent with structure A for the coumarin.

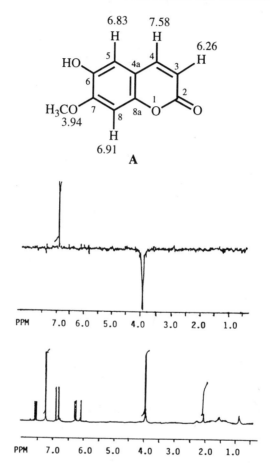

✧ *5.32*

The NOESY spectrum of buxatenone shows four cross-peaks, A–D. Cross-peak B represents the dipolar coupling between the most upfield C-19 cyclopropyl proton (δ 0.68) with the most downfield olefinic proton (δ 6.72). This could be possible only when the double bond is located either between C-1 and C-2 or between C-11 and C-12. The possibility of placing a double bond between C-11 and C-12 can be excluded on the basis of chemical shift considerations, since conjugation with the carbonyl group leads to a significant downfield shift of the olefenic protons β- to the carbonyl group. Cross-peaks A, C, and D are due to couplings between C-1H/C-2H, C-16αH/C-16βH, and C-19αH/C-19βH, respectively.

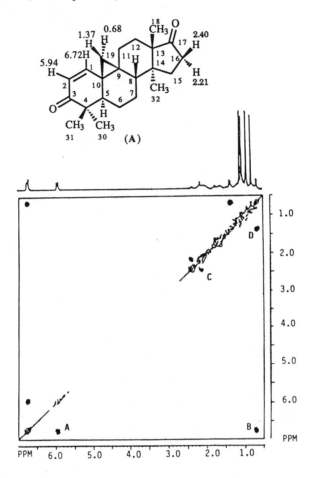

(A)

⬩ *5.33*

The NOESY spectrum of 7-hydroxyfrullanolide reveals the spatial proximities between the various protons. Cross-peak D arises from the geminal coupling between the exomethylenic geminal protons (δ 5.71 and 6.06). Dipolar interaction between the C-6 proton (δ 4.97) and the allylic methyl protons (δ 1.64) is inferred from cross-peak C. This interaction is possible only when the C-6 proton is α-oriented. The C-1β and C-2β protons (δ 1.31 and 1.45, respectively) exhibit cross-peaks

A and B with the angular methyl protons resonating at δ 0.94. This establishes that the C-10 methyl is β-oriented. Important NOESY interactions are presented around the structure.

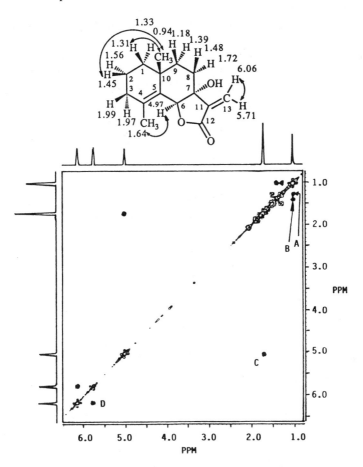

✧ *5.34*

The ROESY spectrum of podophyllotoxin exhibits a number of cross-peaks (A–D) representing interactions between dipolarly coupled (space coupling) hydrogens, which can be helpful to determine the stereochemistry at different asymmetric centers. For example, based on the assumption that the C-1 proton (δ 4.53) is β-oriented, we can trace out the stereochemistry of other asymmetric centers. Cross-peak "B" represents dipolar coupling between the C-1 proton (δ 4.53) and the C-2 proton (δ 2.8), thereby confirming that the C-2 proton is also β-oriented. The C-4 proton resonating at δ 4.7 also exhibits cross-peak "C" with the C-2 proton (δ 2.8), thereby indicating its β-stereo-chemistry in space. The β-position of the C-1 proton establishes the α-stereochemistry of the attached substituted aromatic ring D. The C-15α proton resonating at δ 4.5 shows a cross-peak with the C-15β proton (cross-peak D) as well as vicinal interactions with the C-3α proton (δ 2.7) (cross-peak A). These interactions could only be possible if the hydroxyl group at the C-4 position possesses α-orientation. The homo-nuclear transient nOe interactions based on the ROESY spectrum are shown on the structure.

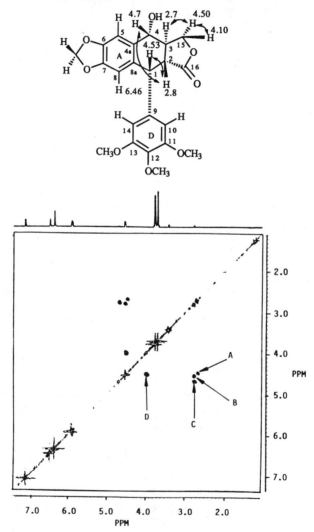

❖ 5.35

The homonuclear 2D *J*-resolved spectrum of ethyl acrylate shows signals for all five magnetically distinct protons. The most downfield signal, A, represents the C-3H$_a$ vinylic proton, which is coupled to both C-3H$_b$ and C-2H$_x$; and it therefore resonates as a double doublet at δ 6.3 with the coupling constants *J* = 1.5 Hz and 17.0 Hz. The coupling constants are determined by measuring the distance between the centers of different components of multiplets. This is more accurately done after plotting the corresponding projections so that the centers of the peaks can be determined. The vertical axis has a *J* scale in hertz. The C-3H$_b$ and C-2H$_x$ protons represented by the signals B and C, respectively, in the spectrum also resonate as double doublets at their respective positions (δ 5.7 and 6.0, respectively). Signals D and E appear as a quartet and triplet for the methylene (δ 4.1) and methyl (δ 1.2) protons, with approximate coupling constant of *J* = 7.4 Hz.

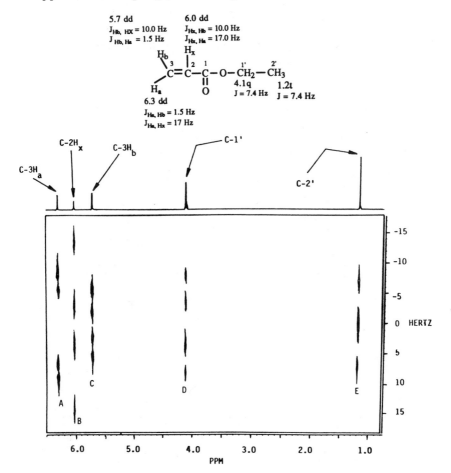

✧ *5.36*

In order to establish the carbon–carbon connectivities, a horizontal line is drawn between satellite peaks that lie at equal distances on either side of the diagonal. For example, a horizontal line is drawn from satellite peak A, which corresponds to the carbon resonating at δ 17.9 (signal "a") to another cross-peak, B, at δ 34.0 (signal "b"), indicating that the two carbons are connected with each other by a C–C bond. A vertical line is next drawn from B downward to the satellite B_1. B_1 has another pair of satellites, C, on the same horizontal axis, thereby establishing the connectivity between the carbons at δ 34.0 and δ 34.7 (signal "c"). Similarly, satellite peak B_2 is connected to another satellite, E, thereby establishing the connectivity between the carbons resonating at δ 34.0 and 74.7 (signal "e"). By dropping a vertical line from satellite peak C, we reach C_1. Its "mirror image" partner satellite on the other side of the diagonal line is D, thereby establishing the connectivity between the carbon at δ 34.7 with the carbon at δ 67.6 (signal "d"). These ^{13}C–^{13}C connectivities are presented around the structure of methyl tetrahydrofuran.

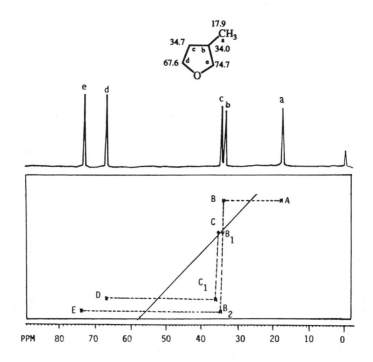

✧ *5.37*

The 2D INADEQUATE spectrum provides carbon–carbon connectivity information and allows the entire carbon framework to be built up. The best strategy for the interpretation of a complex INADEQUATE spectrum is to start with the most downfield satellite carbon resonance and to trace the subsequent ^{13}C–^{13}C connectivities. Using this strategy, the signal at δ 169.4 for the carbonyl carbon, which appeared as satellite peak A in the spectrum, was chosen as the starting point. A horizontal line from satellite peak A to another peak, B (δ 144.4), connects the C-12 and C-11 carbons. A vertical line is next drawn upward to the satellite peaks B′ and B″, which have other pairs of satellites peaks, C and D, at mirror image positions across the diagonal on the horizontal axis, thereby establishing the connectivity of the C-11 olefinic quaternary carbon (δ 144.4) with C-13 and C-7 (δ 120.5 and 75.1), respectively. Similarly, satellite peaks D′ and D″ on the same vertical axis show connectivities with other satellites, E and K, respectively, which correspond to the C-6 and C-8 carbons (δ 80.9 and 30.7), respectively. By drawing a line vertically upward from peak K to K′ and then a horizontal line across the diagonal connecting the cross-peaks K′ and L, the interaction between the C-8 and C-9 carbons (δ 30.7 and 34.3, respectively) can be traced out. The C-9 methine carbon does not show any coupling interaction with the C-10 carbon (δ 32.0) in the spectrum. This serves to illustrate the point that while the presence of a double-quantum peak in the 2D INADEQUATE spectrum provides concrete evidence of a bond, the absence of such a peak does not necessarily prove that a bond is absent. By considering these interactions, fragment **1** could be obtained.

A horizontal line between satellite peaks E′ and F establishes the connectivity between the C-6 and C-5 carbons (δ 80.9 and 126.4), respectively. The mirror image partners of satellites F′ and F″ appearing on the same horizontal axis are M and G, thereby establishing the connectivity of the C-5 carbon with the C-10 and C-4 carbons resonating at δ 32.0 and 139.5, respectively. Signals G′ and G″, which correspond to the C-4 carbon, can be similarly connected to satellites H and I, so connectivities of the C-4 carbon with the C-3 and C-14 carbons resonating at δ 32.5 and 18.7, respectively, can be deduced. The two pairs of satellites (J–W) and (J′–N) establish connectivities of the C-2 carbon (δ 17.6) with the C-3 and C-1 carbons (δ 32.5 and 38.1, respectively). On the other hand, satellites M′ and M″, lying vertically, indicate connectivities of the C-10 carbon (δ 32.0) with the C-1 (M′–N′) and C-15 (M″–Q) carbons (δ 38.1 and 25.5, respectively). These connectivities lead to a fragment **2**.

By joining fragments **1** and **2** together, the following structure for 7-hydroxyfrullanolide was deduced.

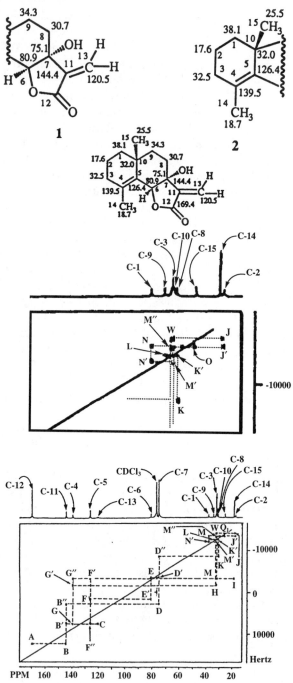

REFERENCES

Atta-ur-Rahman, Pervin, A., Feroz, M., Perveen, S., Choudhary, M. I., and Hasan, M. (1992). *Mag. Reson. Chem.* **29,** 1077.

Bauer, C., Freeman, R., and Wimperis, S. (1984). *J. Magn. Reson.* **58,** 526–532.

Bax, A. (1983). *J. Magn. Reson.* **53,** 517–520.

Bax, A., and Freeman, R. (1981). *J. Magn. Reson.* **44,** 542–561.

Becker, E. D., Ferretti, J. A., and Farrar, T. C. (1969). *J. Am. Chem. Soc.* **91,** 7784–7785.

Garbow, J. R., Weitekamp, D. P., and Pines, A. (1982). *Chem. Phys. Lett.* **93** (5), 504–509.

Gidley, M. J., and Bociek, S. M. (1985). *J. Chem. Soc. Chem. Commun.* (4), 220–222.

Kessler, H., Griesinger, C., and Lautz, J. (1984). *Angew. Chem. Int. Ed. Engl.* **23**(6), 444–445.

Kessler, H., Griesinger, C., Zarbock, J., and Lossli, H. R. (1984). *J. Magn. Reson.* **57**(2), 331–336.

Kessler, H., Bermel, W., Griesinger, C. (1985). *J. Am. Chem. Soc.* **107**(4), 1083–1084.

Kessler, H., Bermel, W., Griesinger, C., and Kolar, C. (1986). *Angew. Chem. Int. Ed. Engl.* **25**(4), 342–344.

Macura, S., and Brown, L. R. (1985). *J. Magn. Reson.* **62,** 328–335.

Muller, L. (1979). *J. Am. Chem. Soc.* **101,** 4481–4484.

Neuhaus, D., and Williamson, M. (1989). *The nuclear Overhauser effect in structural and confirmational analysis.* VCH Pubs., New York, p. 278.

Quast, M. J., Zektzer, A. S., Martin, G. E., and Castle, R. N. (1987). *J. Magn. Reson.* **71**(3), 554–560.

Rance, M., Sörensen, O. W., Bodenhauser, G., Wagner, G., Ernst, R. R., and Weuthrich, K. (1983). *Biochem. Biophys. Res. Comm.* **117,** 479.

Rutar, V. (1984a). *J. Magn. Reson.* **56,** 87–100.

Rutar, V., and Wong, T. C. (1984). *J. Magn. Reson.* **60,** 333–336.

Rutar, V. (1984b). *J. Magn. Reson.* **58,** 132–142.

Schenker, K. V., and Philipsborn, von W. (1985). *J. Magn. Reson.* **61,** 294–305.

States, D. J., Habackorn, R. A., and Ruben, D. J. (1982). *J. Magn. Reson.* **48,** 286.

Wang, J.-S., and Wong, T. C. (1985). *J. Magn. Reson.* **61,** 59–66.

Wang, J.-S., Zhao, D. Z., Ji, T., Han, X.-W., and Cheng, G. B. (1982). *J. Magn. Reson.* **48**(2), 216–224.

Welti, D. H. (1985). *Magn. Reson. Chem.* **23**(10), 872–874.

Wernly, J., and Lauterwein, J. (1985). *J. Chem. Soc., Chem. Commun.* (18), 1221–1222.

Wimperis, S., and Freeman, R. (1984). *J. Magn. Reson.* **58,** 348–353.

Wimperis, S., and Freeman, R. (1985). *J. Magn. Reson.* **61**(1), 147–152.

Zektzer, A. S., John, B. K., and Martin, G. E. (1987a). *Magn. Reson. Chem.* **25**(9), 752–756.

Zektzer, A. S., John, B. K., Castle, R. N., and Martin, G. E. (1987b). *J. Magn. Reson.* **72**(3), 556–561.

Zektzer, A. S., Quast, M. J., Ling, G. S., Martin, G. E., MacKenny, J. D., Jr., Johnston, M. D., and Castle, R. N. (1986). *Magn. Reson. Chem.* **24,** 1083–1088.

CHAPTER 6

The

Third

Dimension

6.1 BASIC PHILOSOPHY

The basic philosophy of three-dimensional NMR spectroscopy was inherent in triple-resonance experiments, in which a system under perturbation by two different frequencies is subjected to further perturbation by a third frequency (Cohen *et al.*, 1963). With the development of two-dimensional NMR spectroscopy (Aue *et al.*, 1976; Chandrakumar and Subramanian, 1987), it was only a matter of time before 3D NMR experiments were introduced, particularly for resolving heavily overlapping regions of complex molecules. The earlier 3D experiments included 3D *J*-resolved COSY (Plant *et al.*, 1986), 3D correlation experiments (Griesinger *et al.*, 1987b,c), and 3D combinations of shift-correlation and cross-relaxation experiments (Griesinger, *et al.*, 1987c; Oschkinat *et al.*, 1988). These illustrated the potential and power of three-dimensional methods. Three-dimensional NMR experiments can be directed at two main objectives: (a) unraveling complex signals that overlap in 1D and 2D spectra, and (b) establishing connectivity of nuclei via *J*-couplings or their spatial proximity via dipolar couplings or cross-relaxation effects.

Two-dimensional spectroscopy has two broad classes of experiments: (a) 2D *J*-resolved spectra (Müller *et al.*, 1975; Aue *et al.*, 1976), in which no coherence transfer or mixing process normally occurs, and chemical shift and coupling constant frequencies are spread along two different axes,

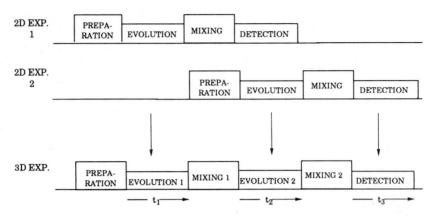

Figure 6.1 Pulse sequence for 3D experiments obtained by merging two different 2D sequences. (Reprinted from *J. Mag. Reson.* **84,** C. Griesinger, *et al.,* 14, copyright (1989), with permission from Academic Press, Inc.)

ω_1 and ω_2, and (b) 2D shift-correlation spectra, involving either coherent transfer of magnetization [e.g., COSY (Aue *et al.,* 1976), hetero-COSY (Maudsley and Ernst, 1977), relayed COSY (Eich *et al.,* 1982), TOCSY (Braunschweiler and Ernst, 1983), 2D multiple-quantum spectra (Braunschweiler *et al.,* 1983), etc.] or incoherent transfer of magnetization (Kumar *et al.,* 1980; Machura and Ernst, 1980; Bothner-By *et al.,* 1984) [e.g., 2D cross-relaxation experiments, such as NOESY, ROESY, 2D chemical-exchange spectroscopy (EXSY) (Jeener *et al.,* 1979; Meier and Ernst, 1979), and 2D spin-diffusion spectroscopy (Caravatti *et al.,* 1985)].

In three-dimensional experiments, two different 2D experiments are combined, so three frequency coordinates are involved. In general, the 3D experiment may be made up of the preparation, evolution (t_1), and mixing periods of the first 2D experiment, combined with the evolution (t_2), mixing, and detection (t_3) periods of the second 2D experiment. The 3D signals are therefore recorded as a function of *two* variable evolution times, t_1 and t_2, and the detection time t_3. This is illustrated in Fig. 6.1.

Three-dimensional NMR spectra, like 2D NMR spectra, may be broadly classified into three major types: (a) 3D *J-resolved spectra* (in which the

Figure 6.2 Pulse sequences for some common 3D time-domain NMR techniques. Nonselective pulses are indicated by filled bars. Nonselective pulses of variable flip angle are shown by the flip angle β. Frequency-selective pulses are drawn with diagonal lines in the bars. (Reprinted from *J. Mag. Reson.* **84,** C. Griesinger, *et al.,* 14, copyright (1989), with permission from Academic Press, Inc.)

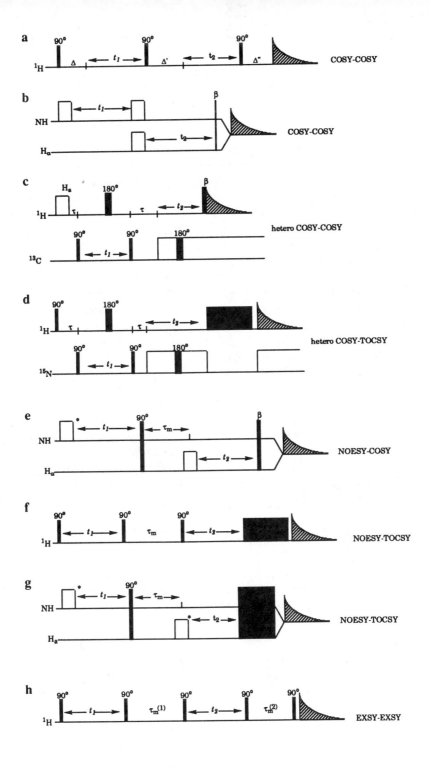

chemical shift frequencies and homonuclear or heteronuclear coupling frequencies are resolved in three different dimensions, no coherence transfer or mixing process being normally involved); (b) *3D shift-correlated spectra*, in which three resonance frequencies are correlated by two coherent or incoherent transfer processes. The two transfer processes may be homonuclear or heteronuclear and may involve transfer of antiphase coherence (e.g., COSY) or in-phase coherence (e.g., TOCSY), leading to such 3D experiments as COSY-COSY, COSY-TOCSY, and hetero-COSY-TOCSY. It is possible to have combinations of *J*-resolved and shift-correlated spectra, e.g., 3D *J*-resolved COSY in which the 2D COSY spectrum is spread into a third dimension through scalar coupling (Plant *et al.*, 1986; Vuister and Boelens, 1987); (c) *3D exchange spectra*, in which two successive exchanges are recorded. The exchange processes may involve either chemical exchange (EXSY) or cross-relaxation in the laboratory frame (NOESY) or in the rotating frame (ROESY) e.g., NOESY-ROESY or EXSY-EXSY spectra.

In some of the most useful 3D experiments, the coherent transfer of magnetization (e.g., COSY) may be combined with incoherent magnetization transfer (e.g., NOESY) to give such 3D experiments as NOESY–COSY or NOESY–TOCSY (Griesinger *et al.*, 1987c; Oschkinat *et al.*, 1988; Vuister *et al.*, 1988). Such 3D experiments are finding increasing use in the study of biological molecules.

Since there are two time variables, t_1 and t_2, to be incremented in a 3D experiment (in comparison to one time variable to increment in the 2D experiment), such experiments require a considerable data storage space in the computer and also consume much time. It is therefore practical to limit such experiments to certain limited frequency domains of interest. Some common pulse sequences used in 3D time-domain NMR spectroscopy are shown in Fig. 6.2.

6.2 TYPES AND POSITIONS OF PEAKS IN 3D SPECTRA

Five different types of peaks can occur in 3D spectra. These are illustrated in an ABC spin system, in which the Larmor frequencies of the three nuclei are ν_A, ν_B, and ν_C, and the coherence transfers are associated with two different mixing processes, M_1 and M_2:

	M_1	M_2
Cross-peaks	$\nu_A \longrightarrow \nu_B$	$\longrightarrow \nu_C$
Cross-diagonal peaks ($\omega_1 = \omega_2$)	$\nu_A \longrightarrow \nu_A$	$\longrightarrow \nu_B$
Cross-diagonal peaks ($\omega_2 = \omega_3$)	$\nu_A \longrightarrow \nu_B$	$\longrightarrow \nu_B$
Back-transfer peaks	$\nu_A \longrightarrow \nu_B$	$\longrightarrow \nu_A$
Diagonal peaks	$\nu_A \longrightarrow \nu_A$	$\longrightarrow \nu_A$

The *cross-peaks* arise when two different mixing processes lead to transfer of magnetization. Cross-peaks correlate three Larmor frequencies and hence provide information on connectivity or spatial proximity of sets of three nuclei. In a COSY-COSY spectrum, for instance, they provide information on three nuclei coupled with one another. In the case of *cross-diagonal peaks,* only one mixing process is involved in the magnetization transfer, while *diagonal peaks* involve no magnetization transfer at all. *Back-transfer peaks* arise by the retransfer of magnetization to the original nucleus,

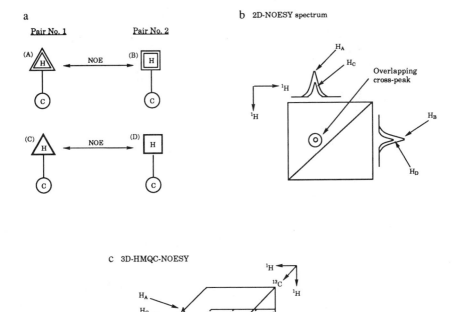

Figure 6.3 Schematic representation of the resolution advantages of 3D NMR spectroscopy. (a) Both pairs of protons have the same resonance frequency. (b) Due to the same resonance frequency, both pairs exhibit overlapping cross-peaks in the 2D NOESY spectrum. (c) When the frequency of the carbon atoms is plotted as the third dimension, the problem of overlapping is solved, since their resonance frequencies are different. The NOESY cross-peaks are thus distributed in different planes.

rather than its being passed on to the third nucleus; hence, two magnetization transfer processes are again involved. Back-transfer peaks provide interesting information in spectra in which unequal mixing processes are involved. For instance, in a NOESY–TOCSY spectrum they would allow us to identify protons that are spatially close and part of the same spin system (Oschkinat *et al.,* 1989).

Increasing the number of dimensions from two to three may result in a reduction in the signal/noise (S/N) ratio. This may be due either to the distribution of the intensity of the multiplet lines over three dimensions or to some of the coherence transfer steps being inefficient, resulting in weak 3D cross-peaks.

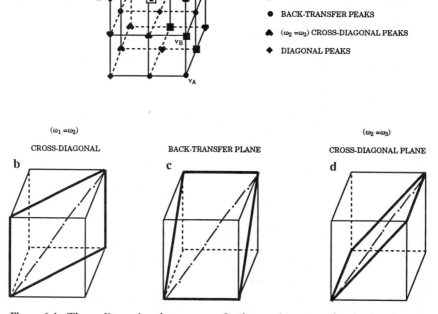

Figure 6.4 Three-dimensional spectrum of a three-spin system showing peak types appearing in a three-dimensional space. Three diagonal peaks, six ($\omega_1 = \omega_2$) and six ($\omega_2 = \omega_3$) cross-signal peaks, six back-transfer peaks, and six cross-peaks are present in the cube. (a) The cubes (b–d) represent three planes in which cross-diagonal peaks and the back-transfer peaks appear on their respective ($\omega_1 = \omega_2$), ($\omega_2 = \omega_3$), and ($\omega_1 = \omega_3$) planes. (Reprinted from *J. Mag. Reson.* **84,** C. Griesinger, *et al.,* 14, copyright (1989), with permission from Academic Press, Inc.)

The major advantage of 3D spectra is that they can enhance resolution of overlapping cross-peaks in 2D spectra. This is presented in schematic form in Fig. 6.3. Let us consider two pairs of protons attached to four different carbons, and assume that each pair has the same resonance frequency. Let us also assume that proton A is spatially close to proton B and gives rise to a corresponding cross-peak in the 2D NOESY spectrum. Similarly, proton C is spatially close to proton D and also gives rise to a second cross-peak in the 2D NOESY spectrum. However, since protons A and C have the same resonance frequencies, as do protons B and D, the two cross-peaks would overlap and only a single cross-peak would actually be observed. If we could somehow introduce a third dimension, say, the ^{13}C chemical shift, then the two overlapping cross-peaks would be separated, since the carbon atoms to which protons B and D are attached have differing chemical shifts. Thus the 3D HMQC–NOESY spectrum would contain two well-separated cross-peaks that had overlapped in the NOESY spectrum.

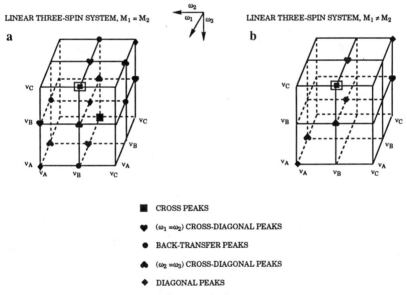

■ CROSS PEAKS

♥ ($\omega_1 = \omega_2$) CROSS-DIAGONAL PEAKS

● BACK-TRANSFER PEAKS

▲ ($\omega_2 = \omega_3$) CROSS-DIAGONAL PEAKS

◆ DIAGONAL PEAKS

Figure 6.5 (a) Schematic representation of a 3D spectrum of a linear spin system ABC with identical mixing processes M_1 and M_2. In a "linear" spin system, the transfer of magnetization between A and C is forbidden for both M_1 and M_2. (b) Schematic representation of a 3D spectrum of a linear spin system ABC, where transfer via M_1 is possible only between A and B and transfer via M_2 occurs only between B and C. (Reprinted from *J. Mag. Reson.* **84**, C. Griesinger, *et al.*, 14, copyright (1989), with permission from Academic Press, Inc.)

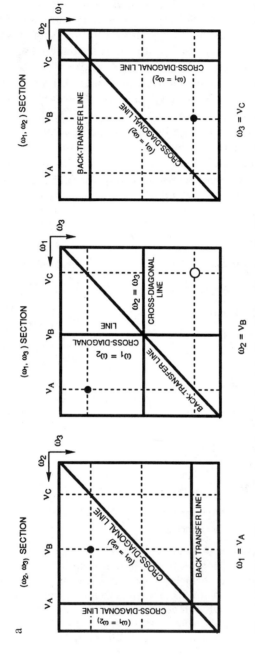

Figure 6.6 Three 2D cross-sections through the A → B → C cross-peaks of a 3D spectrum taken at constant ω_1, ω_2, and ω_3 frequencies, respectively. The open circle indicates a symmetry-related peak with respect to the back-transfer diagonal line. (Reprinted from *J. Mag. Reson.* **84**, C. Griesinger *et al.*, 14, copyright (1989), with permission from Academic Press, Inc.)

Figure 6.4 shows the general features of a 3D-spectrum, which can be visualized as peaks appearing in a three-dimensional space, such as a cube. The cube can be subdivided by two different cross-diagonal planes ($\omega_1 = \omega_2$ and $\omega_2 = \omega_3$) and one back-transfer plane. A diagonal line connecting two opposite corners of the cube is the "space diagonal"; diagonal peaks ($v_A \rightarrow v_A \rightarrow v_A$) would appear on this line. The cross-diagonal peaks and the back-transfer peaks appear on their respective planes.

A 3D spectrum of a three-spin subsystem in which all the nuclei are coupled to one another, such as $C(H_A)(H_B)-C(H_C)$, will lead to 27 peaks, comprising six cross-peaks, 12 cross-diagonal peaks (six at $\omega_1 = \omega_2$ and the other six at $\omega_2 = \omega_3$), six back-transfer peaks, and three diagonal peaks. However, in the case of a linear three-spin network (e.g., $CH_A-CH_B-CH_C$), the number of peaks will depend on whether two equal (e.g., COSY-COSY or NOESY-NOESY) or unequal (e.g., COSY-NOESY) mixing processes are

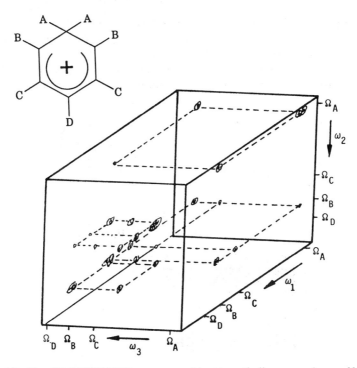

Figure 6.7 The 3D EXSY-EXSY spectrum of heptamethylbenzenonium sulfate in H_2SO_4 representing $\omega_1 = \omega_3$ reflection symmetry. Cross-peaks lying on identical planes (ω_1, ω_3) are connected by lines. (Reprinted from *J. Mag. Reson.* **84**, C. Griesinger, *et al.*, 14, copyright (1989), with permission from Academic Press, Inc.)

involved. If the mixing processes are equal, then the forward and reverse transfers, such as:

$$\nu_A \rightleftarrows \nu_B \rightleftarrows \nu_C$$
$$\nu_C \rightleftarrows \nu_B \rightleftarrows \nu_C$$

produce 17 peaks distributed as two cross-peaks, eight cross-diagonal peaks (four at $\nu_1 = \nu_2$ and four at $\nu_2 = \nu_3$), four back-transfer peaks, and three diagonal peaks (Fig. 6.5a).

In the case of a COSY-NOESY spectrum having an unequal mixing, let us consider two nuclei (say, A and B) that are spatially close but belong to different coupling networks, and nuclei B and C, which have scalar coupling with each other but are spatially distant. The only transfers allowed in this situation are:

$$\mathbf{M}_1 \qquad \mathbf{M}_2$$
$$\nu_A \rightleftarrows \nu_B \rightleftarrows \nu_C$$

So a simplified 3D spectrum is obtained having eight peaks, i.e., one cross-peak, four cross-diagonal peaks (two at $\nu_1 = \nu_2$ and two at $\nu_2 = \nu_3$), three diagonal peaks, and no back-transfer peaks (Fig. 6.5b).

Three-dimensional spectra can be conveniently represented by 2D cross-sections taken at appropriate points and perpendicular to one

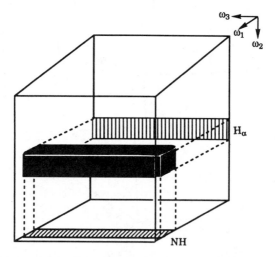

Figure 6.8 Volume chosen for a selective 3D NOESY-TOCSY spectrum of a peptide with NH region in ω_1, H_α in ω_2, and all protons in ω_3. (Reprinted from *J. Mag. Reson.* **84,** C. Griesinger *et al.*, 14, copyright (1989), with permission from Academic Press, Inc.)

Table 6.1

A Selection of 3D and 4D NMR Techniques Used in Protein
Structure Determination*

Experiment	Information/linkage obtained
(a) Unlabeled proteins—homonuclear experiments	
3D-TOCSY-TOCSY	^{1}H spin system/amino acids
3D-TOCSY-NOESY	Sequential assignment of the spin systems, identification of nOes
3D-NOESY-NOESY	Identification of nOes
(b) ^{15}N Labeled proteins	
3D-TOCSY-^{1}H–^{15}N-HMQC/HSQC	^{1}H spin systems, ^{15}N shifts
3D-NOESY-^{1}H–^{15}N-HMQC/HSQC	Sequential assignment of the spin system, identification of nOes with amide protons
3D-HNHB	$H_N(i)-N(i)-H_\beta(i)$, stereospecific assignment of methyl protons
(c) ^{13}C–^{15}N labeled proteins	
3D-TOCSY-^{1}H–^{13}C-HMQC/HSQC	^{1}H spin systems, ^{13}C shifts
3D-NOESY-^{1}H–^{13}C-HMQC/HSQC	Sequential assignment of the spin systems, in particular, identification of nOes between side chains
3D-HCCH-COSY	^{1}H spin systems, ^{13}C shifts
3D-HCCH-TOCSY	^{1}H spin systems, ^{13}C shifts
3D-HC(C)NH-TOCSY	Connection of H_N and side-chain ^{1}H and ^{13}C
3D-H(CCO)NH	Correlation of all side-chain ^{1}H and ^{13}C of amino acid i with $H_N(i + 1)$
H(CA)NH	$H_\alpha(i)-N(i)-H_N(i)$
3D-HCACO	$H_\alpha(i)-C_\alpha(i)-CO(i)$
3D-HNCA	$H_N(i)-N(i)-C_\alpha(i)/C_\alpha(i-1)$
3D-HNCACB	$H_N(i)-N(i)-C_\alpha(i)/C_\alpha(i-1)/C_\beta(i)/C_\beta(i-1)$
3D-HN(CA)NNH	$HN(i)-N(i+1)-H_N(i+1)$
3D-HA(CA)NNH	$H_\alpha(i)/H_\alpha(i-1)-N(i)H_N(i)$
3D-HN(NCA)NNH	$H_N(i)-N(i+1)-H_N(i+1)$
3D-HCA(CO)N	$H_\alpha(i)-C_\alpha(i)-N(i+1)$
3D-HNCO	$H_N(i)-N(i)-CO(i-1)$
3D-HN(CA)HA	$H_N(i)-N(i)-H_\alpha(i)/H_\alpha(i-1)$
3D-HN(CO)CA	$H_N(i)-N(i)-C_\alpha(i-1)$
3D-HN(COCA)HA	$H_N(i)-N(i)-H_\alpha(i-1)$
3D-H(N)CACO	$H_N-C_\alpha(i)/C_\alpha(i-1)-CO(i)/CO(i-1)$

continued

Table 6.1 Continued

Experiment	Information/linkage obtained
3D H(C)(CO)NH TOCSY	Correlation of side-chain ^{1}H and ^{13}C with $H_N(i,i-1)$ and $N(i,i-1)$
4D HCANNH	$H_\alpha(i)-C_\alpha(i)-N(i)-H_N(i)$
4D HCA(CO)NNH	$H_\alpha(i)-C_\alpha(i)-N(i+1)-H_N(i+1)$
4D HNCAHA	$H_N(i)-N(i)-C\alpha(i)-H_\alpha(i)$
4D HCCH TOCSY	^{1}H spin systems, ^{13}C shifts
4D ^{1}H–^{13}C HMQC-NOESY ^{1}H–^{13}C HMQC	Identification of nOes between side chain
4D ^{1}H–^{15}N HMQC-NOESY ^{1}H–^{13}C HMQC	Identification of nOes with amide protons
$^2J_{COH\alpha}$–E. COSY	$^2J(CO, H_\alpha)$
3D HNCA	$^3J(HN, H_\alpha)$, ϕ angle
3D HMQC J_{HH} TOCSY	$^3J(HN, H_\alpha)$ ϕ angle
3D H,C COSY ct-C,C COSY-β-INEPT	Vicinal ^{1}H–^{1}H couplings
Soft HCCH COSY	$^3J(H_\beta, CO)$
Soft HCCH E. COSY	$^3J(H_\alpha, H_\beta)$

[*]Reproduced with permission from T. Müller *et al.* (1994) *Angew. Chem. Int. Ed. Engl.* **33**, 277.

of the three axes. In such cross-sections, the cross-diagonal planes and back-transfer planes appear as lines running across the 2D spectrum. Two-dimensional cross-sections may be taken so they cut across the cross-peak due to the magnetization transfer $v_A \rightarrow v_B \rightarrow v_C$ in a three-spin system. Three such cross-sections are illustrated in Fig. 6.6, taken perpendicular to the three different axes v_1, v_2, and v_3. The cross-section (v_2, v_3), taken perpendicular to v_1 at the chemical shift frequency of nucleus A ($v_1 = v_A$), shows those cross-peaks that arise from the transfer of magnetization from nucleus A to the other spins during the two transfers. The 2D cross-section (v_1, v_2), taken perpendicular to v_2 at $v_2 = v_B$, represents the magnetization reaching and leaving nucleus B. This cross-section may contain not only the cross-peak (v_A, v_B, v_C) but also its symmetry equivalent peak (v_C, v_B, v_A) (hollow circle) if such a pathway is allowed. The third cross-section (v_1, v_2), taken perpendicular to v_3 at $v_3 = v_C$ provides information about all the magnetization arriving at v_C.

The peak shapes in 3D spectra can be obtained from the phases of the corresponding signals in the two 2D experiments from which the 3D spectrum is derived. This, if the two 2D spectra have pure phases, e.g., absorption signals, then the 3D cross-peaks will also be in the pure-absorption mode; and if the two 2D spectra from which the 3D spectrum is constituted have dispersive-mode signals, then the 3D spectrum will also have dispersive peak shapes. Mixed peak shapes in 2D spectra lead to mixed peak shapes in the corresponding 3D spectrum.

Symmetric 2D spectra result from a mixing sequence that is time- and phase-reversal (TPR) symmetric (Griesinger *et al.*, 1987a). Three-dimensional correlation spectra constituted from two identical symmetric mixing sequences, such as COSY-COSY or ROESY-ROESY, are themselves symmetric with respect to the plane $\omega_1 = \omega_3$. If unequal mixing processes are involved in the creation of 3D spectra, e.g., COSY-NOESY or NOESY-TOCSY, then such spectra lack global symmetry. The 3D EXSY-EXSY spectrum of heptamethyl benzenonium sulfate in sulfuric acid shown in Fig. 6.7 serves as an example of a 3D spectrum having $\omega_1 = \omega_3$ reflection symmetry. Peaks lying in the same plane ($\omega_1 = \omega_3$) are shown connected by dotted lines and have $\omega_1 = \omega_3$ reflection symmetry (i.e., symmetry about the diagonal line connecting the two opposite crosses, or in other words, drawn with $\omega_1 = \omega_3$). The exchange process giving rise to the EXSY-EXSY spectrum are due to the 1,2-shifts of the methyl groups (A $\leftrightarrow$ B, B $\leftrightarrow$ C, C $\leftrightarrow$ D).

To record a 3D spectrum that covers the entire spectral range in all three dimensions with a resolution comparable to that obtained in 2D spectra, an enormous amount of data must be processed, which not only may be beyond the storage capabilities of most computer systems, but may

Table 6.2

NMR Experiments and the Types of Labeling Suited for Protein
Structure Determination*

	Normal (without isotopic labeling)	With ^{15}N labeling	With ^{15}N and ^{13}C labeling
Assignment of spin system	TOCSY-TOCSY	TOCSY-HMQC TOCSY-TOCSY	3D/4D HCCH-TOCSY HCCH-COSY HC(C)NH-TOCSY
Sequential assignments	TOCSY-NOESY	NOESY-HMQC TOCSY-NOESY	HCACO HNCA HNCO HCA(CO)N
Identification of nOes	NOESY-NOESY TOCSY-NOESY	NOESY-HMQC NOESY-NOESY TOCSY-NOESY	3D NOESY-HMQC 4D HMQC-NOESY HMQC

*Reproduced with permission from Müller *et al.* (1994) *Angew. Chem. Int. Ed. Engl.* **33**, 277.

also take weeks to accumulate. It is therefore usual to record partial 3D spectra in only a certain region of interest. For instance, in Fig. 6.8 the volume chosen for the 3D NOESY-TOCSY spectrum of a peptide covers the NH region in ω_1, H_α in ω_2, and all protons in ω_3.

Three-dimensional NMR experiments are proving to be of particular use in assignments to amino acids in polypeptides. Some of the 3D and 4D NMR techniques used for the structure determination of proteins are presented in Table 6.1. The NMR experiments and the types of labeling suited for protein structure determination are given in Table 6.2. The development of 3D NMR techniques (Griesinger *et al.*, 1987; Vuister *et al.*, 1988; Fesik and Zuiderweg, 1988; Oschkinat *et al.*, 1988; Kay *et al.*, 1989; Zuiderweg and Fesik, 1989) and 4D NMR techniques (Kay *et al.*, 1990) has allowed protein structures of molecular weights of up to 35 kilodaltons to

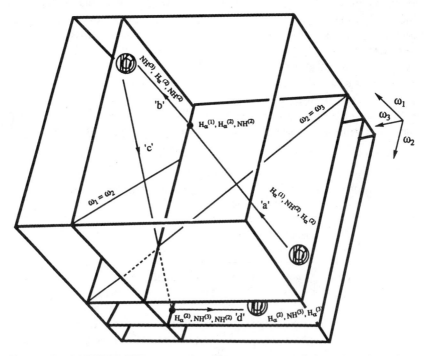

Figure 6.9 NOESY-TOCSY spectrum and its sequence analysis. ω_2, ω_3 planes intersect ω_1, ω_2 planes. Cross-peak ($H_\alpha^{(1)}$, $HN^{(2)}$, $H_\alpha^{(2)}$), when reflected across the ω_2, ω_3 diagonal, yields cross-peak $H_\alpha^{(1)}$, $H_\alpha^{(2)}$, $NH^{(2)}$. Along the line $\omega_1^{(2)}$, cross-peak $NH^{(3)}$, $H_\alpha^{(2)}$, $NH^{(2)}$ is found. By reflection at line $\omega_1 = \omega_2$ in the ω_1, ω_2 section, the latter peak is converted into point $H_\alpha^{(2)}$, $NH^{(3)}$, $NH^{(2)}$, from which cross-peak $H_\alpha^{(2)}$, $NH^{(3)}$, $H_\alpha^{(3)}$ is found by a search paralleled to ω_3. (Reprinted from *J. Mag. Reson.* **84**, C. Griesinger *et al.*, 14, copyright (1989), with permission from Academic Press, Inc.)

be investigated. Three-dimensional experiments that combine incoherent and coherent transfers of magnetization (NOESY-COSY, NOESY-TOCSY, hetero-COSY-NOESY, and ROESY-TOCSY) would give cross-peaks in which each cross-peak correlates three different nuclei. Two of these nuclei would belong to the same amino acid and will be correlated through a coherent magnetization transfer step (e.g., through COSY or TOCSY); the third nucleus would belong to an adjacent amino acid, thereby establishing the sequential connectivity of the two amino acids. Once a cross-peak has been assigned, we proceed to the next cross-peak by keeping two of the frequency coordinates fixed and searching for the next cross-peak along the third frequency domain of the three-dimensional space.

Different assignment strategies can be employed, depending on whether selective or nonselective pulses have been used in recording 3D spectra. A homonuclear 3D NOESY-TOCSY spectrum in which the NH/H_α region has been recorded is presented in schematic form in Fig. 6.9.

a

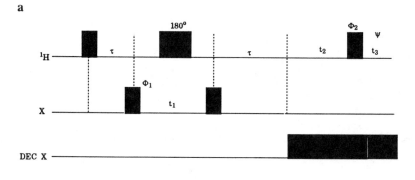

b

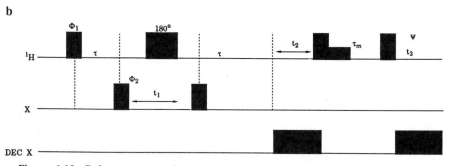

Figure 6.10 Pulse sequences for (a) the HMQC-COSY experiments and (b) the HMQC-NOESY experiments. (Reprinted from *J. Mag. Reson.* **78,** S. W. Fesik and E. R. P. Zuiderweg, 588, copyright (1988), with permission from Academic Press, Inc.)

The superscripts in brackets indicate the number of the amino acid in a given sequence. For example, the cross-peak due to ($H_\alpha^{(1)}$, $NH^{(2)}$, $H_\alpha^{(2)}$) represents the TOCSY interaction between the α protons of the first amino acid (H_α) and the NH proton of the second amino acid (NH) and the NOESY interactions between this $NH^{(2)}$ proton and the α proton of the second amino acid, $H_\alpha^{(3)}$. Having located the first cross-peak due, say, to $H_\alpha^{(1)}$, $NH^{(2)}$, $H_\alpha^{(3)}$, we reflect this cross-peak across the diagonal line (the

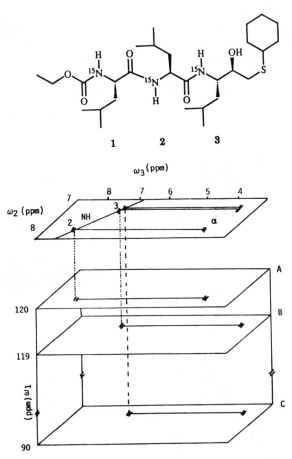

Figure 6.11 A 3D heteronuclear HMQC-COSY spectrum of a tripeptide. The ω_1-axis represents ^{15}N chemical shifts, whereas the ω_2- and ω_3-axes exhibit proton chemical shifts. The assignment pathways are indicated in the top spectrum for reference purposes, not as part of the 3D experiment. (Reprinted from *J. Mag. Reson.* **78**, S. W. Fesik and E. R. P. Zuiderweg, 588, copyright (1988), with permission from Academic Press, Inc.)

$\omega_2 = \omega_3$ diagonal) within the (ω_2, ω_3) section containing this cross-peak (line "a" in Fig. 6.9). This takes us to a position in the (ω_2, ω_3) section that is determined by $\omega_3 = \delta_{NH}^{(2)}$ and $\omega_2 = H_\alpha^{(2)}$. The peak at this position may be very weak or even absent due to the weak nOe expected between $H_\alpha^{(1)}$ and $H_\alpha^{(2)}$. Next we pass a line "b" through this peak and parallel to

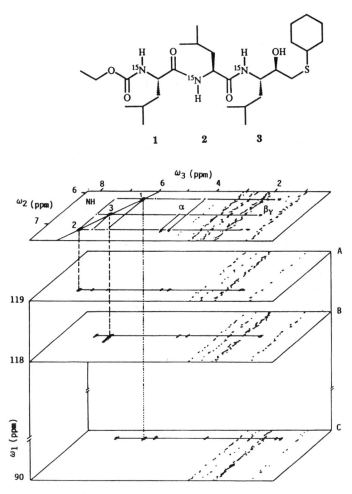

Figure 6.12 A 3D heteronuclear HMQC-NOESY spectrum of a tripeptide. The ω_1-axis represents ^{15}N chemical shifts, whereas ω_2- and ω_3-axes exhibit proton chemical shifts. The assignment pathways are indicated in the top spectrum for reference purposes, not as part of the 3D experiment. (Reprinted from *J. Mag. Reson.* **78,** S. W. Fesik and E. R. P. Zuiderweg, 588, copyright (1988), with permission from Academic Press, Inc.)

the ω_1-axis. This takes us to the $(NH^{(3)}, H_\alpha^{(2)}, NH^{(2)})$ cross-peak. Reflection of this cross-peak across the $\omega_1 = \omega_2$ diagonal takes us along line "c" to position $(H_\alpha^{(2)}, NH^{(3)}, NH^{(2)})$. This cross-peak will not appear, since there is no J-coupling between $NH^{(2)}$ and $NH^{(3)}$. However, by following line "d" from this point parallel to the ω_3-axis, we arrive at the $(H_\alpha^{(2)}, NH^{(3)}, H_\alpha^{(3)})$ cross-peak. By repeating this process we can assign the amino acids sequentially.

The pulse sequences for HMQC-COSY and HMQC-NOESY experiments are presented in Fig. 6.10. The 3D HMQC-COSY spectrum of a N^{15} labeled tripeptide is shown in Fig. 6.11. Since the coherence transfer involved in this experiment is $^{15}N(t_1) \rightarrow {}^{15}N\underline{H}(t_2) \rightarrow {}^{15}N\underline{H}(t_3) + C_\alpha\underline{H}(t_3)$, it is expected that the ^{15}N resonance will appear in ω_1, that the $^{15}N\underline{H}$ resonance will appear in ω_2, and that $^{15}N\underline{H}$ and $C_\alpha H$ resonance will appear in ω_3. The HMQC-NOESY spectrum shown in Fig. 6.12 has been edited with respect to ^{15}N frequencies in ω_1. The spectrum shows that the amide protons of residues 1 and 3 have close chemical shifts but that the ^{15}N nuclei to which they are coupled have well-separated chemical shifts (90 and 119 ppm). Similarly, the near overlap in the ^{15}N dimension (Leu-2 at 120 ppm and leu-3 at 119 ppm) is well resolved in the spectrum. Readers are directed to two excellent articles for a more detailed discussion on the theory and applications of various types of 3D NMR experiments (Griesinger *et al.*, 1989; Oschkinat *et al.*, 1989, 1994).

REFERENCES

Aue, W. P., Bartholdi, E., and Ernst, R. R. (1976). *J. Chem. Phys.* **64**, 2229.
Aue, W. P., Karhan, J., and Ernst, R. R. (1976). *J. Chem. Phys.* **64**, 4226.
Braunschweiler, L., and Ernst, R. R. (1983). *J. Magn. Reson.* **53**, 521.
Braunschweiler, L., Bodenhausen, G., and Ernst, R. R. (1983). *Mol. Phys.* **48**, 535.
Bothner-By, A. A., Stephens, R. L., Lee, J., Warren, C. D., and Jeanloz, R. W. (1984). *J. Am. Chem. Soc.* **106**, 811.
Caravatti, P., Neuenschwander, P., and Ernst, R. R. (1985). *Macromolecules* **18**, 119.
Chandrakumar, N., and Subramanian, S. (1987). *Modern techniques in high resolution FT-NMR.* Springer, New York.
Cohen, A. D., Freeman, R., and McLauchlan, K. A. (1963). *Mol. Phys.* **7**, 45.
Eich, G. W., Bodenhausen, G., and Ernst, R. R. (1982). *J. Am. Chem. Soc.* **104**, 3731.
Ernst, R. R., Bodenhausen, G., and Wokaun, A. (1987). *Principles of NMR in one and two dimensions.* Clarendon, Oxford.
Fesik, S. W., and Zuiderweg, E. R. P. (1988). *J. Magn. Reson.* **78**, 588.
Griesinger, C., Gemperle, C., and Ernst, R. R. (1987a). *Mol. Phys.* **62**, 295.
Griesinger, C., Sörensen, O. W., and Ernst, R. R. (1987b). *J. Magn. Reson.* **73**, 574.
Griesinger, C., Sörensen, O. W., and Ernst, R. R. (1987c). *J. Am. Chem. Soc.* **109**, 7227.
Griesinger, C., Sörensen, O. W., and Ernst, R. R. (1989). *J. Magn. Reson.* **84**, 14.
Jeener, J., Meier, B. H., Bachmann, P., and Ernst, R. R. (1979). *J. Chem. Phys.* **71**, 4546.
Kay, L. E., Marion, D., and Bax, A. (1989). *J. Magn. Reson.* **84**, 72.
Kay, L. E., Clore, G. M., Bax, A., and Gronenborn, A. M. (1990). *Science* **249**, 411.

Kumar, A., Ernst, R. R., and Wüthrich, K. (1980). *Biophys. Res. Commun.* **95,** 1.

Machura, S., and Ernst, R. R. (1980). *Mol. Phys.* **41,** 95.

Maudsley, A. A., and Ernst, R. R. (1977). *Chem. Phys. Lett.* **50,** 368.

Meier, B. H., and Ernst, R. R. (1979). *J. Am. Chem. Soc.* **101,** 6441.

Müller, L., Kumar, Anil, and Ernst, R. R. (1975). *J. Chem. Phys.* **63,** 5490.

Oschkinat, H., Cieslar, C., Holak, T. A., Clore, G. M., and Gronenborn, M. (1989). *J. Magn. Reson.* **83,** 450.

Oschkinat, H., Griesinger, C., Kraulis, P. J., Sörensen, O. W., Ernst, R. R., Gronenborn, A. M., and Clore, G. M. (1988). *Nature (London)* **332,** 374.

Oschkinat, H., Müller, T., and Diekmann, T. (1994). *Angew. Chem. Int. Ed. Engl.* **33,** 277.

Plant, H. D., Mareci, T. H., Cockman, M. D., and Brey, W. S. (1986). 27th experimental NMR conference. Baltimore, Maryland, April 13–17.

Vuister, G. W., and Boelens, R. (1987). *J. Magn. Reson.* **73,** 328.

Vuister, G. W., Boelens, R., and Kaptein, R. (1988). *J. Magn. Reson.* **80,** 176.

Zuiderweg, E. R. P., and Fesik, S. W. (1989). *Biochemistry* **28,** 2387.

CHAPTER 7

Recent Developments in NMR Spectroscopy

7.1 SELECTIVE PULSES IN MODERN NMR SPECTROSCOPY

While the majority of NMR pulse sequences involve nonselective pulses, the use of selective pulses has become popular in recent years, not only in 1D but also in 2D and 3D NMR experiments (Kessler *et al.*, 1986,1989a; Cavanagh *et al.*, 1987a,b; Griesinger *et al.*, 1987a,b,1989). This is because nonselective pulses affect the whole spectral region, whereas selective pulses affect only a limited region of the spectrum in a well-defined manner. The use of selective pulses therefore makes it possible to confine the investigation to an interesting region of the spectrum. In multidimensional NMR experiments this improves digital resolution, reduces data storage requirements, and saves data-processing time.

There are three main requirements of good-quality selective pulses: (a) they should elicit uniform excitation within a limited spectral range; (b) they should exhibit a uniform phase behavior; and (c) the pulse should be as short as possible. Nonselective Rf pulses, called *hard pulses*, are characterized by generally rectangular shape, high power, and short duration. Soft pulses generally are selective pulses, and they usually have a Gaussian shape and are of low power and long duration so they produce excitation within a narrow spectral region. "Soft" multiple-dimensional NMR experiments are those in which only a limited spectral region is excited in one or more dimensions. Some important pulse shapes and their Fourier transforms are shown in Table 7.1.

7.2 ONE-DIMENSIONAL EXPERIMENTS USING SOFT PULSES

A 1D NMR experiment is converted into a 2D NMR experiment by incrementing the delay t_1. When only limited information within a narrow spectral region is required, it is advantageous to use a soft excitation pulse and a constant delay instead of incrementing t_1. Such a spectrum corresponds to a cross-section through the 2D spectrum at the frequency of the selectively excited resonance (Fig. 7.1). The difference between such soft 1D experiments and the classical frequency-selective 1D NMR experiments [e.g., TO1 (Wagner and Wüthrich, 1979), SPI (Pachler and Wessels, 1973)] is that the selective 1D experiments only cause disturbance of populations, while the new frequency-selective 1D NMR experiments, like 2D NMR experiments, utilize the transverse magnetization generated by the soft excitation pulse.

Table 7.1

A Fourier Transform Relationship between Time-Domain and Frequency-Domain Excitation Functions.

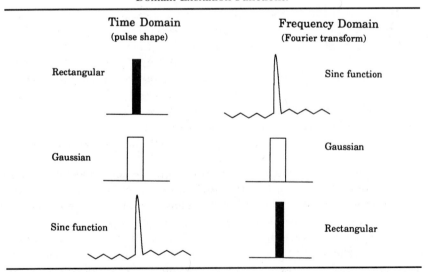

Reproduced with permission, from H. Kessler, S. Mronga and G. Gemmecker, (1991). Multi-Dimensional NMR Experments Using Selective Pulses, *Mag. Reson. Chem.* **29**, 527

One-Dimensional NMR Experiment
(e.g. 1D COSY)

F_2

Two-Dimensional NMR Experiment
(e.g. COSY)

F_2

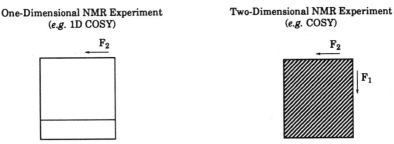

F_1

Figure 7.1 Selective excitation of only one multiplet by a selective pulse transforms a 2D experiment into a 1D technique. A selective pulse generates the transverse magnetization. The result is a "trace" of the corresponding 2D spectrum. (Reprinted from *Mag. Reson. Chem.* **29**, H. Kessler *et al.*, 527, copyright (1991), with permission from John Wiley and Sons Limited, Baffins Lane, Chichester, Sussex PO19 IUD, England.)

7.2.1 1D COSY

A 90° Gaussian pulse is employed as an excitation pulse. In the case of a simple AX spin system, the delay τ between the first, soft 90° excitation pulse and the final, hard 90° detection pulse is adjusted to correspond to the coupling constant J_{AX} (Fig. 7.2). If the excitation frequency corresponds to the chemical shift frequency of nucleus A, then the doublet of nucleus A will disappear and the total transfer of magnetization to nucleus X will produce an antiphase doublet (Fig. 7.3). The antiphase structure of the multiplets can be removed by employing a refocused 1D COSY experiment (Hore, 1983).

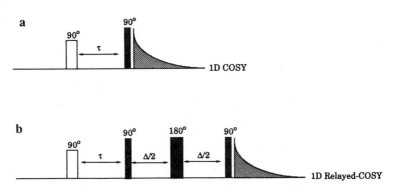

Figure 7.2 Pulse sequences for 1D COSY and 1D relayed COSY. A soft 90° Gaussian pulse serves as an excitation pulse for these experiments. (Reprinted from *Mag. Reson. Chem.* **29**, H. Kessler *et al.*, 527, copyright (1991), with permission from John Wiley and Sons Limited, Baffins Lane, Chichester, Sussex PO19 IUD, England.)

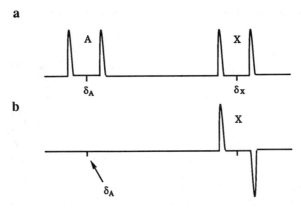

Figure 7.3 One-dimensional COSY spectrum for an AX system. (a) A common 1D spectrum. (b) Selective excitation of spin A leads to a 1D COSY spectrum with antiphase X lines and maximum transfer of magnetization from A to X. (Reprinted from *Mag. Reson. Chem.* **29**, H. Kessler *et al.*, 527, copyright (1991), with permission from John Wiley and Sons Limited, Baffins Lane, Chichester, Sussex PO19 IUD, England.)

7.2.2 1D Relayed COSY

In molecules in which a signal in the NMR spectrum is clearly separated from other signals (e.g., anomeric protons of sugars or NH signals of peptides), relayed NMR experiments can be employed with success, especially when the coupling partner of the nucleus is hidden under a crowded region. The pulse sequence of 1D relayed COSY is shown in Fig. 7.2b. The initial excitation of the selected nucleus (the one clearly separated from other signals) is affected by a 90° Gaussian pulse. Magnetization is then transferred in a step-wise manner to other nuclei in the spin network. In a peptide, for example, the NH signal will be affected by the first 90° Gaussian pulse. The next 90° (hard) pulse will transfer the magnetization to the neighboring H_α. This magnetization is then transferred to H_β. As a result H_α affords in-phase dispersive signals while H_β gives antiphase absorption signals. The delay $\Delta/2$ is adjusted according to the magnitude of the NH/H_α coupling constant so that maximum magnetization transfer occurs.

An interesting feature of the relayed 1D COSY experiment not found in the corresponding 2D version is that directed magnetization transfer can be achieved with more than one relay nucleus, e.g., $A \rightarrow (B_1, B_2) \rightarrow C$. When the coupling constants of A with B_1 and B_2 are different, then the delay Δ can be adjusted to the larger coupling constant so that magnetization transfer occurs to the corresponding nucleus and it is then further transferred to nucleus C. In such a case, nucleus C would show pure phases (antiphases) to only one nucleus B. A 2D relay experiment, however, would have always produced mixed phases at nucleus C because of transfer from both B nuclei.

7.2.3 SELINQUATE

SELINQUATE (Berger, 1988) is the selective 1D counterpart of the 2D INADEQUATE experiment (Bax *et al.*, 1980). The pulse sequence is shown in Fig. 7.4. Double-quantum coherences (DQC) are first excited in the usual manner, and then a selective pulse is applied to only one nucleus. This converts the DQC related to this nucleus into antiphase magnetization, which is refocused during the detection period. The experiment has not been used widely because of its low sensitivity, but it can be employed to solve a specific problem from the ^{13}C–^{13}C connectivity information.

7.2.4 1D NOESY

In the 1D NOESY experiment (Kessler *et al.*, 1986), which is a 1D version of 2D NOESY, a combination of a selective 90° half-Gaussian pulse is followed by a 90° hard pulse (Fig. 7.5). Phase cycling is used to eliminate all nOe effects before the hard pulse so that the mixing time τ_{mix} can be determined exactly. Selective inversion of a single spin causes a change in the longitudinal magnetization of another close-lying spin, and the rate of initial buildup of the longitudinal magnetization gives the cross-relaxation rate, σ, between the two spins.

7.2.5 1D Relayed NOESY

Two-dimensional relayed NOESY experiments (Wagner, 1984; Kessler *et al.*, 1988) give cross-peaks that could be a superposition of nOe's resulting from different relay nuclei. This complicates the extracting of cross-relaxation rates. The 1D NOESY experiment (Kessler *et al.*, 1989a), however, allows the path of the magnetization transfer to be clarified.

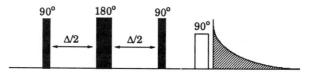

Figure 7.4 Pulse sequence for the SELINQUATE experiment, which is a selective INADEQUATE experiment. (Reprinted from *Mag. Reson. Chem.* **29**, H. Kessler *et al.*, 527, copyright (1991), with permission from John Wiley and Sons Limited, Baffins Lane, Chichester, Sussex PO19 IUD, England.)

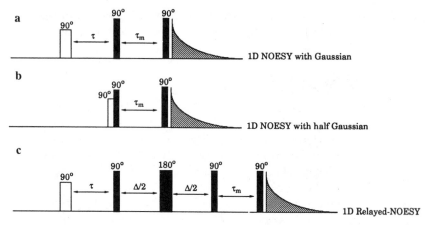

Figure 7.5   Pulses sequences for 1D NOESY and 1D relayed NOESY experiments.
(a) A 1D NOESY sequence with full Gaussian pulses is inferior to the 1D NOESY
sequence (b) with half-Gaussian excitation. (c) A 90° Gaussian pulse in the pulse
sequence of 1D relayed NOESY is appropriate for excitation, since antiphase magnetization is required for the first mixing step. (Reprinted from *Mag. Reson. Chem.*
29, H. Kessler *et al.,* 527, copyright (1991), with permission from John Wiley and
Sons Limited, Baffins Lane, Chichester, Sussex PO19 IUD, England.)

7.2.6 1D TOCSY

The pulse sequence for the 1D TOCSY experiment is shown in Fig.
7.6. The original experiment used a Gaussian pulse, but a half-Gaussian

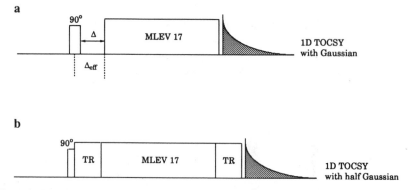

Figure 7.6   (a) A 1D TOCSY experiment with Gaussian excitation. (b) A 1D TOCSY
experiment with half-Gaussian excitation and TR (trim) pulses. (Reprinted from
Mag. Reson. Chem. **29,** H. Kessler *et al.,* 527, copyright (1991), with permission from
John Wiley and Sons Limited, Baffins Lane, Chichester, Sussex PO19 IUD, England.)

pulse has been found to give better results. The 1D TOCSY experiment has no significant advantage over the 2D version, which also has good sensitivity and can be recorded with good resolution.

7.2.7 1D ROESY

The pulse sequence for the 1D ROESY experiment using purged half-Gaussian pulses is shown in Fig. 7.7. The purging is required to remove the dispersive components, since these are not completely eliminated by the weak spin-lock field employed in the 1D ROESY experiment.

7.3 HETERONUCLEAR SELECTIVE 1D NMR EXPERIMENTS

7.3.1 SELINCOR

The SELINCOR experiment is a selective 1D inverse heteronuclear shift-correlation experiment (*i.e.,* 1D H,C-COSY inverse experiment) (Berger, 1989). The last ^{13}C pulse of the HMQC experiment is in this case substituted by a selective 90° Gaussian pulse. Thus the soft pulse is used for coherence transfer and not for excitation at the beginning of the sequence, as is usual for other pulse sequences. The BIRD pulse and the Δ_2 delay are optimized to suppress protons bound to ^{12}C nuclei; Δ_3 is adjusted to correspond to the direct H,C couplings. The soft pulse at the end of the pulse sequence (Fig. 7.8) serves to transfer the heteronuclear double-quantum coherence into the antiphase magnetization of the protons attached to the selectively excited ^{13}C nuclei.

7.3.2 1D HMQC-TOCSY

A 1D analog of the HMQC experiment with TOCSY magnetization transfer has been reported (Crouch *et al.,* 1990). The pulse sequence is

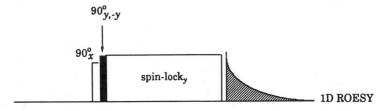

Figure 7.7 A 1D ROESY pulse sequence with purged half-Gaussian excitation. (Reprinted from *Mag. Reson. Chem.* **29,** H. Kessler *et al.,* 527, copyright (1991), with permission from John Wiley and Sons Limited, Baffins Lane, Chichester, Sussex PO19 IUD, England.)

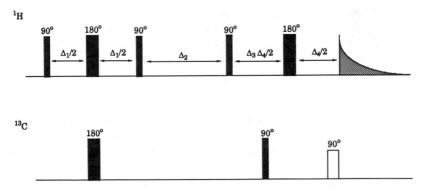

Figure 7.8 SELINCOR (Selective Inverse Correlation) pulse sequence with BIRD presaturation. (Reprinted from *Mag. Reson. Chem.* **29**, H. Kessler *et al.*, 527, copyright (1991), with permission from John Wiley and Sons Limited, Baffins Lane, Chichester, Sussex PO19 IUD, England.)

similar to that employed in SELINCOR, with an additional delay to refocus $^1J_{CH}$ splitting, a TOCSY magnetization transfer, and ^{13}C decoupling during detection (Fig. 7.9).

7.3.3 Selective Heteronuclear J-Resolved Experiment

The selective heteronuclear *J*-resolved experiment (Bax, 1984) is used to determine heteronuclear long-range coupling constants via a selective π pulse that causes splitting of all carbon signals coupled to that proton.

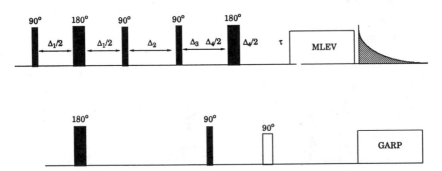

Figure 7.9 A 1D HMQC-TOCSY pulse sequence. The sequence involves a TOCSY transfer and BIRD presaturation. (Reprinted from *Mag. Reson. Chem.* **29**, H. Kessler *et al.*, 527, copyright (1991), with permission from John Wiley and Sons Limited, Baffins Lane, Chichester, Sussex PO19 IUD, England.)

This experiment requires a well-separated proton resonance for excitation; rectangular pulses are used for the soft excitation.

7.3.4 Selective INEPT

The selective INEPT experiment also requires a separated proton resonance for excitation and also uses rectangular soft pulses (Bax and Freeman, 1982). Magnetization is selectively transferred by the soft proton pulses to the corresponding carbons to which the protons are coupled.

7.4 TWO-DIMENSIONAL EXPERIMENTS USING SOFT PULSES

When a desired resonance cannot be excited selectively in a soft 1D experiment, then it is advantageous to use a "semisoft" 2D version, which involves soft excitation in one dimension. This gives higher resolution in the F_1 domain in reasonable time, since only a small spectral region in F_1 is recorded. This semisoft 2D experiment is derived from the corresponding conventional 2D experiment by substituting the nonselective excitation pulse by a soft pulse so that the excitation is limited to a narrow region of the 2D spectrum (Fig. 7.10).

Two-Dimensional Experiment   **Two-Dimensional Semisoft Experiment**

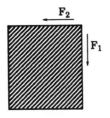   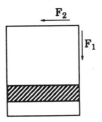

Figure 7.10 A 2D experiment can be transformed into a 2D semisoft experiment by using selective pulses to excite a limited number of multiplets before t_1. The evolution of the magnetization of the multiplets takes place during t_1. The hard pulse immediately after t_1 leads to the recording of a "strip" out of the corresponding 2D spectrum that has reduced spectral range in F_1 and full spectral range in F_2. (Reprinted from *Mag. Reson. Chem.* **29**, H. Kessler *et al.*, 527, copyright (1991), with permission from John Wiley and Sons Limited, Baffins Lane, Chichester, Sussex PO19 IUD, England.)

The pulse sequences employed in semisoft experiments are shown in Fig. 7.11. The last two sequences are preferred, since the baseline distortions and distortions due to large couplings in multiplets are suppressed.

7.4.1 Semisoft COSY

The semisoft COSY experiment (Cavanagh *et al.*, 1987a; Brüschweiler *et al.*, 1988) results in increased resolution in F_1 so that better cross-peak

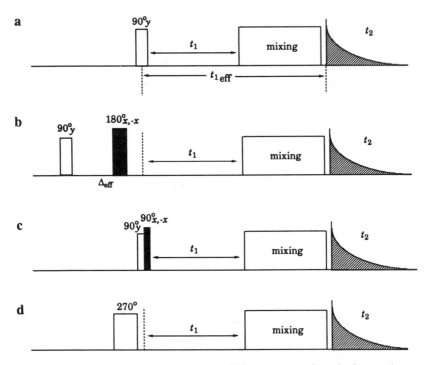

Figure 7.11 Pulse sequences representing different types of semisoft experiments with different excitations. (a) Chemical shift evolution and *J*-coupling starts in the middle of the Gaussian pulse, while the second half of the Gaussian pulse is a part of $t_{1\text{eff}}$ and cannot be adjusted to zero for the first experiment. (b) Chemical shift evolution is refocused at the start of t_1. *J*-coupling evolves during Δ_{eff} and therefore cannot be refocused. (c) The effect of purged half-Gaussian pulses on the evolution of the chemical shift and *J*-coupling during the preparation period. (d) In case of a 270° Gaussian pulse, the t_1 starts at the end of the pulse duration, since the *J*-coupling and chemical shifts that evolve during the soft pulse are refocused. (Reprinted from *Mag. Reson. Chem.* **29**, H. Kessler *et al.*, 527, copyright (1991), with permission from John Wiley and Sons Limited, Baffins Lane, Chichester, Sussex PO19 IUD, England.)

fine structure is obtained. The pulse sequences used for semisoft COSY and its variations are shown in Fig. 7.12. Additional F_1 decoupling may be employed to reduce the size of the cross-peaks and thereby remove overlap between them.

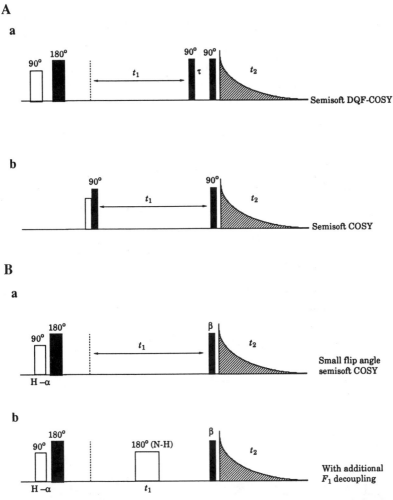

Figure 7.12 (A) Pulse sequences for semisoft COSY experiments. (B) Pulse sequences for small flip-angle semisoft β-COSY. (a) Normal; (b) same but with additional F_1 decoupling. (Reprinted from *Mag. Reson. Chem.* **29,** H. Kessler *et al.*, 527, copyright (1991), with permission from John Wiley and Sons Limited, Baffins Lane, Chichester, Sussex PO19 IUD, England.)

7.4.2 Semisoft NOESY

The semisoft NOESY experiment (Fig. 7.13) (Cavanagh *et al.*, 1987a; Brüschweiler *et al.*, 1988) has been employed successfully for observing nOe effects between NH and H_α in proteins dissolved in water. The H_α are excited by semiselective excitation so that the large t_1 ridge due to the residual water signal, which is problematic in the classical nOe experiment, is suppressed. F_1 decoupling can be employed with advantage in the semisoft nOe experiment. The 180° Gaussian pulse in the middle of the evolution period inverts the H_α spins in proteins, which are thus excited semiselectively. The subsequent 180° hard pulse brings back the H_α spins to their starting points, while other spins are inverted (with the two pulses together acting as a 360° pulse for the H_α spins but behaving as a 180° pulse for the other spins). Zero-quantum coherences have also been suppressed in another modification of this experiment (Otting *et al.*, 1990).

7.4.3 Heteronuclear Multiple-Bond Correlation, Selective (HMBCS)

This is a variation of the proton-detected shift-correlation experiment via long-range couplings proposed by Bax and Summers (Bax and Summers, 1986), with the difference that the first ^{13}C pulse is substituted by a frequency selective pulse (Fig. 7.14) (Bermel *et al.*, 1989; Kessler *et al.*, 1989b,1990). This significantly increases resolution in the F_1 dimension. For example, this can be used to remove the overlap between the cross-peaks of the carbonyl resonances of peptide bonds in proteins that all occur within a

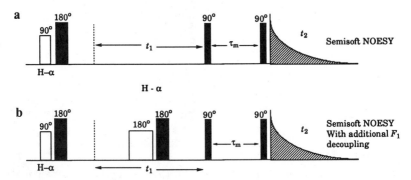

Figure 7.13 Pulse sequences of semisoft NOESY. (a) Normal; (b) same but with additional F_1 decoupling. (Reprinted from *Mag. Reson. Chem.* **29**, H. Kessler *et al.*, 527, copyright (1991), with permission from John Wiley and Sons Limited, Baffins Lane, Chichester, Sussex PO19 IUD, England.)

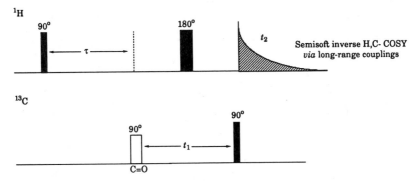

Figure 7.14 Pulse sequence for the HMBCS (heteronuclear multiple-bond correlation, selective) experiment, which uses advantageously a 270° Gaussian pulse for exciting the carbonyl resonances. It is also called the *semisoft inverse COLOC*. (Reprinted from *Mag. Reson. Chem.* **29,** H. Kessler *et al.,* 527, copyright (1991), with permission from John Wiley and Sons Limited, Baffins Lane, Chichester, Sussex PO19 IUD, England.)

6-ppm range. A 270° Gaussian pulse has been used in this experiment to improve resolution, and J_{CH} coupling constants can be determined (Kessler *et al.*, 1990). Decoupling during detection improves sensitivity (Shaka *et al.*, 1985).

7.4.4 Semisoft TOCSY

The pulse sequence involving excitation by a half-Gaussian pulse is shown in Fig. 7.15 (Kessler *et al.*, 1989a). Its use was demonstrated by semiselective excitation of the NH spectral region of a hexapeptide.

7.5 SOFT EXCITATION IN TWO DIMENSIONS

The semiselective pulses in the previously described experiments were employed for either excitation or decoupling, but they were not used for

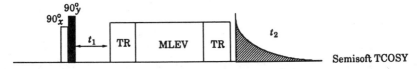

Figure 7.15 Pulse sequence for the semisoft TOCSY experiment. Purged half-Gaussian pulse excitation sequence and trim pulses (TR) are used. (Reprinted from *Mag. Reson. Chem.* **29,** H. Kessler *et al.,* 527, copyright (1991), with permission from John Wiley and Sons Limited, Baffins Lane, Chichester, Sussex PO19 IUD, England.)

Two-Dimensional Experiment   **Soft Two-Dimensional Experiment**

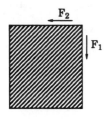
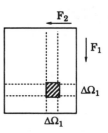

Figure 7.16   When soft pulses are used for excitation and mixing in a 2D experiment, it becomes a 2D "soft" experiment. The spectrum of the 2D soft experiment has reduced frequency ranges in F_1 and F_2. The excitation ranges of the selective pulse depend on the type of experiment. For example, in a soft COSY-COSY experiment, one multiplet is excited, while in the soft NOESY experiment the whole resonance region of a group of signals is excited. (Reprinted from *Mag. Reson. Chem.* **29,** H. Kessler *et al.,* 527, copyright (1991), with permission from John Wiley and Sons Limited, Baffins Lane, Chichester, Sussex PO19 IUD, England.)

mixing, e.g., for transfer of coherence. If semiselective pulses are used for both excitation and mixing, we obtain a 2D spectrum in which the frequency ranges are restricted in both dimensions (Fig. 7.16). This allows the particular segment of the 2D spectrum to be recorded with high resolution.

7.5.1 *Soft H,H-COSY*

The soft H,H-COSY pulse sequence (Brüschweiler *et al.,* 1987) involves three semiselective pulses, one for excitation on the first nucleus, and the other two for mixing on both nuclei (Fig. 7.17). The two selective mixing

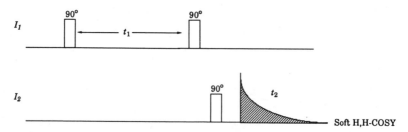

Figure 7.17   Pulse sequence for soft H,H-COSY with two selective pulses exciting the I_1 multiplet and the second mixing pulse exciting the I_2 multiplet of a three-spin system ((I_1, I_2, I_3)). (Reprinted from *Mag. Reson. Chem.* **29,** H. Kessler *et al.,* 527, copyright (1991), with permission from John Wiley and Sons Limited, Baffins Lane, Chichester, Sussex PO19 IUD, England.)

pulses limit the coherence transfer within a specific spectral range. An E. COSY-like multiplet pattern is obtained, with the correlations appearing only between connected transitions, so passive spins remain unaffected. In contrast to E. COSY, the cross-peak components are not eliminated by subtraction, so soft COSY is more sensitive than E. COSY. In a three-spin system, a fourfold sensitivity enhancement results, so there is a 16-fold saving in the recording time.

7.5.2 Soft NOESY

One problem in recording nOe spectra of proteins in aqueous solutions is the presence of a water signal. Soft NOESY produces minimal excitation of the solvent signal. The pulse sequence used is shown in Fig. 7.18 (Oschkinat et al., 1988; Oschkinat and Bermel, 1989).

7.5.3 Soft H,C-COSY

The soft H,C-COSY experiment (Kessler et al., 1988a) allows the heteronuclear long-range couplings in F_2 to be determined. There is good sensitiv-

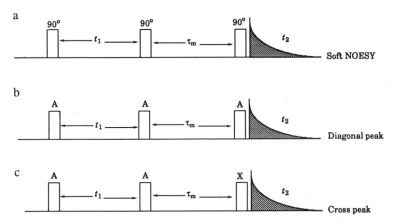

Figure 7.18 The soft NOESY pulse sequence consists of three pulses. (a) The first two pulses have the same excitation frequency (NH resonance in this case). The third and last pulse in the region of NH (F_1)/H $- \alpha(F_2)$ results in a soft NOESY experiment. (b) and (c) sequences of the soft NOESY, with all pulses being multiplet selective. (Reprinted from Mag. Reson. Chem. **29**, H. Kessler et al., 527, copyright (1991), with permission from John Wiley and Sons Limited, Baffins Lane, Chichester, Sussex PO19 IUD, England.)

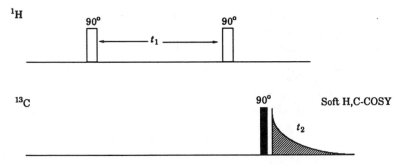

Figure 7.19 Soft H,C-COSY pulse sequence with two soft pulses having the same excitation frequency. (Reprinted from *Mag. Reson. Chem.* **29,** H. Kessler *et al.,* 527, copyright (1991), with permission from John Wiley and Sons Limited, Baffins Lane, Chichester, Sussex PO19 IUD, England.)

ity, since only directly connected transitions are excited. The pulse sequence used is shown in Fig. 7.19.

7.6 THREE-DIMENSIONAL EXPERIMENTS USING SOFT PULSES

Three-dimensional NMR experiments are constructed from a combination of two 2D NMR experiments, with the detection period of the first and the preparation period of the second being omitted. Triple Fourier transformation results in a 3D NMR spectrum in which each individual cross-peak is defined by three chemical shifts in F_1, F_2, and F_3. The large amount of data required poses severe problems in data storage capacity and processing time. The digital resolution also cannot be improved (i.e., by having a higher number of increments) in view of the time constraints. However, the use of selective pulses allows the spectral range in t_1 or t_2 to be reduced so that a subvolume of the total 3D spectrum can be recorded with higher digital resolution (Fig. 7.20).

The frequency selection in the 3D spectrum can be carried out in one, two, or all three dimensions. For instance, the soft COSY-COSY experiment (Griesinger *et al.,* 1987b) employs only soft pulses, so the frequency range is restricted in all three dimensions. The majority of soft 3D spectra are recorded with restricted spectral widths in F_1 and F_2 and full spectral width in F_3, so they are soft in two dimensions and have a high resolution in F_3. The soft pulses are normally applied before the t_1 and t_2 periods, while hard pulses are applied before the detection period. Examples of such experiments are 3D soft COSY-COSY (Griesinger *et al.,* 1987b), 3D soft NOESY-COSY (Griesinger *et al.,* 1987a), 3D soft NOESY -TOCSY (Oschki-

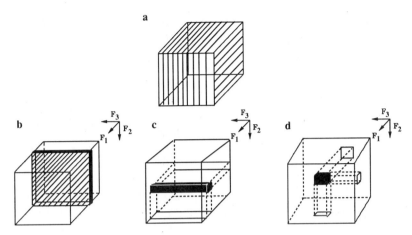

Figure 7.20 Transformation of a 3D experiment into a soft 3D experiment by selective pulses. (a) A typical 3D spectrum. (b) A 3D spectrum with reduced frequency range in F_1 only. (c) A 3D spectrum with reduced frequency range in F_1 and F_2, or (d) in F_1, F_2, and F_3. Spectra (b), (c), and (d) require selective soft pulses preceding t_1, t_1, and t_2, and t_1, t_2, and t_3, respectively. (Reprinted from *Mag. Reson. Chem.* **29**, H. Kessler *et al.*, 527, copyright (1991), with permission from John Wiley and Sons Limited, Baffins Lane, Chichester, Sussex PO19 IUD, England.)

nat *et al.*, 1988b; Oschkinat *et al.*, 1989a), and 3D soft TOCSY-NOESY (Oschkinat *et al.*, 1989a,b; Montelione and Wagner, 1990). A soft inverse hetero-COSY-COSY experiment (Griesinger *et al.*, 1989) has also been reported. A detailed discussion of these methods is beyond the scope of this book, but readers are referred to an excellent review in the area (Kessler *et al.*, 1991).

7.7 FIELD GRADIENTS

Field gradients have been known to NMR spectroscopists for a long time, largely as a nuisance value: They can cause broadening of the spectral lines due to the inhomogeneties in the static magnetic field, B_0. In recent years, however, several important features of field gradients have been used widely by spectroscopists, leading to important developments that extend the role of NMR spectroscopy as a versatile analytical tool and making it a routine technique in medicine. Thus magnetic resonance imaging (MRI), which is now widely used in clinical diagnosis, relies on the properties of field gradients. Magnetic field gradients have also been employed in NMR techniques for studying diffusion and the microscopic structure of heterogeneous systems. Another important development in the last few years has been the modification of some multipulse experiments by using field gradients instead of phase cycling, thereby allowing cleaner spectra to be obtained in shorter measurement times.

At equilibrium, the net magnetization vector of a sample will be parallel to the applied field, B_0, which, by convention, points toward the z-axis. To detect this magnetization in the form of an NMR signal, it must be bent to the $x'y'$-plane, i.e., the original longitudinal magnetization must be converted into transverse magnetization. This is achieved by applying a coherent Rf pulse of a predetermined power, length, and phase. A $90°_x$ pulse, for instance, would bend the magnetization about the x'-axis by $90°$, i.e., bend it from the z-axis to the y'-axis. In a homogeneous magnetic field, each type of magnetically active nucleus (e.g., a particular proton in a molecule) will precess at a single frequency that will decay only due to *transverse* relaxation.* However, in an inhomogeneous magnetic field, the magnetization will become subdivided into a number of smaller vectors, each of which will precess at a slightly differing frequency. These smaller vectors spread out (dephase) during the evolution period, reducing the net transverse magnetization of the sample. This leads to a broadening of the NMR signals and to a more rapid disappearance of the FID.

Magnetic field variations can arise due to field inhomogeneities or due to variations in magnetic susceptibility across the sample. In homogeneous liquids, these variations are particularly large at the solution–air and solution–glass–air interfaces. The axial symmetry of the NMR tubes produces largely symmetrical variations in the magnetic field. These are neutralized by applying compensating fields across the sample using the instrument's shim coils. However, if the sample size is large (as in *in vivo* imaging or NMR spectroscopy of living systems) or when heterogeneous samples are encountered, the compensation measures employed using the spectrometer's shim coils may not succeed. In such cases, echo-based experiments may be used to refocus the spread in magnetization, or we can opt for experiments employing zero-quantum coherence that are unaffected by field gradients.

7.7.1 *Magnetic Resonance Imaging (MRI)*

One of the most important applications of NMR spectroscopy has been its use in medicine for creating images of sections of the human body to detect tumors or other growths. The technique has gained popularity because it provides high-quality images and does not have the damaging effects of X rays used in conventional scanning techniques. Hence it is a noninvasive tool for clinical investigation.

---

* This is because the precessional frequency of a nucleus depends on the magnetic field, and as the magnetic field varies, so will the precessional frequency.

The basic principle underlying the development of images is simple (Lauterber, 1973). Consider a body cavity containing two pools of water in different quantities. In a uniform magnetic field, the NMR spectrum will consist of a single peak, since all the water molecules will precess at the same frequency, irrespective of their spatial location. If, however, a linear field gradient is applied in the x'-direction, the Larmor frequency of the water will increase *linearly* across the sample as a function of the x'-coordinate, thereby creating a one-dimensional profile, or spectrum, of the sample (Fig. 7.21).

To create a two-dimensional image, two gradients are applied along the x- and y-directions, and a series of one-dimensional images recorded in different directions in the xy-plane. A technique known as *back-projection*

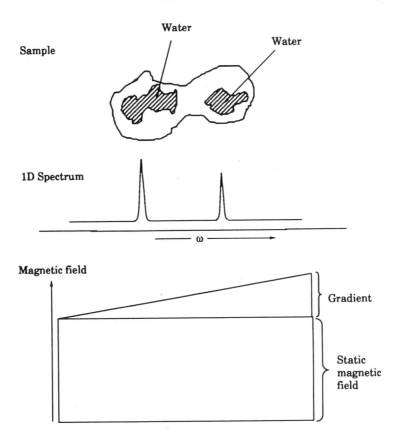

Figure 7.21 One-dimensional NMR imaging. When a magnetic field gradient is applied across a sample, it gives a spectrum that is a profile of the sample in the direction of the gradient.

reconstruction can then be used to combine the various one-dimensional images and create the two-dimensional image of the object (Fig. 7.22). The majority of imaging experiments used presently employ multidimensional Fourier transformation rather than back projection to create the two-dimensional images. The techniques can be extended to three-dimensions. NMR images of a section of a head and of a spinal cord are given in Fig. 7.23. By taking different cross-sections at different depths, it is easy to determine the location, shape, and size of, say, a tumor in the brain or in the spinal cord. The main observed species is water, with the image intensity being dependent on the concentration of water in the sample and also transverse and longitudinal relaxation times, or its diffusion coefficient.

7.7.2 Gradient-Accelerated Spectroscopy (GAS)—An Alternative to Phase Cycling

The pulses in a pulse sequence rotate the magnetization about a specific axis by a specific angle. However, this ideal behavior is, in practice, not

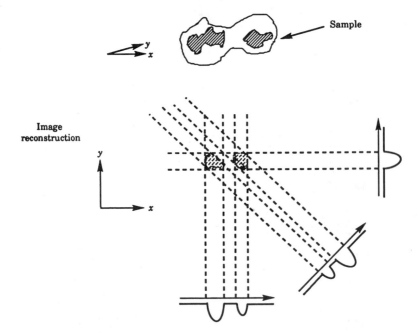

Figure 7.22 The principle of creating a two-dimensional NMR image. A number of profiles of the sample are obtained in different orientations in the presence of magnetic field gradients pointing in different directions (designated by arrows). The x′-gradient yields an x′-profile, and a y′-gradient generates a y′-profile. A combination of these profiles produces a two-dimensional image.

a

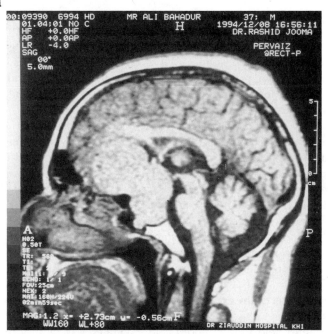

b

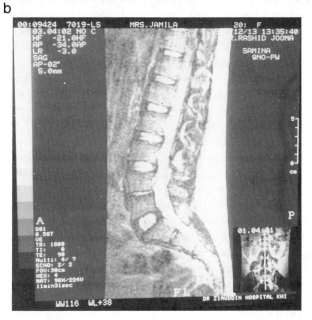

Figure 7.23 A 2D MRI of (a) a cross-section of the human head, (b) a cross-section of the human spinal cord. Different cross-sections can provide valuable information about the sizes and locations of tumors, etc. (Courtesy of Dr. I. H. Bhatti, Director, Jinnah Postgraduate Medical Center, Karachi.)

observed, due to Rf inhomogeneities across the sample, missetting of pulse angles, and resonance offset effects. This can bend the magnetization by a different angle than desired, and it can then follow a number of different coherence transfer pathways. That portion of the net magnetization that follows an unwanted transfer pathway may cause artifact peaks in multi-dimensional experiments. The traditional way of removing these artifact signals is by *phase cycling*. This involves repeating an experiment using different phases of some or all of the pulses (i.e., by applying the pulses along different axes). The resulting FIDs are combined so as to eliminate the signals that arise from undesired coherence transfer pathways.

Phase cycling procedures suffer from a number of drawbacks. The repetition of the experiment lengthens the total experimental time. Intense artifact signals are often not completely eliminated, due to imperfect phase cycling caused by instrumental instability, missetting of pulses, insufficient relaxation period between successive pulses (so that the magnetization does not return completely to its equilibrium value), sample spinning, etc. In 2D NMR experiments, each artifact peak may appear as ridges of "t_1-noise" that may run vertically down the spectrum, thereby concealing underlying cross-peaks.

Magnetic field gradients offer an excellent way of suppressing such artifact peaks. We have already mentioned that field gradients can cause the net magnetization to split into a number of different vectors, leading to the reduction in intensity of the signal and its eventual disappearance. The process can be reversed by applying a refocusing pulse. Both these features of field gradients can be used to advantage for eliminating artifact peaks. For instance, if we apply a 180° pulse at equilibrium, it should ideally cause the z-magnetization to invert and point toward the $-z$-axis with no residual transverse magnetization (i.e., magnetization in the $x'y'$-plane). However, if the 180° pulse is not accurate, then not all of it will be bent to the $-z$-axis, and there may be same residual transverse magnetization. This can be removed by applying a single field gradient pulse (*instead of phase cycling*) during the evolution period, which will cause any unwanted transverse magnetization to dephase and disappear. In an inversion recovery experiment, for instance, it would allow the experiment time to be halved, since only a single transient would be acquired at each τ value.

Field gradients can be employed to select a particular coherence pathway from a number of coherence transfer processes. This was traditionally done by incorporating suitable phase cycling procedures. However, it is possible to exploit the differing sensitivities of different orders of coherence to magnetic field gradients to select a specific coherence transfer process. We can also utilize the fact that a pulse that results in a coherence transfer process also acts as a refocusing pulse, and gives rise to a *transfer echo* (Ernst *et al.*, 1987). Coherence transfer-echos differ from conventional spin-echos in that the coherence that is rephased in the coherence transfer echo is different from what was dephased before it.

A useful feature of coherence transfer echos is that echos arising from different coherence transfer processes occur at different times. This is because different multiple-quantum coherences dephase at different rates, in contrast to single-quantum coherences, all of which dephase at the same rate, irrespective of how they have been generated. This allows a simple trick to be used utilizing magnetic field gredients in order to select the desired coherence transfer process: The gradient is switched off at when the particular coherence transfer echo is generated. Then the only magnetization detected is the one which has followed the coherence transfer process corresponding to the coherence transfer echo.

Let us illustrate this with a practical example. Suppose we wish to record a spectrum containing cross-peaks arising only from zero-quantum coherence. A single gradient pulse will then be applied during t_1, which

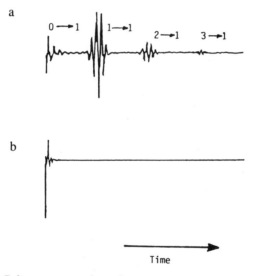

Time

Figure 7.24 (a) Coherence transfer echos formed after a dephasing period of 200 ms and a 90° coherence transfer pulse in the presence of a magnetic field gradient. The echos occurring at 0, 200, 400, and 600 ms in FID are due to the transfer of zero-, single-, double-, and triple-quantum coherences to single-quantum coherence. (b) Conventional free induction decay recorded under the same condition, for comparison. (Reprinted from *Chemical Reviews* **23**, T. J. Norwood, 59, copyright (1994), with permission from The Royal Society of Chemistry, Thomas Graham House, Science Park, Milton Road, Cambridge CB4 4WF, U.K.)

will result in the dephasing (and hence disappearance) of all magnetization except that generated from the zero-quantum coherence process. The spectrum will then contain only signals arising from the zero-to-single quantum coherence transfer process. If, however, we wish to record the double-to-single quantum coherence transfer process, we would have to apply the magnetic field gradient pulse during the detection period, which is twice as long as that applied during the evolution period (Ernst *et al.*, 1987). Figure 7.24 shows coherence transfer echos formed in the presence of a magnetic field gradient at dephasing time intervals of 200 ms. The echos occurring at 0, 200, 400, and 600 ms are due to transfer of zero-, single-, double-, and triple-quantum coherence to single-quantum coherence, respectively.

Figure 7.25 illustrates the power of magnetic field gradient pulses to eliminate unwanted coherences. The double-quantum filtered COSY spec-

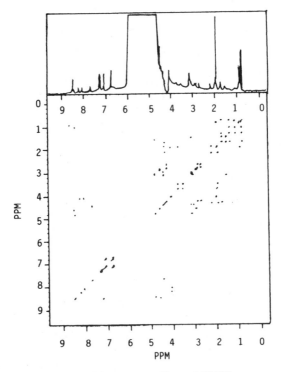

Figure 7.25 Homonuclear double-quantum filtered COSY spectrum (400 MHz) of 8-m*M* angiotensin II in H_2O recorded without phase cycling. Magnetic field gradient pulses have been used to select coherence transfer pathways. (Reprinted from *J. Mag. Reson.* **87,** R. Hurd, 422, copyright (1990), with permission from Academic Press, Inc.)

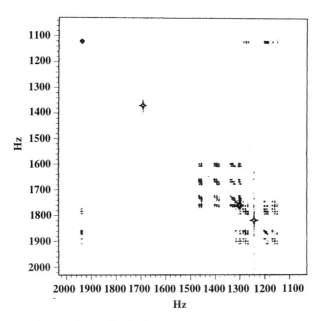

Figure 7.26 Gradient-enhanced TOCSY spectrum of 10 mM sucrose in D_2O recorded at 400 MHz by using modified MLEV-16 pulse sequence.

trum of angiotension II in water contains only cross-peaks arising from spins that can generate double-quantum coherence (i.e., those nuclei coupled through bonds to one or more nuclei). The large water signal present in the 1D spectrum would normally have resulted in a prominent ridge of t_1 noise if a phase cycling procedure had been adopted, which is totally absent, illustrating the striking advantage of magnetic field gradients. Gradient-accelerated NMR spectroscopy (Hurd and Freeman, 1989; Hurd, 1990a,b) thus provides an improved signal-to-noise ratio and cleaner spectra in a number of NMR experiments.

Gradient-enhanced 2D TOCSY spectrum of 10 mM of sucrose in D_2O is shown in figure 7.26. The clean spectrum obtainable without any t_1 noise and without the necessity of any phase cycling illustrates the power of this new technique in modern NMR spectroscopy.

REFERENCES

Atta-ur-Rahman, (1989). *One- and two-dimensional NMR spectroscopy*. Elsevier Science Publishers, Amsterdam.

Bax, A. (1984). *J. Magn. Reson.* **57**, 314.

Bax, A., and Freeman, R. (1982). *J. Am. Chem. Soc.* **104**, 109.

Bax, A., and Summers, M. F. (1986). *J. Am. Chem. Soc.* **108**, 209.

Bax, A., Freeman, R., and Kempsell, S. P. (1980). *J. Am. Chem. Soc.* **102**, 4849.

Berger, S. (1988). *Angew. Chem. Int. Ed. Engl.* **27**, 1196.

Berger, S. (1989). *J. Magn. Reson.* **81**, 561.

Bermel, W., Wagner, K., and Griesinger, C. (1989). *J. Magn. Reson.* **83**, 223.

Brüschweiler, R., Griesinger, C., Sörensen, O. W., and Ernst, R. R. (1988). *J. Magn. Reson.* **78**, 178.

Brüschweiler, R., Madsen, J. C., Griesinger, C., Sorensen, O. W., and Ernst, R. R. (1987). *J. Magn. Reson.* **73**, 380.

Cavanagh, J., Waltho, J. P., and Ernst, R. R. (1987a). *J. Magn. Reson.* **74**, 386.

Cavanagh, J., Waltho, J. P., and Keeler, J. (1987b). *J. Magn. Reson.* **74**, 386.

Crouch, R. C., Shockar, J. P., and Martin, G. E. (1990). *Tetrahedron Lett.* **31**, 5273.

Ernst, R. R., Bodenhausen, G., and Wokaun, A. (1987). *Principles of nuclear magnetic resonance in one and two dimensions.* Clarendon Press, Oxford.

Griesinger, C., Sörensen, O. W., and Ernst, R. R. (1987a). *J. Am. Chem. Soc.* **109**, 7227.

Griesinger, C., Sörensen, O. W., and Ernst, R. R. (1987b). *J. Magn. Reson.* **73**, 574.

Griesinger, C., Sörensen, O. W., and Ernst, R. R. (1989). *J. Magn. Reson.* **84**, 14.

Hore, P. J. (1983). *J. Magn. Reson.* **55**, 283.

Hurd, R. E., and Freeman, D. M. (1989). *Proc. Natl. Acad. Sci. U.S.A.* **86**, 4402.

Hurd, R. E. (1990a). *J. Mag. Reson.* **87**, 422.

Hurd, R. E. (1990b). *J. Magn. Reson.* **87**, 422.

Hurd, R. E., and Freeman, D. M. (1989). *Proc. Natl. Acad. Sci. U.S.A.* **86**, 4402.

Kessler, H., Anders, U., and Gemmecker, G. (1988a). *J. Magn. Reson.* **78**, 382.

Kessler, H., Anders, U., Gemmecker, G., and Steuernagel, S. (1989a). *J. Magn. Reson.* **85**, 1.

Kessler, H., Gemmecker, G., Haase, B., and Steuernagel, S. (1988b). *Magn. Reson. Chem.* **26**, 919.

Kessler, H., Mronga, S., and Gemmecker, G. (1991). *Magn. Reson. Chem.* **29**, 527–557.

Kessler, H., Mronga, S., Will, M., and Schmidt, U. (1990). *Helv. Chim. Acta* **73**, 25.

Kessler, H., Oschkinat, H., Griesinger, C., and Bermel, W. (1986). *J. Mag. Reson.* **70**, 106.

Kessler, H., Schmieder, P., Köck, M., and Kurz, M. (1990). *J. Magn. Reson.* **88**, 615.

Kessler, H., Will, M., Antel, J., Beck, H., and Sheldrich, G. M. (1989b). *Helv. Chim. Acta* **72**, 530.

Lauterber, P. (1973). *Nature* **242**, 190.

Montelione, G. T., and Wagner, G. (1990). *J. Magn. Reson.* **87**, 183.

Norwood, T. J. (1994) *Chem. Revs.* **23**, 59.

Oschkinat, H., and Bermel, W. (1989). *J. Magn. Reson.* **81**, 220.

Oschkinat, H., Cieslar, C., Gronenborn, A. M., and Clore, G. M. (1989a). *J. Magn. Reson.* **81**, 212.

Oschkinat, H., Cieslar, C., Holak, T. A., Clore, G. M., and Gronenborn, A. M. (1989b). *J. Magn. Reson.* **83**, 450.

Oschkinat, H., Clore, G. M., and Gronenborn, A. M. (1988a). *J. Magn. Reson.* **78**, 371.

Oschkinat, H., Griesinger, C., Kraulis, P. J., Sörensen, O. W., Ernst, R. R., Gronenborn, A. M., and Clore, G. M. (1988b). *Nature (London)* **332**, 374.

Otting, G., Orbons, L. P. M., and Wüthrich, K. (1990). *J. Magn. Reson.* **89**, 423.

Pachler, K. G. R., and Wessels, P. L. (1973). *J. Magn. Reson.* **12**, 337.

Shaka, A. J., Barker, P. B., and Freeman, R. (1985). *J. Magn. Reson.* **64**, 547.

Wagner, G. (1984). *J. Magn. Reson.* **57**, 497.

Wagner, G., and Wüthrich, K. (1979). *J. Magn. Reson.* **33**, 675.

CHAPTER 8

Logical Protocols for Solving Complex Structural Problems

The following is a procedure recommended for elucidating the structure of complex organic molecules. It uses a combination of different NMR and other spectroscopic techniques. It assumes that the molecular formula has been deduced from elemental analysis or high-resolution mass spectrometry. Computer-based automated or interactive versions of similar approaches have also been devised for structural elucidation of complex natural products, such as SESAMI (systematic elucidation of structures by using artificial machine intelligence), but there is no substitute for the hard work, experience, and intuition of the chemist.

(a) The first step in elucidating structure is to find a secure molecular formula by high-resolution mass spectrometry and to calculate the degree of unsaturation or double-bond equivalents (DBEs). An acyclic saturated hydrocarbon has the formula C_nH_{n+2}. Each double bond or ring represents a degree of unsaturation, while triple bonds represent two degrees of unsaturation. The presence of oxygen or other divalent elements bound by a single bond does not affect the value of DBEs. One proton is subtracted for each trivalent atom (such as nitrogen), while a monovalent atom, such as chlorine, can be treated as a proton for purposes of calculation. A simple formula to deduce the double-bond equivalents is $(C + 1) - H/2 + W/2$, where W is nitrogen.

(b) Inspect the infrared and ultraviolet spectra to get preliminary clues about the functional groups and conjugation that might be present in the molecule.

(c) A cursory examination of the ^{1}H- and ^{13}C-NMR spectra gives an idea about the ratio of the aromatic to the aliphatic carbons. The ^{1}H-NMR spectrum can also be used to gain additional evidence to confirm the molecular formula (indirectly by counting the protons), and the ^{1}H- and ^{13}C-NMR spectra contain valuable information about the type of unsaturation (e.g., the number of vinylic carbons and protons).

(d) A good 1D ^{1}H-NMR spectrum is always an important starting point for structure elucidation. First, assign each proton appearing in the spectrum a letter, starting from the left side of the spectrum, each proton being labeled individually as A, B, C, etc. If a peak or a group of overlapping peaks represents more than one proton, then assign combinations of letters. Thus, four *overlapping* protons can be labeled as a group, such as DEFG. The total number of protons labeled on the spectrum should correspond to the number of protons determined from the high-resolution mass measurement of the molecular ion. In case of symmetrical molecules with identical protons, say, in symmetrical halves of the same molecule, a fraction of the actual number of protons present may be observed.

(e) Once it has been established through preliminary NMR and mass spectroscopic studies that the substance is new and requires further work, record the ^{13}C-NMR (BB and DEPT or GASPE), COSY-45°, HMBC (or COLOC, if inverse facilities are not available), HMQC (or HETCOR, if inverse facilities are not available), HOHAHA (20, 40, 80, and 120 ms) (or delayed COSY/relayed COSY, if HOHAHA cannot be performed), and NOE difference spectra.

(f) Label each contour *on the diagonal* line of the COSY-45° spectrum with the same letters as used in the 1D ^{1}H-NMR spectrum, and deduce the ^{1}H–^{1}H connectivities from the corresponding cross-peaks.

(g) Next, label the HMQC spectrum with the same letters at the ^{1}H chemical shifts as those already assigned in the ^{1}H-NMR spectrum. This establishes the one-bond ^{1}H/^{13}C connectivities. This should be done on the one-bond hetero-COSY (HETCOR) spectrum if the HMQC experiment cannot be performed due to instrumental constraints. Note that if there are two cross-peaks at any particular carbon chemical shift in the HMQC spectrum, then these two protons must be the two *geminal* protons attached to that particular carbon. Prepare a table of chemical shifts of the ^{13}C nuclei and their corresponding attached ^{1}H nuclei. By deleting the *geminal coupling interactions* from the COSY spectrum, we arrive at the vicinal ^{1}H/^{1}H connectivities (caution: Some weaker long-range couplings, such as "W" coupling interactions, may also appear in the COSY spectrum). Since we already know the direct (one-bond) carbon–hydrogen connectivities of adjacent protonated carbons, *the carbon–carbon connectivities of protonated carbons* can be deduced. This is the "pseudo-INADEQUATE" information that has been derived.

(h) Now select one unambiguously identifiable proton (preferably in an unclustered region of the spectrum) that shows coupling in the COSY spectrum with other protons, and work out the $^1H/^1H$ connectivities from this "end of the spin network chain." Repeat this process, starting from some other proton in another coupled spin network in the molecule, in order to arrive at other "spin networks" of protons and their vicinal $^1H/^1H$ connectivities (cross-peaks due to geminal couplings have already been identified and deleted from consideration in step g).

(i) Spin networks can also be readily determined from the HOHAHA (120-ms) spectrum, in which all the protons in a coupled network (*irrespective of whether they are directly coupled to each other or not*) appear on the same horizontal lines. HOHAHA spectra therefore differ from long-range COSY spectra. Since in long-range COSY spectra contain cross-peaks only if the protons are coupled to one another, whereas in HOHAHA spectra their presence in the same spin network is all that is required for cross-peaks to appear. The relative intensities of cross-peaks within a spin network may provide some information about the relative distances of the coupled protons within that spin network. This can be done by comparing the HOHAHA spectra recorded with different mixing delays (say, 20, 40, 80, 120 ms). Delayed COSY or relayed COSY spectra may be recorded if instrumental constraints do not allow measurement of HOHAHA spectra, and some information regarding long-range $^1H/^1H$ couplings gained from them.

(j) Delete the direct connectivities (as determined by the COSY spectrum) from the connectivities found in the HOHAHA spectrum to arrive at the *distant connectivities* (this procedure is better than the conclusions reached on the basis of the "delayed COSY" spectrum, since it gives you *all* the protons in each coupled network). Thus, by combining the COSY and HOHAHA data, we can obtain several groups of coupled protons. Hence, by taking into consideration the $^1H/^{13}C$ one-bond connectivities obtained from the HMQC spectrum, we can arrive at protonated carbon connectivities. These fragments must be joined at their ends either to quaternary carbons or to heteroatoms (e.g., N, O, S).

(k) In order to connect these fragments to complete the structure, we can join them together on the basis of long-range $^1H/^{13}C$ connectivities obtained in the HMBC spectrum (or COLOC spectrum). These fragments should be joined together in a way that explains the observed $^2J_{CH}$, $^3J_{CH}$, and $^4J_{CH}$ interactions.

Alternatively, the fragments (comprising protonated carbons) thus obtained can be joined together on the basis of nOe difference measurements. NOE difference measurements of protons in different protonated carbon fragments can help us to determine the spatial proximities of protons if the fragments lie close to each other, and hence help in joining the fragments together. If large sample amounts are available and when other methods have failed, the 2D INADEQUATE experiment may be used to obtain the C–C connectivity information directly.

The following is a summary of the protocol just described.

(i) Obtain a secure molecular formula, and calculate the degree of unsaturation.
(ii) Record a 1D ¹H-NMR spectrum, and assign each proton a letter.
(iii) Record the HMQC spectrum (or a one-bond HETCOR spectrum), and label each at the respective chemical shifts with the corresponding letters from the 1D spectrum.
(iv) Label the protons along the diagonal line of the COSY-45° spectrum with the same letters as in the ¹H-NMR spectrum. Deduce vicinal ¹H/¹H connectivities from the cross-peaks in the COSY-45° spectrum by deleting all the coupling interactions from geminal protons (geminal protons are readily identified from the HMQC or HETCOR spectrum).
(v) Identify the various spin networks obtained from the HOHAHA spectra (20, 40, 60, 120 ms). Alternatively, deduce them from the COSY and long-range HETCOR spectra.
(vi) Using HMBC (or long-range HETCOR) and NOE difference spectra, connect the fragments that this procedure has created.

The following are examples of structure elucidation of natural products carried out in our lab by using this procedure.

8.1 3α-HYDROXYLUPANINE (1)

Our first example is 3α-hydroxylupanine (**1**), a new lupine alkaloid isolated by us from *Leontice leontopetalum* (Taleb *et al.*, 1991). Each proton (or sets of close-lying protons) of 3α-hydroxylupanine (**1**) were first assigned letters (or sets of letters, if more than one proton overlapped in an area), on the ¹H-NMR spectrum (Fig. 8.1). Thus, the farthest downfield proton is assigned letter A, and protons upfield to proton A are designated B, C, D, etc. In the case of overlapping multiplets, e.g., at δ 1.86 (integration 4H), a group of letters (KLMN) was assigned. When all the protons in the ¹H-NMR spectrum had been labeled, we examined the hetero-COSY spectrum (Fig. 8.2), in order to determine which protons were attached directly to which carbon atoms by observing the one-bond C/H interactions.

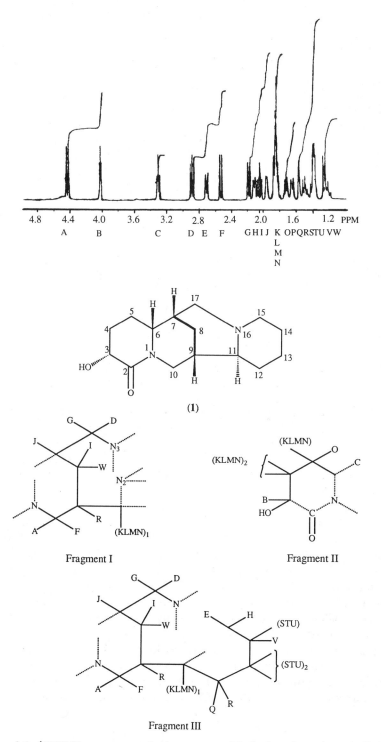

Figure 8.1 ¹H-NMR spectrum and substructures of 3α-hydroxylupanine, an alkaloid isolated from *Leontice leontopetalum*. The fragments, I, II, and III are labeled with the alphabetic assignment of every proton or set of close-lying protons.

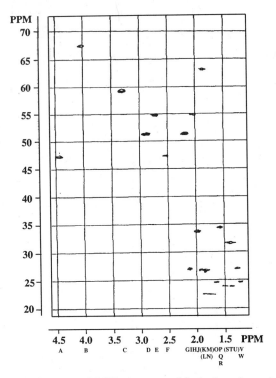

Figure 8.2   Hetero-COSY spectrum of 3α-hydroxylupanine.

The hetero-COSY results were then tabulated by writing the letters and ^{13}C-NMR values together, with the data for CH, CH$_2$, and CH$_3$ groups being placed in *separate* columns (Table 8.1).

Table 8.1 therefore provides us the direct connectivities between the protons and the carbons to which they are attached in one glance. For instance the carbon which resonates at δ 47.23 (C-10) is attached to the protons A (δ 4.52) and F (δ 2.53), whereas the carbon resonating at δ 33.66 (C-9) is coupled to proton J (δ 1.96). Similarly, the carbon at δ 51.31 (C-17) shows cross-peaks with protons D (δ 2.89) and G (δ 2.26), whereas the carbon at δ 34.27 (C-9) is coupled to proton R (δ 1.58).

Once all the direct C/H connectivity relationships had been determined, we examined the COSY-45° spectrum (Fig. 8.3). Since all of the protons of methylene groups in 3α-hydroxylupanine (**1**) were already identified earlier from the hetero-COSY spectrum, the corresponding *geminal* coupling interactions could readily be determined. By deleting such geminal coupling interactions, the COSY-45° spectrum can be simplified greatly,

Table 8.1

Correlations of Methylene and Methine Protons with
Attached Carbons

CH Protons (δ_c of CH)	CH$_2$ Protons (δ_c of CH$_2$)
B (67.5)	A, F (47.2)
C (59.2)	D, G (51.3)
	Q, P (51.3)
J (33.6)	STU—2H (23.8)
R (34.3)	E, H (54.8)
	STU—1H,W (24.5)
KLMN—1H (63.2)	K, V (26.9)
	KLMN—2H (36.7)
	KLMN—1H,O (22.4)

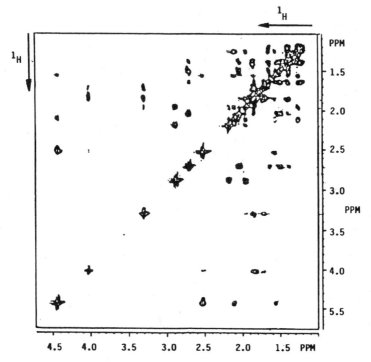

Figure 8.3 COSY-45° spectrum of 3α-hydroxylupanine.

and it allows the vicinal coupling interactions to be seen with greater clarity, since what is of interest at this stage is to deduce C–C connectivity information of protonated carbon atoms from the vicinal proton coupling interactions. Since we had already established the $^1H/^{13}C$ one-bond connectivities from the HETCOR spectrum (Table 8.1), a combination of these two sets of data told us which protonated carbons are bound to one another.

The COSY-45° spectrum (Fig. 8.3) of 3α-hydroxylupanine (1) showed that one of the protons of a CH_2 group (with ^{13}C chemical shift of δ 47.22) — i.e., proton A—was coupled to proton R (δ 1.58). The latter was attached to the methine carbon having a chemical shift of δ 34.27, indicating that the CH_2 carbon (with $δ_C$ = 47.22) was adjacent to a CH carbon (with $δ_C$ = 34.27). Similarly, proton R showed cross-peaks with protons W (δ 1.28) and I (δ 2.12) as well as with one of the protons present in the 4H multiplet (KLMN) at δ 1.88, leading to fragment I. Other important cross-peaks were shown by proton J (δ 1.96) with protons D (δ 2.89), G (δ 2.26), I (δ 2.12), and W (δ 1.28), confirming the substructure of fragment I (Fig. 8.1).

Similarly, proton B (δ 4.02) was coupled to one of the protons in the 4H multiplet labeled KLMN (δ 1.85), whereas proton C (δ 3.31) was coupled to proton O (δ 1.72) as well as to one of the protons present in the 4H multiplet KLMN (δ 1.85). The various coupling interactions of 3α-hydroxylupanine (1) obtained by considering both hetero-COSY and COSY-45° data led to deducing the two fragments II and III (Fig. 8.1).

The spin systems in 3α-hydroxylupanine were investigated further by applying the HOHAHA experiment (Fig. 8.4) recorded, with a mixing time of 100 ms. The various coupling connectivities present in the HOHAHA spectra not only confirmed their assignments but also helped to join fragments I and II together (fragment III). The spectrum obtained, with a mixing time of 20 ms, resembled closely the COSY-45° spectrum, showing mainly direct connectivities, while with longer mixing intervals the magnetization was seen to spread to more distant protons within individual spin systems. This spreading of the magnetization with the increase of the mixing delay can be seen in the projections of the HOHAHA spectra at δ 4.52 and 3.39 (Fig. 8.5). These considerations led to structure **I** for 3α-hydroxylupanine.

8.2 (+)-BUXALONGIFOLAMIDINE (2)

A steroidal alkaloid, (+)-buxalongifolamidine (2) was isolated from the leaves of *Buxus longifolia* Boiss of Turkish origin (Atta-ur-Rahman *et al.*, 1993). The HREIMS of 2 showed the molecular ion at m/z 578.3533 corresponding to the molecular formula $C_{35}H_{50}N_2O_5$ representing twelve

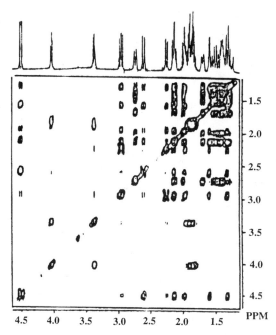

Figure 8.4 HOHAHA spectrum of 3α-hydroxylupanine recorded at 100 ms.

degrees of unsaturation in the molecule. The compound showed a peak at m/z 563, resulting from the loss of a methyl group from the M⁺ ion.

The ¹H-NMR spectrum (Figure 8.6) of **2** revealed the presence of three tertiary methyl groups at δ 0.74, 0.93, and 1.16, and the secondary methyl

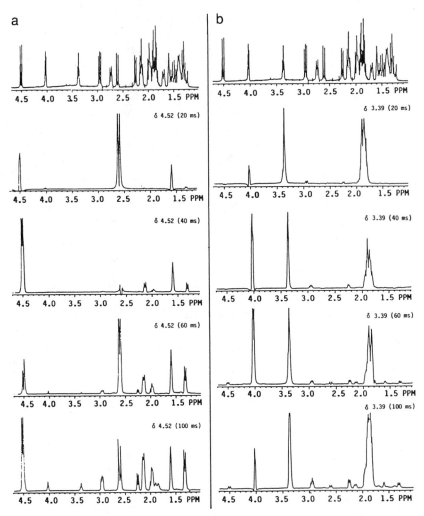

Figure 8.5 The slice plots taken at (a) δ 4.52 and (b) δ 3.39 of HOHAHA spectra of 3α-hydroxylupanine.

group resonated as a doublet at δ 1.25 ($J_{20,21}$ = 6.6 Hz). These observations favor a cycloartenol-type triterpenoidal skeleton, as present in all *Buxus* alkaloids. An AB doublet resonating at δ 3.65 and 3.70 ($J_{31\alpha,31\beta}$ = 8.7 Hz) was assigned to the C-31 hydroxymethylenic protons. A doublet at δ 5.85 ($J_{1,2}$ = 10.3 Hz) due to the C-1 vinylic proton showed coupling with the C-2 vinylic proton resonating at δ 5.87 as a doublet of a doublet ($J_{1,2}$ = 10.3 Hz, $J_{2,3}$ = 3.3 Hz). The C-2 vinylic proton, on the other hand, showed

COSY cross-peaks with the C-3 methine proton. The C-3 proton resonated as a doublet of double doublets at δ 4.80 ($J_{3,NH}$ = 10.0 Hz, $J_{3,2}$ = 3.3 Hz, $J_{3,5}$ = 1.7 Hz), showing cross-peaks with the C-2 vinylic, benzamidic NH, and W coupling with the C-5 methine proton in the COSY-45° spectrum (Fig. 8.7). These observations suggested the presence of fragment I, including ring A of the skeleton.

I

The C-19 methylene protons appeared as a broad singlet at δ 2.72 and showed weak allylic interactions with the vinylic C-11 proton that appeared at δ 5.45 as a multiplet and itself showed vicinal coupling with the C-12 methylene protons. This trisubstituted olefinic bond when tailored into ring C of the skeleton gives rise to fragment II.

A multiplet centered at δ 4.90 was assigned to the C-16 proton, which exhibited COSY interactions with the C-15 methylene and C-17 methine protons. Since an acetoxy group was inferred from IR absorptions and the presence of a singlet at δ 1.90 in the 1D ^{1}H-NMR spectrum, C-16 seemed to be a plausible site of its substitution in ring D of the skeleton (fragment III).

II   III

Other major peaks in the ^{1}H-NMR spectrum were a six-proton singlet at δ 2.03 assigned to the N(CH$_3$)$_2$ protons and groups of three- and two-proton multiplets centered at δ 7.50 and 7.85 due to the aromatic protons of the benzamidic moiety substituted on C-3 of ring A.

The ^{13}C-NMR spectra (CDCl$_3$) (broad-band and DEPT) of compound 2 showed resonances for all 35 carbons. A notable feature of the ^{13}C-NMR

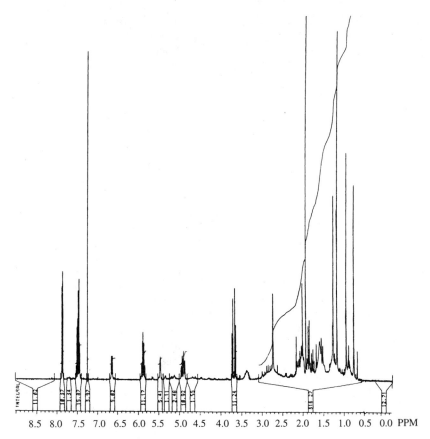

Figure 8.6 ¹H-NMR spectrum of buxalongifolamidine.

broad-band decoupled spectrum was the appearance of downfield signal for a quaternary carbon at δ 79.5, which was assigned to C-10, bearing a hydroxy group. The chemical shifts for various carbons are listed in Table 8.2. One-bond correlations between ¹H- and ¹³C-nuclei were established by inverse ¹H-detected HMQC experiment (Fig. 8.8); the results are presented in Table 8.2.

The ¹H/¹³C long-range coupling information obtained from the inverse HMBC experiment (Fig. 8.9) allowed the various fragments to be connected together. The proton at δ 3.65 (C-31αH) showed $^2J_{CH}$ interaction with C-4 (δ 45.1) and $^3J_{CH}$ interactions with C-3 (δ 56.1) and C-5 (δ 53.6). The C-31β proton (δ 3.70) showed $^2J_{CH}$ interaction with

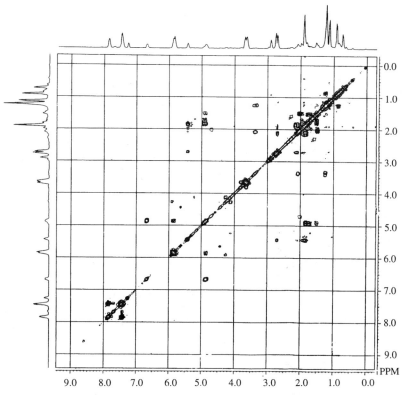

Figure 8.7 COSY 45° spectrum of buxalongifolamidine.

Table 8.2

$^1H/^{13}C$-NMR Connectivities (from HMQC) for **2**

Carbon	Chemical shift (δ)	Multiplicity (DEPT)	$^1H/^{13}C$ Connectivity δ (J = Hz)
1	132.5	CH	5.85 (d, J = 10.3)
2	131.7	CH	5.87 (dd, J = 10.3, 3.3)
3	56.1	CH	4.80 (ddd, J = 10.0, 3.3,1.7)
4	45.1	C	—
5	53.6	CH	2.68 (m)
6	29.7	CH_2	1.25 (m)
7	34.3	CH_2	1.81 (m)
8	41.9	CH	2.00 (m)
9	133.6	C	—
10	79.5	C	—
11	124.0	CH	5.45 (m)
12	37.3	CH_2	1.85 (m)
13	46.2	C	—
14	47.7	C	—
15	43.8	CH_2	1.50 (m)
16	75.1	CH	4.90 (m)
17	56.5	CH	2.05 (m)
18	16.8	CH_3	0.93 (s)
19	44.3	CH_2	2.72 (s)
20	65.4	CH	3.40 (m)
21	11.7	CH_3	1.25 (d, J = 6.6)
30	19.9	CH_3	1.16 (s)
31	78.6	CH_2	3.65 (d, J = 8.7) 3.70 (d, J = 8.7)
32	17.3	CH_3	0.74 (s)
N_b-CH_3	35.8	CH_3	2.03 (s)
N_b-CH_3	44.1	CH_3	2.03 (s)
1'	135.3	C	—
2'	127.4	CH	7.85 (m)
3'	128.5	CH	7.48 (m)
4'	130.3	CH	7.50 (m)
5'	128.5	CH	7.48 (m)
6'	127.4	CH	7.85 (m)
Ph-CO-N	167.5	C	—
CO-$\underline{C}H_3$	21.2	CH_3	1.90 (s)
$\underline{C}$O-CH_3	171.2	C	—

Table 8.3
Long Range ^{1}H/^{13}C Connectivities (from HMBC) for
Buxalongifolamidine (2) in CDCl$_3$

^{1}H (δ)	$^2J_{CH}$ (δ)	$^3J_{CH}$ (δ)	$^4J_{CH}$ (δ)
7.85 (H-2',6')	128.5 (C-3')	167.5 (amide C=O)	
5.87 (H-2)		79.5 (C-10)	
		45.1 (C-4)	
5.85 (H-1)	79.5 (C-10)	53.6 (C-5)	
		44.3 (C-19)	
4.90 (H-16)	171.2 (C-CH$_3$)	46.2 (C-13)	
	56.5 (C-17)		
4.80 (H-3)	131.7 (C-2)		79.5 (C-10)
3.70 (H-31βH)	45.1 (C-4)	56.1 (C-3)	79.5 (C-10)
		19.9 (C-30)	
3.65 (H-31αH)	45.1 (C-4)	56.1 (C-3)	
		53.6 (C-5)	
2.72 (H-19)	133.6 (C-9)	41.9 (C-8)	
	79.5 (C-10)	53.6 (C-5)	
1.81 (H-7)	41.9 (C-8)	17.3 (C-32)	
		133.6 (C-9)	
1.50 (H-15)	75.1 (C-16)	56.6 (C-17)	
	47.7 (C-14)	41.9 (C-8)	
1.25 (H-21)	65.4 (C-20)	56.6 (C-17)	
1.16 (H-30)	78.6 (C-31)	56.1 (C-3)	
	45.1 (C-4)	53.6 (C-5)	

C-4, $^3J_{CH}$ interactions with C-3 (δ 56.1) and C-30 (δ 19.9), and $^4J_{CH}$ interaction with C-10 (δ 79.5). The protons at δ 2.72 (C-19) showed $^2J_{CH}$ coupling with C-9 (δ 133.6) and C-10 (δ 79.5) and $^3J_{CH}$ connectivities between C-5 (δ 53.6) and C-19 protons were also observed. The C-2 protons (δ 5.87) showed $^3J_{CH}$ couplings with C-10 (δ 79.5) and C-4 (δ 45.1). The proton at δ 1.16 (C-30) showed $^3J_{CH}$ connectivities with C-3 (δ 56.1). Other connectivities are shown in Table 8.3. HOHAHA Spectrum (100 ms) (Fig. 8.10) of buxalongifolamidine further confirmed the ^{1}H-^{1}H connectivities within each spin-system.

On the basis of these studies, structure 2 was assigned to (+)-buxalongi-folamidine.

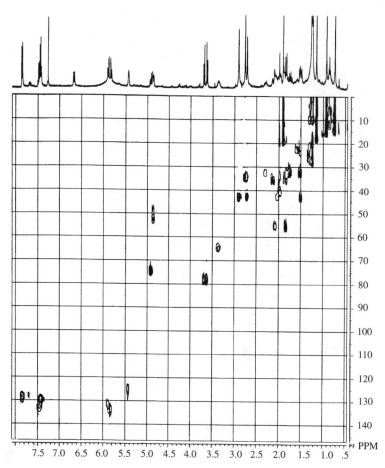

Figure 8.8   HMQC spectrum of buxalongifolamidine.

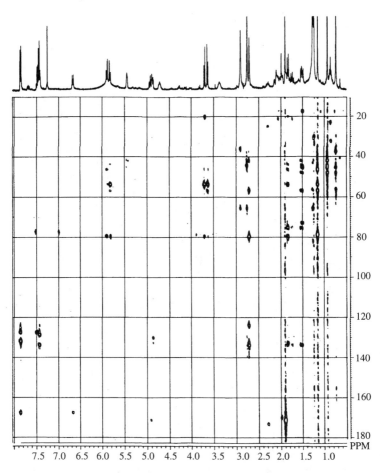

Figure 8.9 HMBC spectrum of buxalongifolamidine.

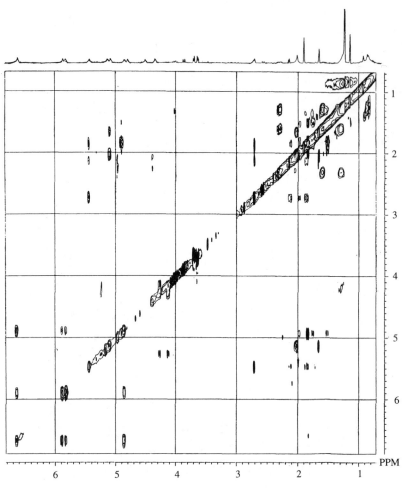

Figure 8.10   HOHAHA spectrum (100 ms) of buxalongifolamidine.

REFERENCES

Atta-ur-Rahman, Noor-e-ain, F., Ali, R. A., Choudhary, M. I., Pervin, A., Turkoz, S., and Sener, B. (1993). *Phytochemistry* **32**(4), 1059–1063.

Gray, N. A. B. (1986). *Computer-assisted structure elucidation.* Wiley, New York.

Taleb, H. Al-Tel., Sabri, S. S., Abu Zarga, M. H., Pervin, A., Shah, Z., Atta-ur-Rahman, and Rycroft, D. S. (1991). *Phytochemistry* **30**(7), 2393.

Glossary

Absolute-value-mode spectrum: The spectrum is produced by recording the square root of the sum of the squares of the real (R) and imaginary (I) parts of the spectrum ($R^2 + I^2$).

Absorption-mode spectrum: The spectrum in which the peaks appear with Lorentzian line shapes. NMR spectra are normally displayed in absolute-value mode.

Acquisition time: The time taken to digitize the free induction decay. In 2D NMR experiments this takes place during time t_2 when the FID is "acquired".

Aliasing: False peaks produced that originate outside the spectral window, when the digitization rate is less than twice the Nyquist frequency.

Analog-to-digital converter: The computer hardware that converts the voltage from the detector into a binary number, usually 12 or 16 bits long.

Antiphase: The phases opposite to that of an individual line in a multiplet.

Apodization: Applying weighting functions for enhancing sensitivity or resolution.

Artifacts: False peaks, noise, or other unwanted signals in the NMR spectrum.

Band-pass filter: A filter used for selecting the desired coherence order, p.

Carrier frequency: The transmitter frequency that consists of high-frequency pulses.

Chemical shift: The difference between the nuclear precession frequency and the carrier frequency.

Coherence: A condition in which nuclei precess with a given phase relationship and can exchange spin states via transitions between two eigenstates. Coherence may be zero-quantum, single-quantum, double-quantum, etc., depending on the ΔM_z of the transition corresponding to the coherence. Only single-quantum coherence can be detected directly.

Coherence order, p: The difference in the magnetic quantum number, M_z of the two energy levels connected by the same coherence.

Coherence pathway selection: Choosing a corresponding phase cycle so that undesired resonances are suppressed and only the desired magnetization is observed.

Coherence transfer: The transfer of coherence between two different types of nuclei.

Composite pulse: A composite "sandwich" of pulses that replaces a single pulse; employed to compensate for B_1 field inhomogeneities, phase errors, or offset effects.

Constant-time experiment: An NMR experiment in which the total duration of t_1 is kept constant.

Continuous wave: A method of recording an NMR spectrum in which the field B_1 is applied continuously and either the magnitude of B_0 or the radiofrequency is varied so that the nuclei are brought successively into resonance.

Contour plot: The most widely applied method for recording 2D spectra. The peaks appear as concentric circular lines (*contours*), with the two axes of the 2D spectrum representing chemical shifts or coupling constants.

Convolution: The random noise in a spectrum can be smoothed out by the application of convolution functions. The NMR spectrum is sampled at regular frequency intervals, and each ordinate in the resulting new spectrum is multiplied by the corresponding ordinate on the convoluting function. The sum of the products is normalized and plotted.

Convolution difference: To emphasize the narrow peaks in the spectrum of a macromolecule and to suppress the broad resonances, we can apply a severe broadening function that is calculated to affect the narrow lines much more than the broad lines. This causes the narrow peaks to broaden, and the resulting spectrum is subtracted from the original spectrum to give the convolution difference spectrum, in which each narrow line appears on shallow depressions.

Correlation time (τ_c): A parameter related to the mean time during which a molecule maintains its spatial geometry between successive rotations. For internuclear vectors, τ_c is approximately equal to the average time for it to rotate through an angle of 1 radian.

COSY (H–H correlation spectroscopy): An important 2D experiment that allows us to identify the protons coupled to one another.

Cross-peaks: The off-diagonal peaks in a 2D experiment that appear at the coordinates of the correlated nuclei.

Cross-relaxation: The mutual intermolecular or intramolecular relaxation of magnetically equivalent nuclei, e.g., through dipolar relaxation. This forms the basis of nOe experiments.

CYCLOPS: A four-step phase cycle that corrects *dc* imbalance in the two channels of a quadrature detector system.

Dead time: A very short delay introduced before the start of acquisition that allows the transmitter gate to close and the receiver gate to open.

Density matrix: A description of the state of nuclei in quantum mechanical terms.

DEPT (distortionless enhancement by polarization transfer): A one-dimensional ^{13}C-NMR experiment commonly used for spectral editing that allows us to distinguish between CH_3, CH_2, CH, and quaternary carbons.

Detectable magnetization: The magnetization precessing in the $x'y'$-plane induces a signal in the receiver coil that is detected. Only single-quantum coherence is directly detectable.

Detection period: The FID is acquired in the last segment of the pulse sequence. This is the detection period, and, in 2D experiments, this is the t_2 domain.

Diagonal peaks: Cross-sections of peaks that appear on or near a diagonal line in a 2D spectrum. The projection produces the 1D spectrum. They give no shift-correlation information.

Diamagnetism: Substances that have no unpaired electrons are diamagnetic. When placed in an applied magnetic field, their induced magnetic fields oppose the applied magnetic field.

Digital resolution: The distance, in hertz, between successive data points; determines the extent of peak definition by the data points.

Digitization: The process of converting voltage by an analog-to-digital converter (ADC) to a corresponding binary number for digital processing; used to represent free induction decay in digital form.

Digitization rate: The rate that represents the number of analog-to-digital conversions per second during data acquisition; must be twice the highest signal frequency.

Dipolar coupling: The direct through-space coupling interaction between two nuclei. It is responsible for nOe and relaxation, and represents the main mechanism for spin-lattice relaxation of spin ½ nuclei in diamagnetic liquids.

Dispersion mode: A Lorentzian line shape that arises from a phase-sensitive detector (which is 90° out of phase with one that gives a pure-absorption-mode line). Dispersion-mode signals are dipolar in shape and produce long tails. They are not readily integrable, and we need to avoid them in a 2D spectrum.

Double-quantum coherence: Coherence between states that are separated by magnetic quantum numbers of ± 2. This coherence cannot be detected directly, but must be converted to single-quantum coherence before detection.

Dynamic range: Range determined by the number of bits in the output of the analog-to-digital converter; represents the ratio between the largest and smallest signals observable in the digitized spectrum.

Editing: Obtaining a given set of subspectra to supply some desired information, e.g., the multiplicity of signals.

Equilibration delay: A time period introduced between successive scans to allow the nuclei to relax to their respective equilibrium states (i.e., to become realigned along the z-axis).

Ernst angle: The angle by which the nuclear magnetization vector must be tipped in order to give the best signal-to-noise ratio for a given combination of spin-lattice relaxation time and the time between successive pulses.

Evolution period: The delay time in the second segment of a pulse sequence, during which the nuclei precess under the influence of chemical shift and/ or spin coupling; incremented successively in 2D experiments, thereby producing the t_1 domain.

F_1 axis: The axis of a 2D spectrum resulting from the Fourier transformation of the t_1 domain signal.

F_2 axis: The axis of a 2D spectrum resulting from the Fourier transformation of the transposed digitized FIDs (t_2 domain).

Fast Fourier transform (FFT): A procedure for carrying out Fourier transformation at high speed, with a minimum of storage space being used.

Field frequency lock: The magnitude of the field B_0 is stabilized by locking onto the fixed frequency of a resonance in the solution, usually ^2H of the solvent.

Field gradients: In an inhomogenous magnetic field, the magnetization will become subdivided into a number of smaller vectors, each of which will precess at a slightly differing frequency. These smaller vectors spread out (dephase) during the evolution period, thereby reducing the net transverse magnetization of the sample. Such field gradients broaden the NMR signals and lead to a more rapid disappearance of the FID.

First order multiplet: First order multiplets are those in which the difference in chemical shifts of the coupled protons, in hertz (Hz), divided by the coupling constant between them is about 8 or more ($\Delta v/_J > 8$). These multiplets have a symmetrical disposition of the lines about their midpoints. The distance between the two outermost peaks of a first order multiplet is the sum of each of the coupling constants.

Flip angle: The angle by which a vector is rotated by a pulse.

Fourier transformation: A mathematical operation by which the FIDs are converted from time-domain data to the equivalent frequency-domain spectrum.

Free induction decay: A decay time-domain beat pattern obtained when the nuclear spin system is subjected to a radiofrequency pulse and then allowed to precess in the absence of Rf fields.

Frequency spectrum: A plot of signal amplitude versus frequency, produced by the Fourier transformation of a time-domain signal.

Gated decoupling: The decoupler is "gated" during certain pulse NMR experiments, so spin decoupling occurs only when the decoupler is switched on and not when it is switched off; used to eliminate either $^1H-^{13}C$ spin-coupling or nuclear Overhauser effect in a 1D ^{13}C spectrum, and employed as a standard technique in many other ^1H-NMR experiments, such as APT and J-resolved.

Heteronuclear correlation spectroscopy (HETCOR): Shift-correlation spectroscopy in which the chemical shifts of different types of nuclei (e.g., ^1H and ^{13}C) are correlated through their mutual spin-coupling effects. These correlations may be over one bond or over several bonds.

High-pass filter: A filter used to select a coherence order $\geq p$.

INADEQUATE: Correlation spectroscopy in which the directly bonded ^{13}C nuclei are identified.

Increment: A short, fixed time interval by which time interval t_1 is repeatedly increased in a 2D NMR experiment.

INEPT (insensitive nuclei enhanced by polarization transfer): Polarization transfer pulse sequence used to record the NMR spectra of insensitive nuclei, e.g., ^{13}C, with sensitivity enhancement; may be used for spectral editing.

In-phase: A term applied to a multiplet structure when the individual lines of the multiplet have the same phase.

Inverse experiments: Heteronuclear shift-correlation experiments in which magnetization of the less sensitive heteronucleus (e.g., ^{13}C) is detected through the more sensitive magnetization (e.g., ^1H).

Inversion recovery: A pulse sequence used to determine spin-lattice relaxation times.

J-scaling: A method for preserving a constant scaling factor for all ^{13}C multiplets.

J-spectroscopy: A 2D experiment in which the chemical shifts are plotted along one axis (F_2 axis) while the coupling constants are plotted along the other axis (F_1 axis). This is an excellent procedure for uncoding overlapping multiplets.

Laboratory frame: The Cartesian coordinates (x, y, and z) are stationary with respect to the observer, in contrast to the rotating frame, in which they rotate at the spectrometer frequency.

Larmor frequency: The nuclear precession frequency about the direction of B_0. Its magnitude is given by $\gamma B_0/2\pi$.

Longitudinal magnetization: This is the magnetization directed along the z-axis parallel to the B_0 field.

Longitudinal operators: Orthogonal operators, designated as I_z.

Lorentzian line shape: The normal line shape of an NMR peak that can be displayed either in absorption or dispersion mode.

Low-pass filter: A filter used for selecting coherence orders $\geq p$.

Magnetic field gradients: See *Field gradients*.

Magnetic quantum number (m_I): The number that defines the energy of a nucleus in an applied field B_0. It is quantized, i.e., has values of $-I$, $-I + 1, \ldots, + I$. In the case of ^1H or ^{13}C nuclei, the spin quantum number is $\pm\frac{1}{2}$, corresponding to the alignment of the z-component of the nuclear magnetic moment with or against the applied magnetic field, B_0.

Magnetic resonance imaging (MRI): A technique, based on magnetic field gradients, that is widely used as an aid to clinical diagnosis. MRI has proved to be a promising tool for tumor detection.

Magnetization vector (M): The resultant of individual magnetic moment vectors for an ensemble of nuclei in a magnetic field.

Magnetogyric ratio (γ): The intrinsic property of a nucleus that determines the energy of the nucleus for specific values of the applied field, B_0. It can be either positive (e.g., ^1H, ^{13}C) or negative (e.g., ^{15}N, ^{29}Si).

Matched filter: The multiplication of the free induction decay with a sensitivity enhancement function that matches exactly the decay of the raw signal. This results in enhancement of resolution, but broadens the Lorentzian line by a factor of 2 and a Gaussian line by a factor of 2.5.

Mixing period: The third time period in 2D experiments, such as NOESY, during which mixing of coherences can occur between correlated nuclei.

Modulation: The variation in amplitude and/or phase of an oscillatory signal by another function, e.g., modulation of the nuclear precession frequency of one nucleus by the nuclear precession frequency of a correlated nucleus in COSY spectra.

Multiple-quantum coherence: When two or more pulses are applied, they can act in cascade and excite multiple-quantum coherence, i.e., coherence between states whose total magnetic quantum numbers differ by other than ± 1 (e.g., 0, ± 2, ± 3). Such coherence is not directly detectable but must be converted to single-quantum coherence to be detected.

Multiple-quantum filtration: A pulse sequence applied for preferentially selecting transitions involving multiple-quantum coherences; may be used to simplify 2D spectra by eliminating signals due to single-quantum coherence or to attenuate diagonal peaks in a COSY spectrum.

Multiple-pulse experiment: An experiment in which more than one pulse is applied to the nuclei.

Nonselective pulse: A pulse with wide frequency bandwidth that excites all nuclei of a particular type indiscriminately.

Nuclear magnetic moment (μ): The consequence when a nucleus has both a charge and an angular momentum. The magnitude of the observed magnetic moment is given by $m_I = h\gamma/2\pi$ (where m_I = magnetic quantum number and γ = magnetogyric ratio).

Nuclear Overhauser effect (nOe): The change in intensity in the signal of one nucleus when another nucleus lying spatially close to it is irradiated, with the two nuclei relaxing each other via the dipolar mechanism.

Nuclear Overhauser effect correlation spectroscopy (NOESY): A 2D method for studying nuclear Overhauser effects.

Nyquist frequency: The highest frequency that can be characterized by sampling at a given rate; equal to one-half the rate of digitization.

Off-resonance decoupling: Applying an unmodulated radiofrequency field B_2 of moderate intensity in the proton spectrum just outside the spectral window of the carbon spectrum shrinks the multiplet structure of the carbon signals. Some multiplet structure is, however, still left behind, allowing the CH_3 (quartets), CH_2 (triplets), CH (methines), and C (quaternary carbons) to be distinguished.

Paramagnetism: Magnetic behavior of nuclei that contain unpaired electrons. When such substances are placed in a magnetic field, the induced magnetic field is parallel to the applied magnetic field.

Phasing: A process of phase correction that is carried out by a linear combination of the real and imaginary sections of a 1D spectrum to produce signals with pure absorption-mode peak shapes.

Phase cycling: As employed in modern NMR experiments, repeating the pulse sequence with all the other parameters being kept constant and only the phases of the pulse(s) and the phase-sensitive detector reference being changed. The FIDs are acquired and coadded. The procedure is used to eliminate undesired coherences or artifact signals, or to produce certain desired effects (e.g., multiple-quantum filtration).

Phase-sensitive data acquisition: NMR data are acquired in this manner so that peaks are recorded with pure absorption-mode or pure dispersion-mode line shapes.

Phase-sensitive detector: A detector whose output is the product of the input signal and a periodic reference signal, so that it depends on the amplitude of the input signal and its phase relative to the reference.

Polarization: The results of an excess of nuclear magnetic moment vectors (μ_z) aligned with or opposed to the applied magnetic field, B_0. The magnitude of the polarization effect is determined by the magnetogyric ratio, the applied field B_0, and the temperature.

Polarization transfer: Application of certain pulse sequences causes the transfer of the greater-equilibrium polarization from protons to a coupled nucleus, *e.g.,* ^{13}C, which has a smaller magnetogyric ratio.

Precession: A characteristic rotation of the nuclear magnetic moments about the axis of the applied magnetic field B_0 at the Larmor frequencies.

Preparation period: The first segment of the pulse sequence, consisting of an equilibration delay. It is followed by one or more pulses applied at the beginning of the subsequent evolution period.

Presaturation: Selective irradiation of a nucleus prior to application of a nonselective pulse causes it to be saturated, so its resonance is suppressed. This technique may be employed for suppressing solvent signals.

Product operator: A set of orthogonal operators that can be connected together by a product. The product operator approach can be used to analyze pulse sequences by applying simple mathematical rules to weakly coupled spin systems.

Projection: The one-dimensional spectrum produced when one of the two axes of a 2D spectrum is collapsed and the resulting "projection spectrum" recorded.

Pulse sequence: A series of Rf pulses, with intervening delays, followed by detection of the resulting transverse magnetization.

Pulse width: The time duration for which a pulse is applied; determines the extent to which the magnetization vector is bent.

Quadrature detection: A method for detecting NMR signals that employs two phase-sensitive detectors. One detector measures the x-component of the magnetization; the other measures the *y*-component. The reference phases are offset by 90°.

Quadrature images: Any imbalances between the two channels of a quadrature detection system cause ghost peaks, which appear as symmetrically located artifact peaks on opposite sides of the spectrometer frequency. They can be eliminated by an appropriate phase-cycling procedure, e.g., CYCLOPS.

Real and imaginary parts: Two equal blocks of frequency that result from Fourier transformation of the FIDs.

Rotating frame: The frame is considered to be rotating at the carrier frequency, v_0. This allows a simplified picture of the precessing nuclei to be obtained.

Saturation: If the Rf field is applied continuously, or if the pulse repetition rate is too high, then a partial or complete equalization of the populations of the energy levels of an ensemble of nuclei can occur and a state of saturation is reached.

Scalar coupling: Coupling interaction between nuclei transmitted through chemical bonds (vicinal and geminal couplings).

Selective polarization transfer: A polarization transfer experiment in which only one signal is enhanced selectively.

Signal averaging: A procedure for improving the signal-to-noise ratio involving repeating an experiment n times so that the signal-to-noise ratio increases by $n^{1/2}$.

Signal-to-noise ratio: Ratio of signal intensity to the root-mean-square noise level.

Sine-bell: An apodization function employed for enhancing resolution in 2D spectra displayed in the absolute-value mode. It has the shape of the first half-cycle of a sine function.

Single-quantum coherence: Coherence between states whose total quantum numbers differ by ± 1. The only type of coherence that can be observed directly.

Spin-echo: The refocusing of vectors in the *xy*-plane caused by a $(\tau-180°-\tau)$ pulse sequence produces a spin-echo signal. It is used to remove field inhomogeneity effects or chemical shift precession effects.

Spin-lattice relaxation time: Time constant for reestablishing equilibrium, involving return of the magnetization vector to its equilibrium position along the *z*-axis.

Spin-locking: If a continuous B_1 field is applied along the *y*'-axis immediately after a $90°_x$ pulse, then all the magnetization vectors become spin-locked along the *y*'-axis, and negligible free precession occurs.

Stacked plot: The 2D NMR spectrum is presented as a series of 1D spectra parallel to the F_1 axis stacked together at a number of F_2 values (or vice versa), one below the other.

Symmetrization: A procedure for removing artifact signals that appear nonsymmetrically disposed on either side of the diagonal.

t_1 noise: Noise peaks parallel to the F_1 axis that arise due to instrumental factors, such as random variations of pulse angles, phases, B_0 inhomogeneity, or temperature. It is not removed by signal averaging.

Tip angle (α): The angle by which the magnetization vector is rotated by the applied pulse.

Transient: Time-domain signal (FID) acquired in an FT experiment.

Transmitter: Coil of wire and accompanying electronics from which Rf energy is applied to the NMR sample.

Transverse magnetization: Magnetization existing in the $x'y'$-plane.

Transverse operators: Designated as I_x or I_y in the product operator description of NMR spectroscopy.

Zero-filling: A procedure used to improve the digital resolution of the transformed spectrum (e.g., in the t_1 domain of a 2D spectrum) by adding zeros to the FID so that the size of the data set is adjusted to a power of 2.

Zero-quantum coherence: The coherence between states with the same quantum number. It is not observable directly.

Index